P9-DBJ-712

INTRODUCTION TO ENGINEERING

INTRODUCTION TO ENGINEERING

P. THOMAS BLOTTER
UTAH STATE UNIVERSITY

JOHN WILEY & SONS
NEW YORK · CHICHESTER · BRISBANE · TORONTO

Library of Congress Cataloging in Publication Data:

Blotter, P Thomas.
Introduction to engineering.

Includes index
1. Engineering. I. Title
TA145.B55 1980 620 80-25375
ISBN 0-471-04935-2

Printed in the United States of America

10 9 8 7 6 5 4 3 2 1

PREFACE

This book introduces engineering by involving the student in meaningful solutions to engineering problems. Several years of experience in teaching engineers at the entry level suggest that the problem-solving method is a productive and highly motivational approach. Students are easily bored with broad definitions of engineering disciplines and descriptions of the role they might play as a professional engineer. More than words are required to adequately introduce engineering. Methodically applying scientific principles and eventual solutions to problems provide students with an opportunity to feel the inward triumphs associated with engineering. Through the assessment of problems, the application of scientific concepts, and some manipulation of numbers, students obtain a glimpse of the personal satisfaction experienced by professional engineers as they create new and better things. Such feelings of accomplishment cannot be described with words any more realistically than can feelings associated with getting a big raise or winning a ball game.

In the past, engineering students have been expected to survive a barrage of courses in mathematics, physics, chemistry, and the humanities prerequisite to any engineering application. Approximately the first two years were devoted to a very broad preparation. Consequently, some students were disillusioned with engineering and their careers suffered a premature death. I certainly agree that a general and rigorous science background is essential for the engineering graduate; however, all the excitement associated with the application of fundamentals to the solution of problems need not be held sacred until the last two years of an engineering program. This book presents some engineering applications that can be understood by the entering freshman. The broad prerequisite experience is temporarily circumvented, yet sufficient rigor is maintained in the streamlined approach to provide students with specific fundamentals that are necessary to solve some relevant problems.

There are essentially no post-high school prerequisites required for the text. It has been used by entering freshmen with math backgrounds that range from remedial algebra to one or two quarters of calculus. Students deficient in algebra have been disadvantaged and are advised to take remedial work.

In Chapter One, the role of an engineer, engineering specialties, engineering challenges, and the methodology of problem solving are briefly discussed.

Chapter Two presents both customary and SI dimensions and units. Dimensional homogeneity and unit conversions are included. The entire text is bilingual in that problems are available in either customary or SI units.

Chapter Three covers energy and is presented in two parts. Part A presents some fundamental concepts that allow students to solve problems relating to the transformation of energy from one form to another. It introduces a technical basis that is essential in order to comprehend the so-called energy crisis and complete coverage is recommended. Part B presents some relevant topics in fuels, solar energy, and windmills that may be selected as desired.

Chapter Four begins with a discussion of models for structural loads and continues through statics to the point where a student can analyze the forces in a simple truss. A brief discussion of trigonometry is provided in the Appendix that has proved to be sufficient to enable students to resolve forces into components.

Chapter Five deals with materials. In anticipation of the future, discussions and problems extend beyond metals and include ceramics and polymers. Part A presents some mechanical, thermal, and electrical properties. The student is able to solve problems that deal with such concepts as stress and strain. A brief discussion of the microstructure of materials is presented in an attempt to describe why materials behave as they do. Part B includes some topics that may be selected according to particular interests. The section on building materials and heat transfer has been particularly popular with students.

Chapter Six presents both resistance and digital circuits. Students become involved with truth tables, switching algebra, and design some simple logic circuits.

Chapter Seven begins with some fundamental concepts of fluid statics and eventually includes some discussion and problems relating to the flow of ideal fluids.

It is exciting for the instructor and rewarding for the student to experience how rapidly students can apply such fundamentals as the first law of thermodynamics, equilibrium equations, conservation of energy, and conservation of mass. The book is written so that chapters are essentially independent and may be selected as course time permits. However, it is suggested that Chapters One through Four be covered first and in sequence.

A general introductory course that includes multiple areas of engineering may be challenging to a single instructor with a specific expertise. This book will allow an instructor from any engineering department to teach a variety of basic engineering concepts.

I acknowledge the contributions of several colleagues, students and friends. Fellow staff members at the U.S. Air Force Academy helped formulate the initial concept. J. Clair Batty contributed a major portion of Chapter Three. Derle Thorpe, Carl Spear, and others exchanged ideas during the developmental stages over the past 10 years at Utah State University. Jay Bagley helped with the fluids chapter. Several students have provided significant comments and Nan Wangerin persevered with the typing.

P. Thomas Blotter

CONTENTS

INTRODUCTION TO ENGINEERING

CHAPTER ONE

THE ENGINEER

1.1/What Is an Engineer?

Engineers are problem solvers. They apply the fundamental laws of science to obtain solutions to human problems and answer demands to improve the quality of life (Figure 1.1). An engineer is the human link between the scientist and the consumer. Physical scientists discover and mathematically describe laws of the universe, whereas engineers harness these laws to provide useful products.

Engineers replace muscle, toil, boredom, and sweat with devices that efficiently and conveniently perform our work. Early Americans used hand tools for long hours to harvest grain (Figure 1.2). Later, work crews and teams of horses added excitement and reduced the burden of the harvest. Through engineering, a self-propelled combine now harvests the grain to feed the world.

Many engineers design systems that convert raw materials into useful forms of energy. They extract materials from the crust of the earth and build structures that contribute to our comfort and provide the physical skeleton for our industrial prowess. The materials in the bridge in Figure 1.3 were once scattered in an unorganized manner over the earth.

Space and communication systems are designed and constructed that transmit signals, words, and pictures from other planets and around the world. The space shuttle (shown in Figure 1.4) is an engineering triumph. Engineers produce mechanical components for automobiles, appliances, airplanes and even human bodies. In Figure 1.5, rings are placed in the vena cava of a dog to test the clotting of blood in response to a foreign material. Synthetic materials made from sand, by-products, and other

FIG. 1.1. An engineer uses the fundamentals of science as building blocks to create useful products.

abundant elements are used to replace our dwindling natural resources. Assembly lines, weapons of war, sugar beet harvesters, heating and air conditioning systems, computers, highways, toys, razors, and satellites are products of engineering. A popular engineering challenge centers on recreational products, such as the ski lift and slope-grooming equipment shown in Figure 1.6.

FIG. 1.2. (a) The harvesting of grain began with hand tools. (b) Early combines required teams of horses and a crew of people. (c) A modern combine is self-propelled and requires one operator seated comfortably in an air conditioned cab. (Courtesy of Sperry-New Holland)

FIG. 1.3. The Cart Creek Bridge spans one channel of the Flaming Gorge Reservoir. (Courtesy of Utah Department of Transportation)

FIG. 1.4. The Space Shuttle Orbiter Enterprise. (Courtesy of NASA)

The work arena of engineers is extremely broad. Some engineers are professional dreamers who sift through creative ideas long before a product is conceived. Some design products to meet specifications. Others write the specifications for needed technology. Often engineers are involved in either selling or buying of technical gadgets. Some engineers spend their time building hardware, whereas others are trying to destroy it. Experimental testing, research, development, legal consulting on product liability cases, and teaching are other common pursuits that involve engineers.

What is a typical work environment for an engineer? To some

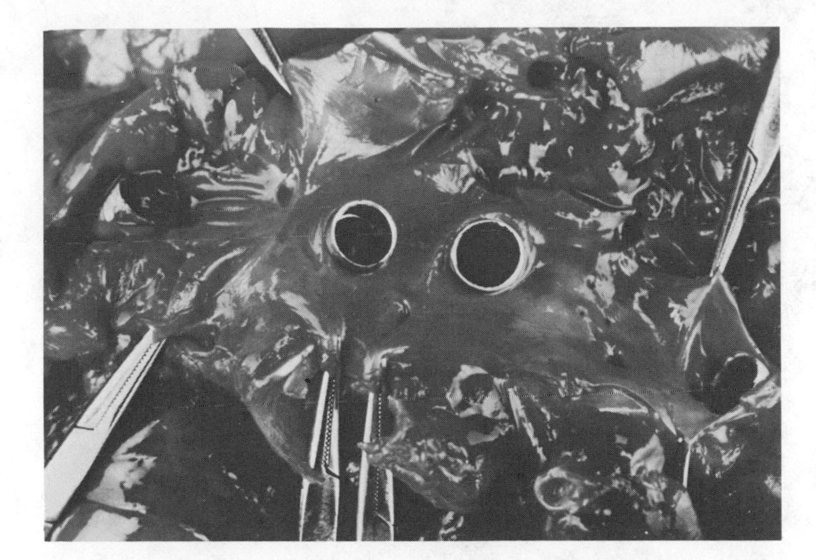

FIG. 1.5. Rings placed in the vena cava of a dog. (Courtesy of UBTL Division of University of Utah Research Institute)

FIG. 1.6. The grooming of a ski slope at Sundance. (Courtesy of De Lorean Mfg. Co./Logan Division)

extent, one can select an appealing environment. Some people wear hard hats, overalls, and heavy shoes as they move around a construction site supervising or inspecting a project. Others spend their day in the company of corporate managers. There are engineers who perform delicate assignments and must wear meticulously clean white coats and work in environmentally controlled rooms. Engineers assigned to foreign projects might wear

FIG. 1.7. An engineer studying the model of an industrial plant. (Courtesy of A. M. Kinney, Inc.)

FIG. 1.8. Engineers are testing materials. (Courtesy of A. M. Kinney, Inc.)

FIG. 1.9. A typical engineering office. (Courtesy of A. M. Kinney, Inc.)

sandals and shorts as they work under a straw shelter in underdeveloped countries. Many engineers are surrounded by physical models of plants being designed (shown in Figure 1.7). Others observe instruments and evaluate data (shown in Figure 1.8). Most engineers, however, work in an office (shown in Figure 1.9), dress in business attire, enjoy some flexibility in their working schedule, and interact with several different kinds of people.

1.2/Engineering Specialties

The field of engineering is frequently divided into some rather broad specialties such as mechanical, electrical, civil, agricultural, chemical, and industrial engineering. An engineer trained only in the theory of electronic circuits obviously could not design bridges, internal combustion engines, and irrigation systems with the same expertise. Engineering is much too broad for any one person to have a competent knowledge of the complete spectrum. Although there is a common core of knowledge required of engineers, the division of engineering into specialties does allow individuals to prepare for particular challenges.

There is considerable overlap between engineering specialties. The boundaries are not clearly defined. For example, courses in fluid mechanics, solid mechanics, and material science may be found in both mechanical and civil engineering departments across the country. Many graduate mechanical engineers are working in assignments typical of electrical and chemical engineers. Some electrical engineers are designing instrumentation systems that would typify mechanical engineering. Most engineers encounter professional assignments that extend their expertise into specialties beyond those prepared for in their college experience.

In this introductory text, the concepts are sufficiently basic to apply to the common core of knowledge anticipated for all engineers. The concepts are not stipulated within the domain of any particular specialty.

1.3/Why Engineers?

The requirements of society such as food, energy, water, transportation, defense, recreation, and shelter filter back to engineers in the form of a cry for additional technology. In the past, technology has developed in response to these cries and engineers have been the captains of the technology teams. Engineers have established an admirable record as implementors of scientific discoveries to the sustenance and even enrichment of our style of living. Improved food-processing systems allow fewer to feed more. Heat and light are available to almost everyone. Water flows from some far-off point through piping networks and into our kitchens and through our showers. Trucks, trains, and airplanes transport our products. Missiles, bombs, and guns defer enemy aggression. Athletic events bounce from earth to satellites and back to earth and appear to take place on our living room television. However, each year the public screams for more and the appetite for additional technology is insatiable. In fact, technology is no longer a luxury. Instead, it is a necessity.

The American worker now has more free time and money to spend than ever before. In 1909, an individual employed in manufacturing worked an average of 51 hours per week and earned $9.74. Corrected for cost-of-living increases, the purchasing power of such a worker existing in 1971 would have been $44 weekly. However, the average wage in 1971 was $142.44 for a 39.9-hour work week. In 62 years, the purchasing power of the worker increased 325 percent and the working time was reduced by 27 percent.[1] Between 1950 and 1978, the average work week of a nonagricultural production worker was reduced by four hours, the productivity measured in output per hour increased 95

[1]J. D. Kemper, *The Engineer and His Profession, Second Edition,* Holt, Rinehart and Winston, New York, 1975, p. 42.

percent, and real compensation (adjusted for inflation) increased 102 percent per hour.[2]

Why worry about the major problems of the world and their solutions through technology? Why not digress to the simple way of life enjoyed by our forebears scattered across the frontiers of America? The answer relates to our increased population. Our present population would have difficulty existing on small family farms. The abundance of land, unlimited natural resources, and a pristine environment of the early frontiers are things of the past. The present demands for food, water, heat, defense, education, medical care and mobility could not be satisfied by a return to the primitive utopia enjoyed by a few of our early ancestors. Society is locked into a progressive pattern that depends on technical innovation in order to continue along its course.

Technology is not a monster that usurps human dignity and leads people along an uncontrollable path. In contrast, technology provides a means of survival and represents a tool box for thinking minds. Technology is the bridge between what people want and what they get. In spite of the emotional pleas promulgated by counterculturalists, the sustenance of quality living rests on the shoulders of innovative technology. Engineers are the pilots of technology and attempt to improve human dignity and provide time over which we might exercise dominion. The common belief is that someone will figure out how to solve our problems. That someone is apt to be you as an engineer.

1.4/Engineering Challenges

If engineers are problem solvers, then what are the problems? Some have been inherited from past generations, others are consequences of solutions to previous problems and some are starting to surface that threaten our quality of living. Only an overview of a few of the major concerns are identified in this section. More detailed analyses follow in subsequent chapters.

The depletion of our petroleum is a serious problem. The total impact of the earth's dwindling fuel tank is affecting our economic stability, defense, employment, productivity, recreation, and social attitudes. The human muscle that pulled the handcarts of the westward migration in the 1850s, as shown in Figure 1.10, has been conveniently replaced with engines mounted in auto-

[2]U.S. Department of Labor, Bureau of Labor Statistics, *Monthly Labor Review,* July 1979, pp. 65, 87.

FIG. 1.10. A handcart pioneer family. (Courtesy of Church Archives. The Church of Jesus Christ of Latter Day Saints)

FIG. 1.11. The Boeing 747 carries an average load of 374 passengers and baggage a distance of 5900 statute miles. (Courtesy Boeing Aircraft)

mobiles, trucks, and airplanes for the 1980s (shown in Figure 1.11). An automobile industry based on the availability of gasoline has developed that is twice the size of the next largest manufacturing industry, constitutes 10 percent of our gross national product (GNP), and is responsible for over 13 million jobs or about one-sixth of the total U.S. employment.[3] The economy withstood

[3]D. G. Harvey and W. R. Menchen, *The Automobile-Energy and the Environment*, Hittman Associates, Inc., Columbia, Maryland, 1974.

the abandonment of the handcarts, but could we survive the loss of the combustion engine? Society is depending on engineers and other members of the technology team to provide continued transportation.

The total energy spectrum is of great concern. The economic growth and stability of developed countries has been highly dependent on the availability and transformation of energy. The United States has the highest gross national product in the world, yet more than 10 times the energy per capita is required as compared to India. Technology must somehow harness greater amounts of energy from the sun, wind, tides, water, atoms, or other resources in addition to the fossil fuels in order to maintain our modern style of living.

The production of food is a problem. In the United States, less than 6 percent of the working population is required to produce the food required. In some underdeveloped countries, at least three out of four individuals are involved in the pursuit of food, and yet many starve. American agriculture is a champion in the area of productivity, yet the challenge continues. Arid lands need to be brought under cultivation as the increasing population of the world absorbs more space. Irrigation systems must be designed and built. Some agricultural machines are almost unbelievable, such as the hay stacker, which harvests over 3000 bales per day untouched by human hands, shown in Figure 1.12. Yet,

FIG. 1.12. A balewagon loads, transports, and stacks hay. (Courtesy of Sperry-New Holland)

better machines to harvest fruit, vegetables, and animal feeds must become available. Processing food, eliminating waste, efficient storing, packaging, and distributing food are in need of improved technology. The American public pays a lower percentage of their income for food than any other country and expects the bonanza to continue.

The protection of our precious environment is a problem. The natural filtering phenomena provided by our ecological system cannot process the tons of waste that it receives continuously. Technology must provide waste disposal systems. As our population increases and space becomes more critical, we cannot simply haul our garbage away to some isolated location. Our streams, rivers, and lakes cannot absorb unlimited amounts of sewage and industrial waste. Sewage treatment plants similar to Figure 1.13 are needed. The outflow is then returned to the natural streams as shown in Figure 1.14. The air that sustains all forms of life must be protected from polluting chemicals that result from the burning process. For example, if the sole source of energy for a typical university of 10,000 students in the Rocky Mountains were coal, over 100 tons of sulfur dioxide would be emitted into the air each year. The delicate balance required by our ecological systems cannot be disregarded. An exposure to air that contains only 1 part of carbon monoxide to 1000 parts of air is sufficient to cause unconsciousness after one hour and death within four hours. Further technology in the area of particle separators, electrostatic precipitators, polarizers, catalytic converters, sewage disposal systems, purifiers, scrubbers, and other

FIG. 1.13. Sewage treatment plant for a small community. (Courtesy of Valley Engineering, Inc., Logan, Utah)

FIG. 1.14. Reclaimed waste water from the sewage treatment system is returned to natural streams. (Courtesy of Valley Engineering, Inc.)

systems must be developed to maintain our ecology, yet allow the required productivity.

Inflation is a major problem. Inflation can be curtailed by increasing our productivity. Engineers improve our productivity through innovative ideas. The growth of an economy as measured through increased productivity is related to a nation's ability to create and develop better ideas. One of the reasons for inflation in the United States is the decline in innovative momentum. A smaller proportion of our total number of people in the United States are involved in technology than ever before.[4] In Figure 1.15 the rate of change of the total number of technologists per GNP shows that the United States is actually on a declining curve as compared to other countries. The capability for technical innovation must increase in order to maintain our world leadership. New and better ways of doing things are more in demand now than perhaps ever before. Nations need more technically trained innovators.

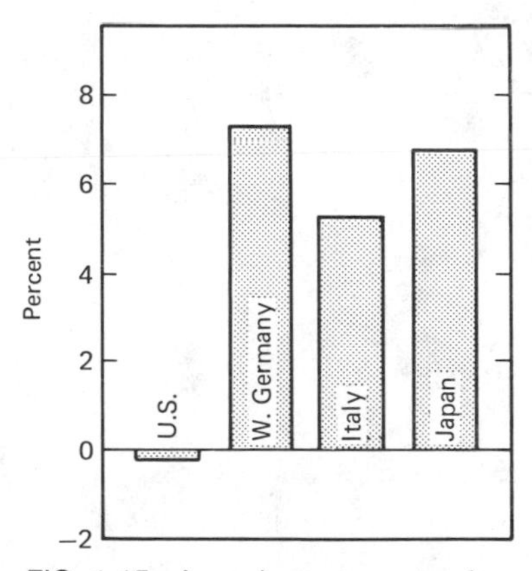

FIG. 1.15. Annual average rate of growth of professional research and development manpower in the United States, West Germany, Italy, and Japan (1970–1973).

Space exploration is a challenge of the future. As our generation reaps the rewards of sacrifices made by early explorers that opened new regions of the earth, we must continue these efforts and unlock the mysteries of outer space. Prior explorations have resulted in multidimensional improvements in our standard of living. In a similar way, space technology overflows the space program to enrich our quality of life. The voice controlled wheelchair is a spinoff from NASA's space program as shown in Figure

[4]Robert C. Dean, "Technical Innovation USA," *Mechanical Engineering*, Vol. 100, No. 11, November 1978.

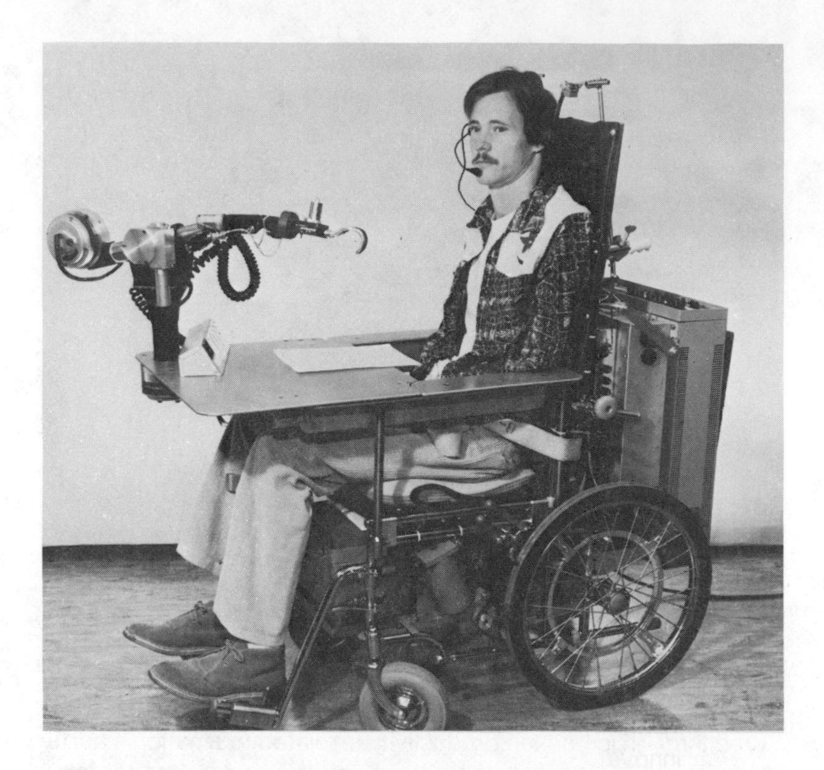

FIG. 1.16. Voice-controlled wheelchair allows a quadraplegic to pick up packages, open doors, turn knobs, and perform other functions. (Courtesy of NASA)

1.16. Other spinoffs from space technology are found in home design, brain surgery, cancer treatment, fire prevention, food products, industrial processes, crop disease control, pollution control, law enforcement, airlines, energy systems, heart pacers, and many others. Space technology is helping to solve many of our domestic problems.

There are numerous other problems that need engineering solutions. Transportation systems, medical equipment, human transplants, dams, aerospace vehicles, computers, communication systems, chemical systems, mining operations, and other challenges await the scrutiny of engineers.

1.5/The Engineering Approach to Problem Solving

An engineer is trained to pursue a solution to a problem in a systematic manner. Whether the answer sought is a rough estimate, a precise number or a judgment decision, a logical step-by-step procedure is followed. Most everyone enjoys some occasional good luck, however, random guesses based on scanty information represent the antipathy of the engineering approach. An organized thought pattern is helpful to obtain answers in both technical and nontechnical situations.

The *first step* is to describe the problem. A problem, such as putting someone on the moon may be very broad and require years of effort by an interdisciplinary team. The problem may be very narrow, such as the design of a lock nut that would securely fasten an instrument to the pallet of the space shuttle and withstand the vibration environment during launch. Perhaps the problem is to determine the speed of an automobile involved in an accident by analyzing the wreckage. One might be challenged to repair an expanding leak in an earth dam. In any event, an engineer begins the solution process by first defining the problem to be solved.

The *second step* is to identify the laws of science that apply to the problem and are necessary to eventually determine a solution. For example, if the question is related to thermal insulation, then an engineer would apply the equations that govern the flow of heat. If a windmill were being developed, engineers would recall equations relating to the transfer of energy. The laws of energy and mass conservation might by applied to fluid flowing through a pipe or gases that are ejected through a nozzle. Laws that relate forces, stresses, and displacements in materials would be applied to structures to predict their behavior when subjected to seismic loads. The design of a minicomputer to control a switch would require the laws that govern the flow of electricity. An engineer must select scientific principles that govern the phenomena of the problem.

The *third step* in the engineering approach is the identification of pertinent known and unknown facts. These facts are usually quantified with numbers and must eventually be substituted into equations. For example, if one applied the law of heat conduction to an insulation problem, then such things as the temperature gradient, material thickness, material conductivity, and surface area would be required to predict the rate of heat flow. A constant deceleration for an aircraft after touchdown could be predicted by knowing the touchdown velocity, the stopping distance and the final velocity. Once an engineer understands the governing equation, then he/she must find the necessary numbers as either knowns or unknowns. This third step is difficult for both practicing engineers and students. In the case of students, they are relatively inexperienced with the governing equations even though the known and unknowns are specified in the problem statement. In the case of professional engineers, the equations are familiar, but the determination of appropriate known values is sometimes difficult. For example, what are the material constants for aluminum inside a cryogenic tank at a temperature of $-450°F$ and zero pressure? What is the spring constant for a small cantilevered wire coated with ice?

The *fourth step* in the engineering approach is the manipulation of numbers. Once numerical values have been assigned to the parameters of an equation, then some mathematical process is initiated to eventually find numbers that describe the unknown quantities. These calculations might be completed with large sophisticated computers, hand-held calculators, slide rules, or even done by hand on the back of an envelope. The logarithmic slide rule as shown in Figure 1.17 was used successfully by engineers for several years. They may be used for such operations as multiplication, division, exponents, and trigonometric functions, yet slide rules cannot add and subtract. Now they have been replaced with electronic calculators. Several calculators with varying capabilities and sophistication are available (a typical hand-held programmable calculator is shown in Figure 1.18). Another hand-held type calculator that interfaces with a desktop printer is shown in Figure 1.19. An engineer without a calculator

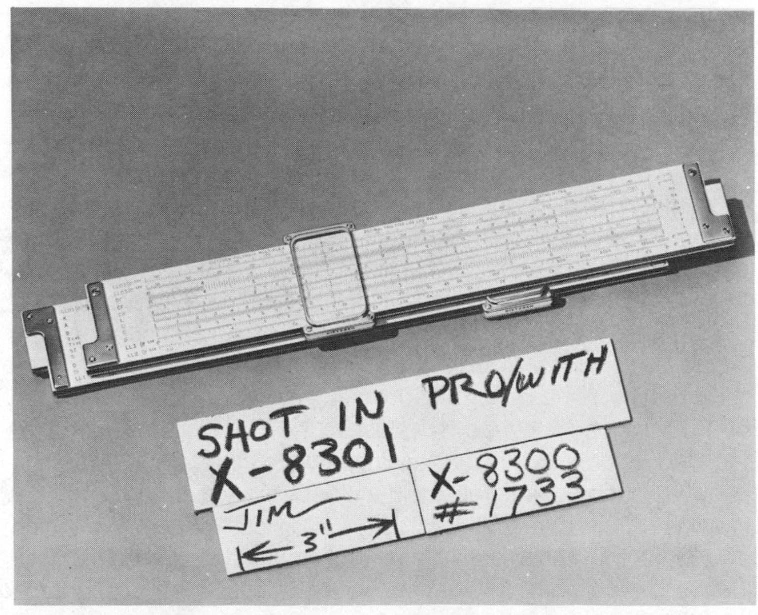

FIG. 1.17. The logarithmic slide rule was used in the past to make engineering calculations. (Photo Courtesy E. Dietzgen Company)

FIG. 1.18. A typical hand-held calculator. (Photo courtesy of the Hewlett-Packard Company)

would be similar to a cowboy without a horse, a dentist without a drill, or a painter without a brush. Some additional discussion of calculators and some practice calculations are given in Appendix A.

The *fifth and final step* of the engineering approach is to evaluate the result. Does it make sense? Additional testing, experimentation, other design iterations, impact statements, or other courses of action might follow in order to understand, verify, or even accept the result as a solution. A solution to one problem may easily be the source of another.

Engineers cannot afford to make many mistakes. The public

FIG. 1.19. A calculator and desk-top printer. (Photo courtesy of Texas Instruments)

cannot tolerate dams that wash away, structures that fall down, engines that won't run, computers that give erroneous numbers, or products that are unreasonably dangerous. Therefore, it is important that the problem-solving steps be clearly documented. Other engineers frequently check solutions and designs. It is important that the solution format is neat and that each major step is identified so that someone else can check the work. The problem statement must be clearly stated. The physical laws applied must be identified and written down. Any simplifying assumptions must be explained. The knowns and unknowns must be written down. Sufficient steps in the manipulation of numbers must be included to allow a check of the procedure. In this text, example problems suggest a solution format that might be followed.

CHAPTER TWO

DIMENSIONS AND UNITS

If you were to listen in on the conversation of a team of engineers as they worked in the calculation phase of the engineering approach to solutions, you might hear some rather strange statements. For example, a structural engineer may be asking as to whether the dead and live loads were given in newtons, kilonewtons, pounds, or tons. Another might specify the energy available from coal in British thermal units or joules. Someone else might ask about the dimensional homogeneity of an equation. Others may desire some unit conversion factors to convert inches to meters. Dimensions and units are very important to engineers and must be understood.

The word *dimension* refers to a physical quantity such as length, mass, or temperature. Some of these dimensions are difficult to define since they are so basic. How would one define the dimension of length? One couldn't simply say that length is a measure of "how long" something is, since the width of a bar, the thickness of a plate, the depth of a well, the radius of a circle, and the distance between the earth and the sun are also length dimensions. What is the definition of time? Perhaps time is synonymous with duration, yet it would seem strange to ask someone wearing a wristwatch for "the duration." Time is a dimension that we intuitively accept without a precise definition. Mass is another dimension that is rather confusing. When we think of mass, such things as a large chunk of lead, a paper weight, our pencil, a human body, the moon, or other objects might flash through our mind. What's an acceptable definition of temperature? One might say that temperature relates to "how hot" or "how cold" a body is compared to the freezing point of water, the temperature in

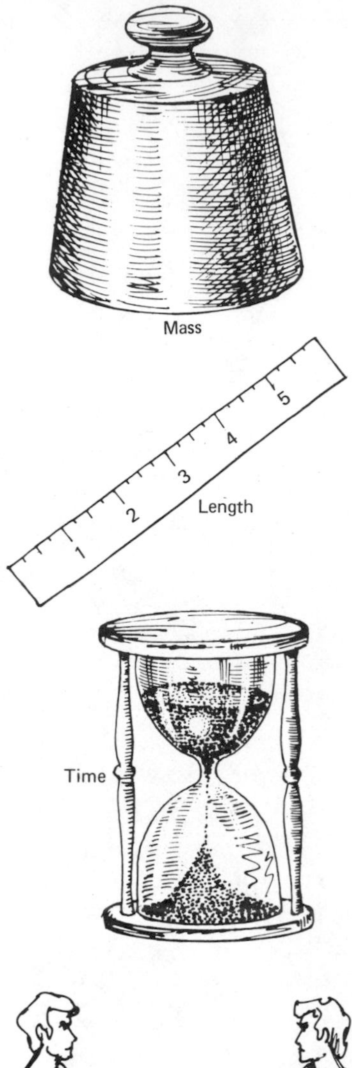

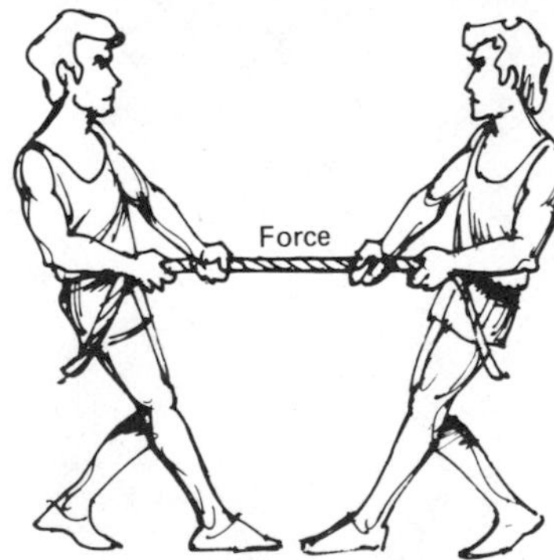

FIG. 2.1. Some commonly used dimensions.

black space, or perhaps the boiling point of aluminum. The dimension of force is generally accepted as a push or pull. These basic concepts of length, time, mass, temperature, and force are typical physical quantities that are called dimensions, some of which are depicted in Figure 2.1.

Dimensions are frequently identified as either fundamental or derived. A limited number of independent dimensions that form a basic set for the definition of all other physical quantities are called *fundamental dimensions. Derived dimensions* are combinations of the fundamental dimensions.

There are several different systems of dimensions that have evolved over the years. This is not surprising, since one could define different sets of fundamental dimensions and then derive other dimensions as necessary. For example, in a mechanical system, the fundamental dimensions might be selected as length (L), time (T), and force (F). Since the average velocity (V) is defined as distance divided by time ($V =$ distance/time), the dimensions of velocity are derived in terms of the fundamental dimensions L and T as $V = L/T$.

The principle of dimensional homogeneity states that an equation expressing a physical relationship must be dimensionally homogeneous. The dimensions of both sides of the equation must be equivalent. In other words, one cannot specify distance as length L, time as T and then arbitrarily assign velocity as something other than L/T.

Dimensions and units are not the same thing. Units are used to measure or quantify a dimension. Units are recognizable standards to which one can relate. The dimension of length (L) might be measured in units of feet, inches, meters, kilometers, miles, cubits, hands, fingers, or whatever else we choose. The dimension of time may be expressed in units of milliseconds, seconds, minutes, hours, years, decades, centuries, millenia, winks, shakes, or even a jiffy. Temperature is a dimension frequently expressed in units of degrees—Fahrenheit or Celsius. The dimension of force could be measured in units such as ounces, pounds, newtons, or tons.

2.1/Newton's Law—Systems of Units

Over 300 years ago, Isaac Newton observed that force and acceleration are proportional. He stated that an object of mass M subjected to a net force F would have an acceleration a, shown in Figure 2.2. Newton's law, $F = Ma$, is accepted as an absolute truth in engineering. It is valid in or out of gravity fields. The equation is true on the surface of the earth, on the moon, in outer space, or in any reference frame as defined by Newton.

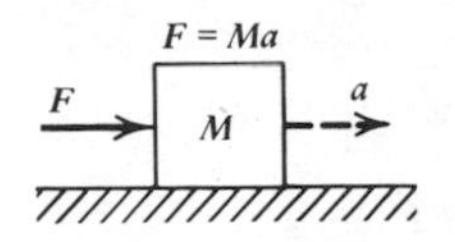

FIG. 2.2. Newton's second law states that force and acceleration are proportional.

The dimensions force, mass, length, and time, along with their units, can be related through Newton's law. For example, one might specify mass, length, and time as fundamental dimensions and on substitution into Newton's law, force would be determined as a derived dimension. Furthermore, the units used to quantify the dimensions would also have to satisfy the equation, that is, the units of the left-hand side must equal the units of the right-hand side.

A dimension or unit system is said to be either absolute or gravitational. An *absolute* system does not depend on gravitational effects of the earth or any other planet. The dimensions and units of a *gravitational* system do depend on the gravitational attraction.

In the absolute systems, the dimensions mass, length, and time are specified as fundamental and their units are often called fundamental units. Force is a dimension derived from Newton's law and has so-called derived units. Since the units of force were derived by substituting the units of mass, length, and time into Newton's law, obviously the equation is dimensionally homogeneous. In this text, dimensional homogeneity implies compatibility of both dimensions and units.

The *British absolute system* is commonly used by scientists. The fundamental unit of mass is called the pound mass (lbm). Since the word "pound" in this case refers to the mass and not the force, the term pound mass is used. The fundamental units of length and time are feet (ft) and seconds (s), respectively. Upon substitution into Newton's law, the units of force are derived as

$$F = Ma = \text{lbm} \cdot \text{ft/s}^2$$

and the unit $\text{lbm} \cdot \text{ft/s}^2$ is called a poundal.

The *metric system* (mks) is absolute. The fundamental dimensions are again mass, length, and time. Mass has units of kilograms (kg), length is measured in meters (m), and time is measured in seconds (s). The derived units of force are found by substitution as

$$F = Ma = \text{kg} \cdot \text{m/s}^2$$

One kilogram meter per second squared ($1\ \text{kg} \cdot \text{m/s}^2$) is called a newton.

In the past, English engineers have not used absolute systems. They have preferred to define force as one of the fundamental dimensions. Such a system is said to be gravitational since the magnitude of force depends on the acceleration due to gravity through Newton's law. Both the British gravitational system and the American engineering system will be described.

In the *British gravitational system,* force, length and time are defined as fundamental dimensions. Mass is a derived dimension based on Newton's law. Force has units of pound force (lbf). Since pounds refer to force in this system and not mass, the term pound force is used and abbreviated as lbf. Length is measured in feet (ft) and time has units of seconds (s). The units of the derived dimension, mass, are found by substitution into Newton's law as

$$M = \frac{F}{a} = \text{lbf} \cdot \text{s}^2/\text{ft}$$

One $\text{lbf} \cdot \text{s}^2/\text{ft}$ is called a slug. If one looks at Newton's law in the form

$$F = Ma$$

$$1 \text{ lbf} = (1 \text{ slug})(1 \text{ ft/s}^2)$$

one could say that one pound force (1 lbf) will accelerate one slug at 1 ft/s^2. Also, someone standing on the surface of the earth is subjected to an acceleration due to gravity of approximately 32.2 ft/s^2. If this person weighs 200 lbf, then the person's mass would be

$$M = \frac{F}{a} = \frac{200 \text{ lbf}}{32.2 \text{ ft/s}^2} = 6.21 \text{ slugs}$$

It is important to note that Newton's law is dimensionally homogeneous if the British gravitational units are substituted as follows:

$$F = Ma$$

$$\text{lbf} = (\text{slug})(\text{ft/s}^2) = (\text{lbf} \cdot \text{s}^2/\text{ft})(\text{ft/s}^2) = \text{lbf}$$

The *American engineering system* is a gravitational system where force is again specified as a fundamental dimension. Furthermore, the dimensions mass, length and time are also specified as fundamental. By definition, one pound force (1 lbf) will accelerate one pound mass (1 lbm) at the rate of 32.174 ft/s^2, which is the acceleration due to gravity at sea level. Using Newton's law, it follows that

$$F = Ma$$

$$1 \text{ lbf} = (1 \text{ lbm})(32.174 \text{ ft/s}^2) = 32.174 \text{ lbm} \cdot 1 \text{ ft/s}^2$$

Comparing the American engineering system with the English gravitational system, it is apparent that the units of force are both pound force and the units of acceleration are also identical. However, the units of mass are different. In the English gravitational system, mass is measured in slugs, while in the American engineering system mass is measured in pound mass. The masses are related since

$$1 \text{ lbf} = (1 \text{ slug})(\text{ft/s}^2) = (32.174 \text{ lbm})(\text{ft/s}^2)$$

and it follows that

$$1 \text{ slug} = 32.174 \text{ lbm}$$

There is a problem with Newton's law and dimensional homogeneity in the American engineering system. In the other systems, only three of the four units of mass, force, length, and time were declared as fundamental. The units of the only derived dimension were defined by substitution into $F = Ma$ and obviously the equation would be dimensionally homogeneous. However, in the American engineering system all four dimensions are declared as fundamental and, on substitution, the equation $F = Ma$ is not dimensionally homogeneous, that is,

$$F = Ma$$

$$(1 \text{ lbf}) \neq (1 \text{ lbm})(\text{ft/s}^2)$$

An additional constant of proportionality must be added to satisfy Newton's law in the American engineering system. In order to understand this constant, recall that Newton's law states that force is proportional to acceleration. The porportionality constant in all the systems discussed except the American engineering system is the mass M. One could write that

$$F \propto a \qquad \text{or} \qquad F = Ma$$

In the American engineering system, one writes that

$$F \propto a \qquad \text{or} \qquad F = \frac{Ma}{g_c}$$

where M/g_c is now the proportionality constant. The units of mass are lbm and g_c is a constant defined as

$$g_c = 32.174 \frac{\text{lbm} \cdot \text{ft}}{\text{lbf} \cdot \text{s}^2}$$

In the American engineering system, Newton's law is expressed as

$$F = \frac{Ma}{g_c}$$

TABLE 2.1. Systems of Units

	British Absolute	*Metric Absolute*	*English Gravitational*	*Engineering Gravitational*
Fundamental Dimensions				
Force (F)	—	—	lbf	lbf
Mass (M)	lbm	kg	—	lbm
Length (L)	ft	m	ft	ft
Time (T)	s	s	s	s
Derived Dimensions				
Force (F)	$\frac{\text{lbm} \cdot \text{ft}}{\text{s}^2}$ (poundal)	$\frac{\text{kg} \cdot \text{m}}{\text{s}^2}$ (newton)	—	—
Mass (M)	—	—	$\frac{\text{lbf} \cdot \text{s}^2}{\text{ft}}$ (slug)	
Velocity $\left(\frac{L}{T}\right)$	ft/s	m/s	ft/s	ft/s
Acceleration $\left(\frac{L}{T^2}\right)$	ft/s^2	m/s^2	ft/s^2	ft/s^2
Density $\left(\frac{M}{L^3}\right)$	lbm/ft^3	kg/m^3	slug/ft^3	lbm/ft^3
Pressure $\left(\frac{F}{L^2}\right)$	poundal/ft^2	newton/m^2 (pascal)	lbf/ft^2	lbf/ft^2
Newton's Law	$F = Ma$	$F = Ma$	$F = Ma$	$F = \frac{Ma}{g_c}$

where the units of force are pound force (lbf), the units of mass are pound mass (lbm), and acceleration is feet per second per second (ft/s^2). If the constant g_c is confusing, then don't feel too badly. It has troubled beginning students for years. That is one reason for the change to the metric system.

A summary of the unit system is given in Table 2.1.

2.2/Weight Force or Mass

There is often some confusion as to whether weight is a force or a mass. Unfortunately, the unit of pound has been used loosely throughout our economy to describe both force and mass. As evident from Newton's law, weight cannot be both force and mass since they are different dimensions and have different units.

One might visualize mass as a chunk of matter. If this mass, or chunk of matter, is accelerated, then there is a force equal in magnitude to the product of the mass and acceleration. On or near the surface of the earth, matter or mass is subjected to the acceleration due to gravity which is commonly called g. Unsupported bodies are accelerated toward the earth as they "fall" through the air. Supported bodies also are subjected to the acceleration of gravity even though they remain at rest. The force required to support a mass and prevent it from falling is equal to the product *Mg*. The gravitational acceleration near the surface of the earth is given in Table 2.2.

To help understand the weight problem, let's consider a large ripe tomato. The tomato is weighed by placing it on the platform scale as shown in Figure 2.3. Initially the platform deflects slightly until the force pushing up on the tomato by the scales is equal to the force pulling down on the tomato and the platform is at rest. The grocer then reads the calibrated scale and says that the tomato weighs 1 lb. The weight of 1 lb in this case is a measure of the force pushing on the platform produced by the product of the tomato mass times the acceleration due to gravity. When you eat the tomato, however, you eat the mass and not the force. Even with an abundance of salt and pepper, it would be rather ridiculous to try and eat dimensions such as velocity, acceleration, and force.

When a young man observes a young lady, or vice versa, they admire the mass and not the force. A car has mass and pushes down on the road with a force. The weight force on a space craft on the earth is different than the weight force on the same space craft on the moon, yet the mass is the same. The so-called dead weight of a structure is a force and would change if the gravitational acceleration changed; however, the mass is invariant.

TABLE 2.2. Acceleration Due to Gravity at Sea Level and at Various Latitudes

Latitude Degrees	*Acceleration Due to Gravity* m/s^2	*Acceleration Due to Gravity* ft/s^2
0	9.780 39	32.0878
10	9.780 78	32.0891
20	9.786 41	32.1076
30	9.793 29	32.1302
40	9.801 71	32.1578
Stand.	9.806 60	32.1739
50	9.810 71	32.1873
60	9.819 18	32.2151
70	9.826 08	32.2377
80	9.830 59	32.2525
90	9.832 17	32.2577

To correct for altitude, subtract $3.086 \times 10^{-6} m/s^2$ for each meter of altitude or subtract 3.086×10^{-6} ft/s^2 for each foot of altitude.

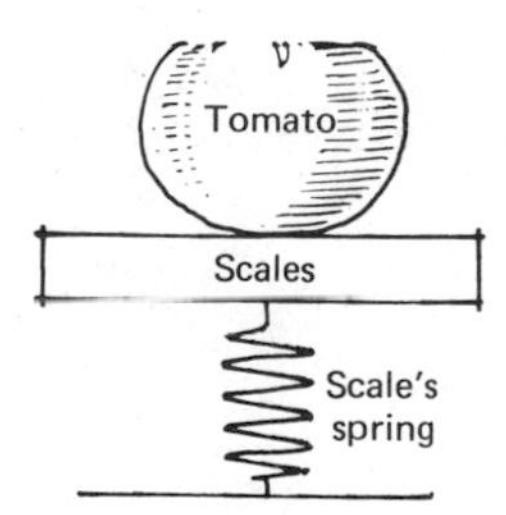

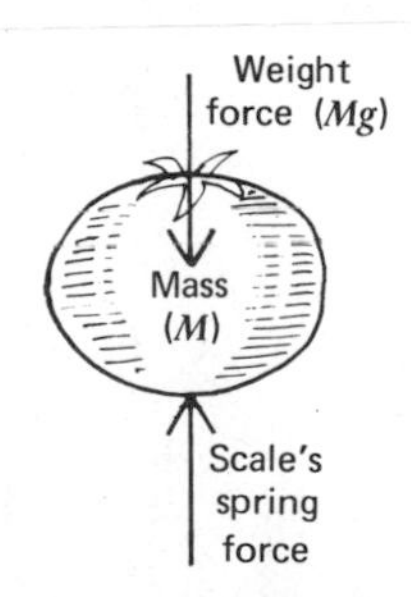

FIG. 2.3. Gravity attracts the mass *M* of the tomato. The product *Mg* is a force that is often called weight.

EXAMPLE PROBLEM 2.1

A certain object weighs 2000 lbf at a location where the acceleration due to gravity is $32.10\ ft/s^2$. Determine the mass of the object in both slugs and pound mass.

Solution

The mass in slugs can be determined using Newton's law in the British gravitational system.

$$F = Ma$$

$$M = \frac{F}{a} = \frac{2000\ lbf}{32.10\ ft/s^2} = 62.305\ lbf \cdot s^2/ft$$

$$M = 62.305\ slugs$$

The mass in terms of pound mass can be determined by recalling that

$$1 \text{ slug} = 32.174 \text{ lbm}$$

then

$$M = 62.305 \text{ slugs} = 62.305 \text{ slugs} \left(\frac{32.174 \text{ lbm}}{1 \text{ slug}}\right)$$

$$M = 2004.60 \text{ lbm}$$

Note that the mass in units of pound mass can also be found by using the definition of force in the American engineering system, that is,

$$1 \text{ lbf} = 32.174 \text{ lbm} \cdot \text{ft/s}^2$$

Since the force in the problem statement is 2000 lbf, then

$$F = (2000)(1 \text{ lbf}) = (2000)(32.174 \text{ lbm} \cdot \text{ft/s}^2)$$

and from Newton's law,

$$M = \frac{F}{a} = \frac{(2000)(32.174 \text{ lbm} \cdot \text{ft/s}^2)}{32.10 \text{ ft/s}^2}$$

$$M = 2004.6 \text{ lbm}$$

Instead of using the defining equation of force in the American engineering system, the alternative form of Newton's law could be used as $F = Ma/g_c$. Then,

$$M = Fg_c/a = (2000 \text{ lbf})/(32.174 \text{ lbm} \cdot \text{ft/lbf} \cdot \text{s}^2)/(32.10 \text{ ft/s}^2)$$

$$M = 2004.6 \text{ lbm}$$

EXAMPLE PROBLEM 2.2

A certain football player weighs 265 lb. Determine his mass in slugs if the acceleration due to gravity is 32.14 ft/s^2.

Solution

$$F = Ma$$

$$265 \text{ lb} = M \cdot 32.14 \text{ ft/s}^2$$

$$M = 8.25 \text{ lb} \cdot \text{s}^2/\text{ft} = 8.25 \text{ slugs}$$

EXAMPLE PROBLEM 2.3

Suppose the 70-kg person mentioned in the previous example is riding an earth satellite in a free-fall orbit. Determine the mass and weight of the person in the satellite.

Solution

The mass of an object is independent of location or acceleration. Therefore, the person's mass would be 70 kg in the satellite, on the earth, on the moon, or elsewhere. The weight of the person in the free-falling satellite would be zero because the satellite is not resisting or preventing the acceleration toward the center of the earth.

2.3/Units for Temperature

Temperature is a dimension and has units that must be defined. In order to do so, one must be familiar with the four temperature scales depicted in Figure 2.4. The temperature of a mixture of pure H_2O solid (ice) and liquid (water) at 1 standard atmosphere pressure is called the ice point. The temperature of boiling water at 1 standard atmosphere pressure is called the steam point. On the Celsius scale of temperature, the ice point is assigned the value of 0°C while the steam point is assigned the value of 100°C. Thus, on the Celsius scale, there are 100 divisions or degrees between the ice point and the steam point. Using this "size" division or degree, it may be shown that the absolute zero of temperature is at −273.15°C. On the Kelvin scale the lowest temperature possible or absolute zero is assigned as 0 K. The Kelvin scale of temperature uses exactly the same "size" divisions or degrees as the Celsius scale. Therefore on the Kelvin scale the ice point has the value of 273.15 K and the steam point is 200 divisions higher at 373.15 K. It is easy to convert a temperature from C to K by simply adding 273.15 to the Celsius temperature.

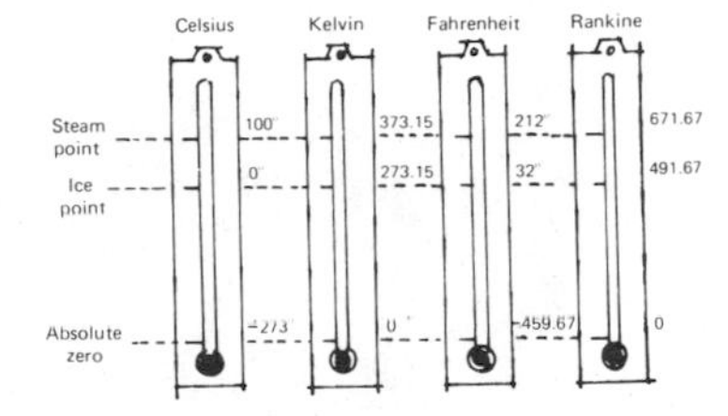

FIG. 2.4. Temperature scales.

On the Fahrenheit scale the ice point is assigned the value of 32.00 F while the steam point is assigned the value 212 F. Thus on the Fahrenheit scale there are 180 divisions between the ice point and the steam point. Using this smaller division or degree it is found that the point of absolute zero of temperature is located at −459.67 F. The Rankine scale of temperature uses exactly the same size division as the Fahrenheit scale. On the Rankine scale, however, the absolute zero of temperature point is assigned the value 0.00. The ice point therefore has the value 491.67 K and the steam point is 180 divisions above at 671.67 K. It is easy to convert from F to R by simply adding 459.67 to the Fahrenheit temperature. It is also easy to convert from K to R by simply multiplying the Kelvin temperature by 1.8. Temperature on the Celsius scale is expressed as a number followed by the symbol °C with no space between the number and the symbol. The small degree symbol is necessary to avoid confusion with the symbol C, which in the SI system represents a quantity of electricity called the coulomb. Temperature on the Kelvin scale is expressed as a number followed by a space followed by the symbol K. Temperatures on the Fahrenheit and Rankine scales are expressed as numbers followed by a space followed by the symbols F and R, respectively.

EXAMPLE PROBLEM 2.4

Express 20°C as Kelvin, Rankine, and Fahrenheit temperatures.

Solution

$$20°\text{C} + 273.15 = 293.15 \text{ K}$$

$$(293.15 \text{ K})(1.8) = 527.67 \text{ R}$$

$$527.67 \text{ R} - 459.67 = 68 \text{ F}$$

For most engineering purposes, it is sufficiently accurate to convert using 273 rather than 273.15 and 460 rather than 459.67.

EXAMPLE PROBLEM 2.5

Express -40 F as Rankine, Kelvin, and Celsius temperatures.

Solution

$$-40 \text{ F} + 460 = 420 \text{ R}$$

$$\frac{420 \text{ R}}{1.8} = 233 \text{ K}$$

$$233 \text{ K} - 273 = -40°\text{C}$$

Note that at -40, the temperatures on the Fahrenheit and Celsius scales are numerically equal.

2.4/SI and America

The early leaders of America had to select a system of dimensions and units. Around 1790, Thomas Jefferson recommended that the United States adopt a unit system based on multiples of 10. Congress feared that the adoption of such a system would hamper trade with Great Britain and so rejected Jefferson's recommendation in favor of the established system of units that started to evolve in England around the thirteenth century. The industrial might of the United States was thus developed using the British system where force is measured in pounds, length in feet, and time in seconds.

Most of the other countries of the world had used the more logical metric or decimal system. In 1960, the International System of Units (the SI system) was recommended for adoption worldwide by the General Conference on Weights and Measures. Finally, on December 23, 1975, President Gerald Ford signed the Metric Conversion Act and declared a national policy to convert to the SI system. The conversion process will be very lengthy and painful since many billions of dollars have been invested in hardware based on the British system. Our factories, tractors, trucks, automobiles, airplanes, appliances, and an immense variety of mechanical parts cannot all be replaced overnight. That process will take many years. Engineers entering the profession during this transition period must be acquainted with systems of the past and the future. The student must become "bilingual" in learning to communicate using engineering units.

2.5/SI Units

In the SI system, all units are separated into three classes—base, supplementary, and derived units. There are seven *base units* that are associated with seven fundamental dimensions, shown in Table 2.3.

TABLE 2.3. SI Base Units

Quantity	*Unit*	*Symbol*
Length	Meter	m
Mass	Kilogram	kg
Time	Second	s
Electric current	Ampere	A
Thermodynamic temperature	Kelvin	K
Amount of substance	Mole	mol
Luminous intensity	Candela	cd

The American National Standards Institute (ANSI) provides authorized translations of the original French definitions of the seven base units and they are given as follows:

Meter/The meter is the length equal to 1 650 763.73 wavelengths in vacuum of the radiation corresponding to the transition between the levels $2p_{10}$ and $5d_5$ of the krypton-86 atom. [Adopted by 11th CGPM (Conférence Générale des Poids et Measures) 1960]

Kilogram/The kilogram is the unit of mass; it is equal to the mass of the international prototype of the kilogram. (adopted by 1st and 3rd CGPM 1889 and 1901)

Second/The second is the duration of 9 192 631 770 periods of the radiation corresponding to the transition between the two hyperfine levels of the ground state of the cesium-133 atom. (adopted by 13th CGPM 1967)

Ampere/The ampere is that constant current, which, if maintained in two straight parallel conductors of infinite length, of negligible circular cross section, and placed one meter apart in vacuum, would produce between these conductors a force equal to 2×10^{-7} newton per meter of length. (adopted by 9th CGPM 1948)

Kelvin/The kelvin, unit of thermodynamic temperature, is the fraction $\frac{1}{273.16}$ of the thermodynamic temperature of the triple point of water. (adopted by 13th CGPM 1967)

Mole/The mole is the amount of substance of a system that contains as many elementary entities as there are atoms in 0.012 kilogram of carbon-12. (adopted by 14th CGPM 1971)

Candela/The candela is the luminous intensity, in the perpendicular direction, of a surface of $\frac{1}{600\,000}$ square meter of blackbody at the temperature of freezing platinum under a pressure of 101 325 newtons per square meter. (adopted by 13th CGPM 1967)

The *supplementary units* in the SI system are identified in Table 2.4 and are defined by ANSI as follows:

Radian/The radian is the plane angle between two radii of

TABLE 2.4. Supplementary SI Units

Quantity	*Name of Unit*	*Symbol*
Plane angle	Radian	rad
Solid angle	Steradian	sr

TABLE 2.5. SI Derived Units with Special Names

	SI Unit		
Quantity	*Name*	*Symbol*	*Expression in Terms of Other Units*
Frequency	Hertz	Hz	s^{-1}
Force	Newton	N	$kg \cdot m/s^2$
Pressure, stress	Pascal	Pa	N/m^2
Energy, work, quantity of heat	Joule	J	$N \cdot m$
Power, radiant flux	Watt	W	J/s
Quantity of electricity, electric charge	Coulomb	C	$A \cdot s$
Electric potential, potential difference, electromotive force	Volt	V	W/A
Capacitance	Farad	F	C/V
Electric resistance	Ohm	Ω	V/A
Conductance	Siemens	S	A/V
Magnetic flux	Weber	Wb	$V \cdot s$
Magnetic flux density	Tesla	T	Wb/m^2
Inductance	Henry	H	Wb/A
Luminous flux	Lumen	lm	$cd \cdot sr$
Illuminance	Lux	lx	lm/m^2
Activity (of ionizing radiation source)	Becquerel	Bq	s^{-1}
Absorbed dose	Gray	Gy	J/kg

a circle that cut off on the circumference an arc equal in length to the radius.

Steradian/The steradian is the solid angle, which, having its vertex in the center of a sphere, cuts off an area of the surface of the sphere equal to that of a square with sides of length equal to the radius of the sphere.

Various combinations of base and supplementary units are used to provide so-called *derived units* for derived dimensions. Any dimension other than those identified as fundamental or base have derived units and may be expressed by multiplying and dividing the base and supplementary units. For example, linear and angular velocity may be expressed as m/s and rad/s, respectively. Some of the derived dimensions that have special names, symbols and units that have been officially approved by CGPM are given in Table 2.5.

Some quantities and their derived SI units that are commonly used in science, yet not officially adopted by CGPM are given in Table 2.6.

In Table 2.7, a set of prefixes is given that may be used to form multiples and submultiples of SI units conveniently. For example, $3(10)^3$ meters may be written as 3 kilometers (3 km) and $3(10)^{-3}$ meters as 3 millimeters (3 mm). The kilogram is the only SI basic unit with a prefix. Since double prefixes should be avoided, the prefixes for mass should be assigned to the gram and not the kilogram.

2.6/SI Language

The names, symbols, and units of SI are international and are identical in all languages. There are SI rules for proper pronunciation, spelling, and punctuation. Some selected guidelines recommended by the American National Metric Council are presented as follows.

1. Unit names, including prefixes, are not capitalized except at the beginning of a sentence or in titles. Note that in degree Celsius the unit "degree" is lowercase. (In the term "degree Celsius," "degree" is considered to be the unit name, modified by the adjective "Celsius," which is always capitalized. The "degree centigrade" is now obsolete).

2. The short forms for metric units are called symbols. They are lowercase except that the first letter is uppercase when the name of the unit is derived from the name of a person.

Examples.

Unit Name	*Unit Symbol*
Gram	g
Newton	N
Pascal	Pa

3. Names of units are plural when appropriate.

   Examples. 1 meter, 100 meters, zero degrees Celsius.

4. Values less than 1 take the singular form of the unit name.

   Examples. 0.5 kilogram or $\frac{1}{2}$ kilogram

5. Symbols of units are the same in singular and plural when abbreviated.

   Examples. 1 m, 100 m

6. A period is *not* used after a symbol, except at the end of a sentence.

   Examples. A current of 15 mA is found . . .
   The field measured 350 × 125 m.

7. Separate digits into groups of three, counting from the decimal marker. A comma should not be used. Instead, a space is left to avoid confusion, because many countries use a comma for the decimal marker. In a four-digit number, the space is not required unless the four-digit number is in a column with numbers of five digits or more.

   Examples. For 4,720,525 write 4 720 525
   For 0.52875 write 0.528 75
   For 6,875 write 6875 or 6 875

8. In symbols or names for units having prefixes, no space is left between letters making up the symbol or the name.

   Examples. kA, kiloampere; mg, milligram

9. When a symbol follows a number to which it refers, a space must be left between the number and the symbol, except when the symbol (such as °) appears in the superscript position. The symbol for degree Celsius may be written either with or without a space before the degree symbol.

   Examples. 455 kHz, 22 mg, 20 mm, 10^6 N, 30°, 20°C or 20 °C

10. When a quantity is used in an adjectival sense, a hyphen should be used between the number and the symbol (except ° and "C").

TABLE 2.6. Some Common Derived Units of SI

Quantity	*Unit*	*Symbol*
Acceleration	Meter per second squared	m/s^2
Angular acceleration	Radian per second squared	rad/s^2
Angular velocity	Radian per second	rad/s
Area	Square meter	m^2
Concentration (of amount of substance)	Mole per cubic meter	mol/m^3
Current density	Ampere per square meter	A/m^2
Density, mass	Kilogram per cubic meter	kg/m^3
Electric charge density	Coulomb per cubic meter	C/m^3
Electric field strength	Volt per meter	V/m
Electric flux density	Coulomb per square meter	C/m^2
Energy density	Joule per cubic meter	J/m^3
Entropy	Joule per kelvin	J/K
Heat capacity	Joule per kelvin	J/K
Heat flux density } Irradiance	Watt per square meter	W/m^2
Luminance	Candela per square meter	cd/m^2
Magnetic field strength	Ampere per meter	A/m
Molar energy	Joule per mole	J/mol
Molar entropy	Joule per mole kelvin	J/(mol·K)
Molar heat capacity	Joule per mole kelvin	J/(mol·K)
Moment of force	Newton meter	N·m
Permeability	Henry per meter	H/m
Permittivity	Farad per meter	F/m
Radiance	Watt per square meter steradian	$W/(m^2 \cdot sr)$
Radiant intensity	Watt per steradian	W/sr
Specific heat capacity	Joule per kilogram kelvin	J/(kg·K)
Specific energy	Joule per kilogram	J/kg
Specific entropy	Joule per kilogram kelvin	J/(kg·K)
Specific volume	Cubic meter per kilogram	m^3/kg
Surface tension	Newton per meter	N/m
Thermal conductivity	Watt per meter kelvin	W/(m·K)
Velocity	Meter per second	m/s
Viscosity, dynamic	Pascal second	Pa·s
Viscosity, kinematic	Square meter per second	m^2/s
Volume	Cubic meter	m^3
Wavenumber	One per meter	1/m

TABLE 2.7. SI Prefixes

Factor	*Prefix*	*Symbol*
10^{18}	Exa	E
10^{15}	Peta	P
10^{12}	Tera	T
10^{9}	Giga	G
10^{6}	Mega	M
10^{3}	Kilo	k
10^{2}	Hecto	h
10^{1}	Deka	da
10^{-1}	Deci	d
10^{-2}	Centi	c
10^{-3}	Milli	m
10^{-6}	Micro	μ
10^{-9}	Nano	n
10^{-12}	Pico	p
10^{-15}	Femto	f
10^{-18}	Atto	a

Examples. It is a 35-mm film; but the film width is 35 mm. I bought a 6-kg turkey; but the turkey weighs 6 kg.

11. Leave a space on each side of signs for multiplication, division, addition, and subtraction.

    Examples. 4 m × 3 m (not 4 m×3 m); kg/m^3, N·m

12. When writing symbols for units such as square meter or cubic

centimeter, write the symbol for the unit, followed by the superscript ^{2} or ^{3}, respectively.

Examples. For 14 square meters, write 14 m^2
For 26 cubic centimeters, write 26 cm^3

13. For a unit symbol derived as a quotient—for example, km/h—do not write k.p.h. or kph because these are understood only in the English language, whereas km/h is used in all languages. The symbol km/h can also be written by means of a negative exponent—for example, $km \cdot h^{-1}$. Do not use more than one slash (/) in any combination of symbols unless parentheses are used to avoid ambiguity.

   Examples. m/s^2, not m/s/s; W/(m·K), not W/m/K

14. For a unit symbol derived as a product, use a product dot—for example, N·m.

15. Do not mix nonmetric units with metric units, except units for time, plane angle, or rotation.

   Example. Use kg/m^3, not kg/ft^3

16. Do not use a mixture of prefixes, unless the difference in size is extreme.

   Examples. Use 40 mm wide and 1500 mm long, not 40 mm wide and 1.5 m long; but 1500 meters of 2-mm diameter wire.

2.7 / Conversion of Units

Engineers frequently substitute numbers into equations and solve problems. In order to obtain a correct solution, one must make certain that the equation is dimensionally homogeneous. The dimensions and units on the left-hand side of the equation must be identical to the right hand side. Furthermore, if an equation adds or subtracts several terms, each term must be identical in both dimensions and units. One cannot add length and time, force and mass, meters and feet, or pounds and seconds. To blindly substitute a number such as 6 into an equation without regards to dimensions and units is like a deposit of six something (pennies, dollars, elephants or chickens) into a bank account. It sounds rather stupid, yet beginning and experienced engineers often overlook homogeneity of units. If you suspect a mistake in a solution, then check for an inadvertent error in units.

In converting numbers from one system of units to another,

it is important that the accuracy is not erroneously reduced or exaggerated. If one measured a distance to the nearest foot, as 12 ft, then it would not be appropriate to convert 12 ft to 3.6576 m. The measurement is not that accurate. The correct conversion from feet to meters would contain the same number of significant digits and be 3.7 m. Conversion factors are often given more accurately than the measurements. When converting units, multiply or divide by the conversion factor first, and then round off the result to keep the same number of significant digits as the original measurement. In the case of multiple measurements, the product or quotient should contain no more significant digits than are contained in the least accurate measurement. For example, the arca of a rectangle with sides of 123.2 m and 1.31 m would be found by taking the product $123.2 \times 1.21 = 161.392$. However, the area should be rounded to 161 m^2. Several unit conversions are given in Appendix E. Some example problems follow.

EXAMPLE PROBLEM 2.6
Determine the number of seconds in a century.

Solution
(100 years/century)(365 days/year)(24 hr/day)
(60 minutes/hour)(60 seconds/minute) =
3 153 600 000 s/century

Notice how the similar units in the numerator and denominator algebraically cancel.

EXAMPLE PROBLEM 2.7
Determine the volume of a room 20 m long by 10 m wide by 3 m high.

Solution
The volume is

$$v = (20 \text{ m})(10 \text{ m})(3 \text{ m}) = 600 \text{ m}^3$$

Notice how the units are multiplied together.

EXAMPLE PROBLEM 2.8
It has been established that a volume of 1 ft^3 is equivalent to 7.48 gallons. How many gallons of water in a swimming pool 60 ft long by 30 ft wide by 5 ft deep?

Solution
The volume of the pool is 9000 ft^3.

$$9000 \text{ ft}^3 (7.48 \text{ gallons/ft}^3) = 67\ 328 \text{ gallons}$$

EXAMPLE PROBLEM 2.9

In a certain engineering calculation, it is required to multiply 80 newtons by 50 meters and divide that product by 20 seconds. What would be the units of the resulting number?

Solution

$$(80 \text{ N})(50 \text{ m})/20 \text{ s} = 200 \text{ N}\cdot\text{m/s}$$

EXAMPLE PROBLEM 2.10

It has been established that 1 horsepower (hp) is equivalent to 550 ft·lbf/s. What is the engine output in horsepower if a certain engine produces 18×10^6 ft·lbf/s?

Solution

$$(18 \times 10^6 \text{ ft}\cdot\text{lbf/s})\left(\frac{1 \text{ hp}}{550 \text{ ft}\cdot\text{lbf/s}}\right) = 32\,727.3 \text{ hp}$$

Again it is instructive for the student to carry out the indicated mathematical operations with the units to be certain the process is clear.

PROBLEMS

2.1. Explain the difference between dimensions and units.

2.2. Attempt a definition of (a) time and (b) length.

2.3. List the proper units for mass, length, time, and force in both the SI, American engineering, and British system of units. Use correct symbols for each unit.

2.4. Using the appropriate prefix and symbol correctly, express the following: (a) one million kilograms, (b) one ten thousandth of a meter, (c) ten meters, (d) one million newtons, (e) one thousandth of a meter, (f) thirty-six hundred seconds, (g) one thousand milliseconds, (h) one millionth of a second.

2.5. Write the following numbers in appropriate SI notation: (a) one thousand, (b) ten thousand, (c) ten million, (d) one tenth (accurate to two decimal places), (e) one hundredth (accurate to three decimal places), (f) one ten thousandth (accurate to four decimal places), (g) one ten thousandth (accurate to five decimal places).

2.6. Explain the difference between weight and mass.

2.7. Determine the weight in N of a 75-kg person at sea level.

2.8. Determine the weight in lbf of a 150-lbm person. Assume the acceleration due to gravity is 32.17 ft/s^2.

2.9. Calculate the net force in N required to accelerate a mass of 1000 kg at 5 m/s^2 in the direction of the net force.

2.10. Calculate the net force in lbf required to accelerate a mass of 1000 lbm at 5 ft/s^2 in the direction of the net force.

2.11. Explain why the weight in lbf of a body on the surface of the earth is approximately numerically equal to the mass in lbm while the weight in N of a body on the surface of the earth is approximately 9.8 times as great as its mass in kilograms.

2.12. Is a temperature of 0 R identical to a temperature of 0 K? Is a temperature of 0°C identical to a temperature of 0 F?

2.13. Which temperature scale has the largest size division or degree?

2.14. A specialized computer is to be housed in a room maintained at 20°C. The thermostats that control the computer room temperature are calibrated in F. What temperature should the thermostats be set at?

2.15. Complete the blanks in the following table:

C	K	R	F
10	____	____	____
____	10	____	____
____	____	10	____
____	____	____	10

2.16. Three railroad trains per day leave Apple County. In each train there are five railroad cars loaded with apples. Each car contains 750 boxes of apples. Each box contains 300 apples. Calculate the number of apples per day leaving Apple County by train. Write the units associated with each number as you make your calculations.

2.17. It has been established that 1 kg is equivalent to 2.204 6 lbm. It has also been established that 1 ft is equivalent to 0.304 8 m. A certain substance has a mass density of 1000 kg/m^3. Using only the given information, express the mass density of this substance in units of lbm/ft^3. Write out units associated with each number as you make your calculations.

2.18. One gallon is equivalent to 3.785 L (liters). One cubic meter is equivalent to 1000 L. One cubic foot is equivalent to

0.028 3 m^3. One cubic foot is equivalent to 7.48 gallons. Using only the given information, determine the number of gallons per cubic meter. Write out all units.

2.19. One acre is equivalent to an area of 43 560 ft^2. One hectare is equivalent to an area of 10 000 m^2. Noting from Problem 2.2 that 1 ft is equivalent to 0.304 8 m determine the number of acres per hectare and the number of hectares per acre. Use only the given information and write all units associated with each other.

2.20. One mile is equivalent to 5280 ft. One hour is equivalent to 3600 s. One ft is equivalent to 0.3048 m. Express a velocity of 55 miles per hour in units of (a) km/h, (b) ft/s, and (c) m/s.

2.21. A certain animal weighs 1200 lb at sea level. Determine the mass of the animal in (a) slugs, (b) kilograms, and (c) pound mass.

2.22. A table of material properties indicates that aluminum weighs 0.098 lb/$in.^3$ at sea level. Determine the total weight of an aluminum cube with sides equal to 25 cm at sea level. Also, determine the mass of the aluminum cube in kilograms.

2.23. A certain material costs $10/lbm. Express this cost in dollars per kilogram.

CHAPTER THREE

ENERGY AND THE ENGINEER[1]

PART A/FUNDAMENTAL ENERGY CONCEPTS

3.1/A Serious Predicament

Life on earth is sustained by a large thermonuclear device traveling the outer fringes of the Milky Way Galaxy. By converting hydrogen to helium, this immense fireball, called the sun, with a surface temperature near 6000 K radiates energy at the rate of 74 000 kilowatts per square meter. Tiny earth, in orbit some 149 million kilometers away, is literally drenched with solar energy. Radiating in all directions, some two billion photons (solar energy bundles) must leave the sun for each one that strikes the earth (Figure 3.1). Of each 100 photons that do strike the earth, about 30 are deflected directly to outer space, approximately 47 are absorbed by the atmosphere, the oceans, and the earth's surface only to be reradiated to deep space. The hydrologic cycle, in which water is evaporated from oceans, lakes, and streams, requires almost 23 of the 100 incident photons. Two are used to drive atmospheric and ocean currents. This energy too is soon dissipated by friction into heat to be lost in space. An even smaller number of the incident solar photons, approximately 0.02 of every 100, are trapped through the miracle of photosynthesis, and eventually stored through sedimentation. Some small fraction of this organic material has been converted to peat, coal, oil, and natural gas.

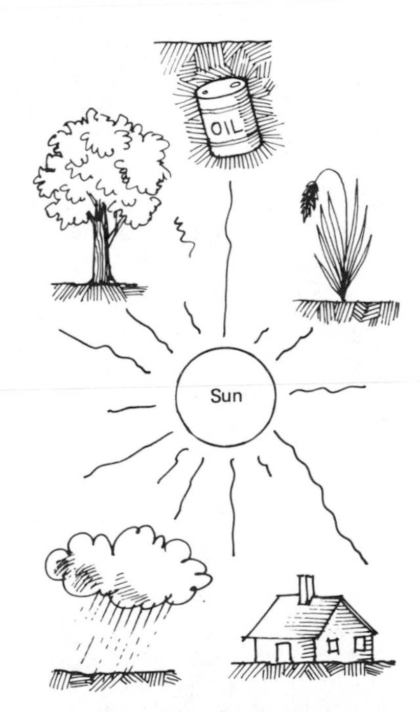

FIG. 3.1. The sun is the primary source of energy for the earth.

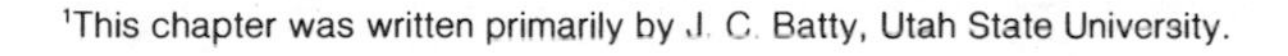
[1]This chapter was written primarily by J. C. Batty, Utah State University.

There are natural energy flows to the earth's surface on which we live that are not traceable directly to the sun (Figure 3.2). Thermal energy from the hot interior of the earth seems to originate from both residual energy left over from the planet's fiery beginnings and ongoing nuclear reactions deep in the earth that continue to release heat. The energy that drives the tides is principally derived from the spinning of the earth.

The energy thus lost through the rising and falling of the tides is inexorably reducing the rotational speed of the earth. The period of rotation is increasing at a less than alarming rate of 1 second every 120 000 years or so.

The magnitude of these nonsolar contributions to the earth's energy budget is minor in comparison with solar sources. For each 100 photons of sunshine striking the earth, the contributions from the earths interior and tidal energies would be equivalent to about 0.018 and 0.0017 photons, respectively.

We have known for a long time how to indirectly utilize the sun's energy. The winds driven by solar energy have been harnessed to drive an almost infinite variety of windmills. Water, lifted from the oceans by the sun, turned innumerable waterwheels and turbines on its way back to the seas. Untold numbers of fires have been kindled from wood to cook our food and keep us warm.

Within the last 200 years or so, we have been clever enough to build fires and arrange things such that the fire would cause a shaft to spin. (See Figure 3.3.) A device that utilizes a fire to spin a shaft is usually called an engine and people who build them are called engineers. The first engines were clumsy, inefficient devices that converted only about 1 percent of the energy stored

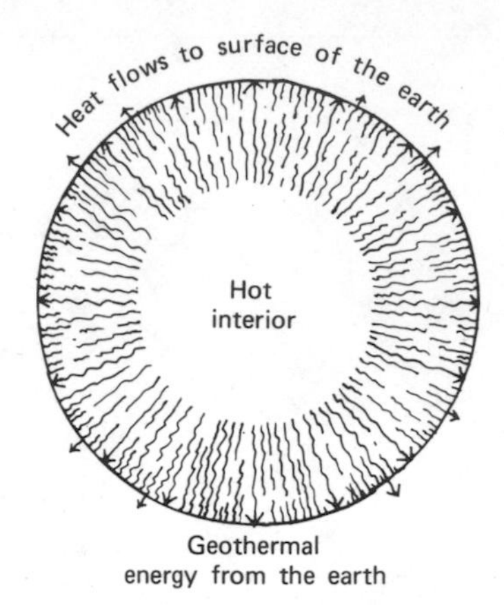

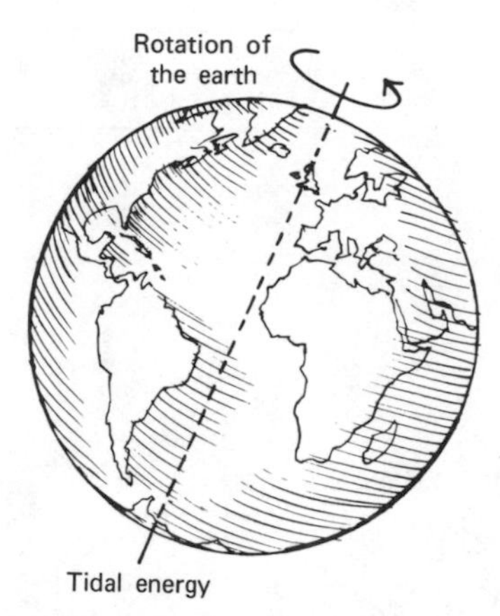

FIG. 3.2. Nonsolar contributions to the earth's energy budget.

FIG. 3.3. The sun's energy is utilized indirectly in the form of wind, falling water, and fire.

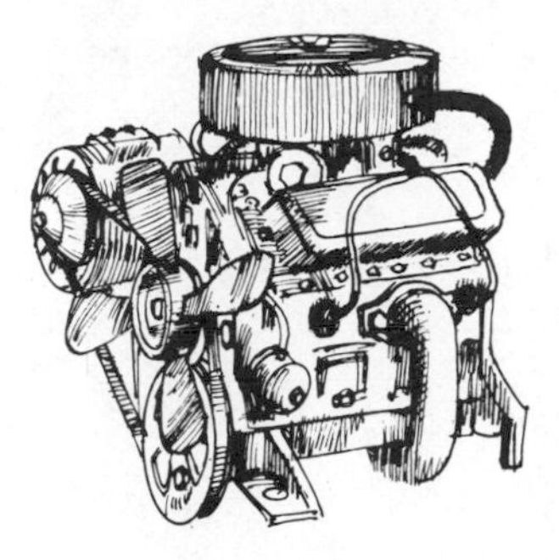

FIG. 3.4. Energy-conversion devices.

within the wood or coal they burned into useful work. But in an incredibly short time, engineers became so skillful in engine building that our entire society was transformed.

Industrialized society's appetite for energy conversion devices and the fuel to burn in them became insatiable. Great industries developed to recover vast deposits of coal and oil and natural gas. In a technological spiral more engines demanded more and better fuels. Better fuels, in turn, spawned more sophisticated energy conversion devices to be demanded in ever increasing numbers by society.

We came to depend on engines and energy conversion devices of all kinds to plow our fields; plant, cultivate, and harvest our crops; process our food; weave our cloth; wash our clothes; construct our buildings; warm us or cool us as necessary; transport us about; entertain us; brush our teeth; and even do our arithmetic. Our present life-style is almost totally centered about energy-conversion devices. These devices include our electrical power generators, automobiles, tractors, airplanes, washing machines, refrigerators, TVs, and calculators (Figure 3.4). They are our slaves. Almost every elite society has had slaves in one form or another and our society is no exception. Our slaves, machines, and the fuel that drives them have enabled even the average person to live better than royalty of old.

We are apparently running out of the miracle fuels—oil and natural gas. It required untold millenia to deposit the fossil fuels, but it seems probable that 80 percent of all the readily recoverable oil and natural gas in the United States will be burned within the period of our lifetime (Figure 3.5). This then is our predicament. We are hooked on oil and natural gas just as a dope addict is hooked on drugs. But, our supply is limited. Just as the dope addict is willing to go to almost any lengths to obtain drugs, we seem to be willing to do almost anything to keep our oil-burning machines running. Clearly we must develop alternative energy sources and improve our energy conversion devices.

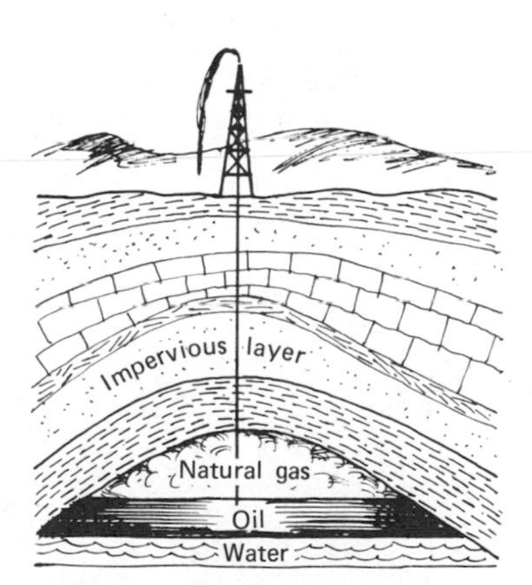

FIG. 3.5. Natural gas and oil are deposited beneath the earth's crust.

We seem to have the following five basic choices for energy development.

1. Solar energy used directly.
2. Solar energy used indirectly (including fossil fuels, biomass, wind, ocean current, hydropower).
3. Tidal energy.
4. Geothermal energy.
5. Nuclear.

The technical challenges associated with effectively converting these energy sources to useful forms are formidable but tremendously exciting. Meeting the energy challenges will keep an army of engineers occupied for a long time. The best minds available must be put to the task of easing the difficulties associated with our energy predicament. The very existence of our society depends on it.

Later in this chapter we examine in detail some innovative energy conversion concepts. First, however, we must develop some problem-solving capability in energy basics and learn to precisely use some engineering terms.

3.2/Systems, Surroundings, and Interaction

Engineers generally refer to that object upon which their attention is focused as the *system*. Everything not in the system is regarded as the *surroundings*. Thus the system together with the surroundings constitutes the entire *universe!* Engineers generally separate the *system* from the *surroundings* by a *boundary*. The *boundary* is an imaginary surface often represented by a dashed line in a sketch of the system, as shown in Figure 3.6*a*. We describe how the surroundings interacts with the system by closely watching what happens at the boundary. No mass crosses the boundary of a *closed system* (Figure 3.6*b*). Mass may enter or leave an *open system* by crossing the boundary (Figure 3.6*c*). Examples of closed systems include a bouncing ball, a pile of bricks, an oxygen molecule, or the earth (if entering meteorites or leaving rocket ships are ignored). Examples of open systems include an automobile engine, a hydropower dam, a water pump, a gas turbine, a busy escalator, a furnace, or the earth (if entering meteorites and leaving rocket ships are not ignored). Basically the only way

Boundary

System

Surroundings

(*a*)

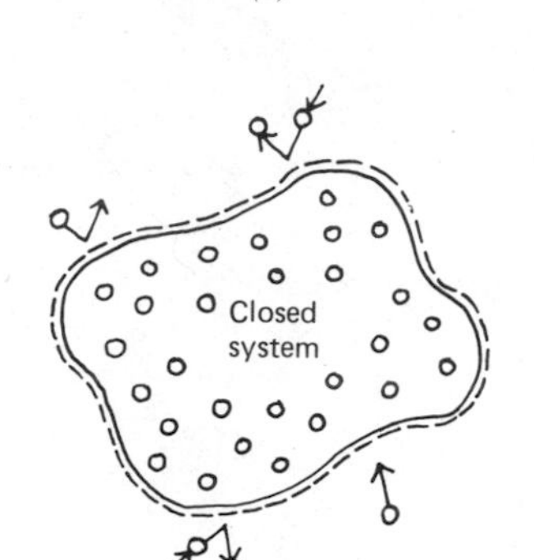

(*b*)

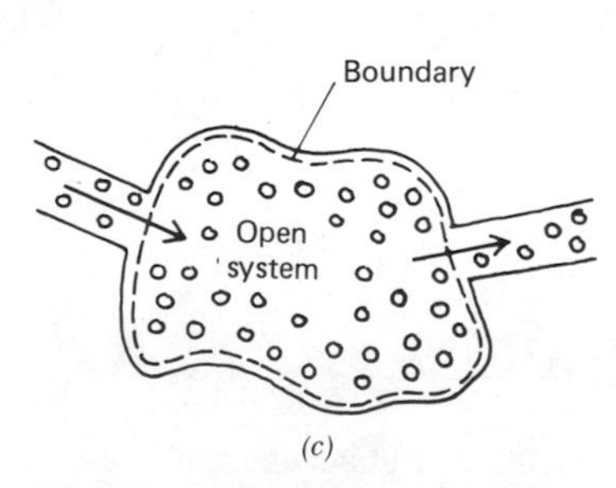

(*c*)

FIG. 3.6. Systems and surroundings.

FIG. 3.7. Interactions between systems and surroundings.

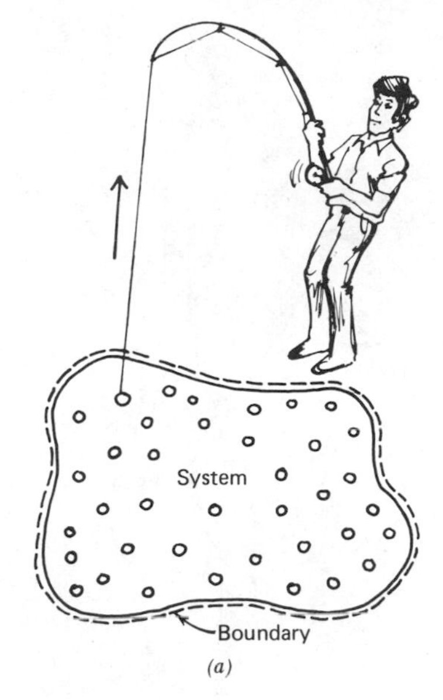

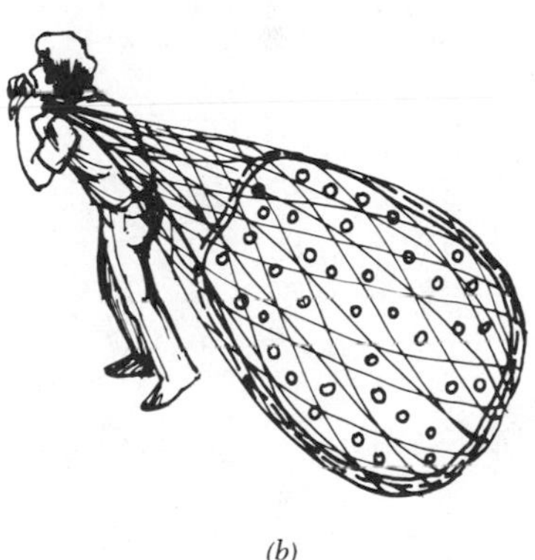

the surroundings can interact with a closed system is to strike it or push it or pull it or tug at it in some fashion (Figure 3.7*a*). Consider the ways in which you, as a system, interact with your surroundings. If a person speaks to you, you are bombarded with sound waves. You see because photons strike your eyes. If you touch something it pushes against you.

Interactions between any system and its surroundings may usually be identified as either work or heat. Interactions that are highly organized such that many molecules are moving in the same direction at the same time are usually called *work interactions.* A hammer striking a nail or your hand lifting a weight are examples of work interactions. Both involve many molecules organized to move in the same direction at the same time (Figure 3.7*b*). On the other hand, if the hits, tugs, or strikes of the interaction are disorganized and on a microscopic scale, then we usually describe it as a *heat interaction.* A lighted match held under an object will not cause that object to rise because the strikes, hits, or pushes of the hot gaseous flame particles lack organization (Figure 3.8). They are not moving in the same direction at the same time. Thus we say interactions between a closed system and its surroundings are classified as work and/or heat.

3.3/Work and Power

When a force is exerted against the moving boundary of a system we generally call that a work interaction. The amount of *work* involved is expressed as the product of the force times the distance that force moves (Figure 3.9).

Work = (force)(distance)

or $W = F(\Delta x)$

where work is denoted by the symbol W. The work is positive if the force and displacement are in the same direction. The work is negative if force and displacement have opposite directions.

EXAMPLE PROBLEM 3.1

Compute the work required to lift a weight of 700 N through a vertical distance of 8 m.

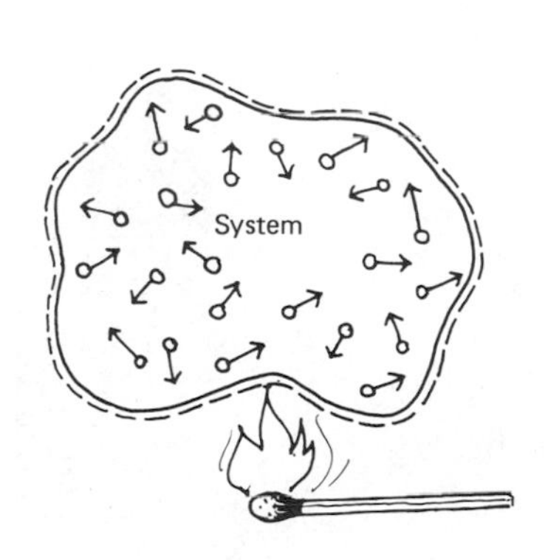

FIG. 3.8. Heat interaction at the boundary of a closed system.

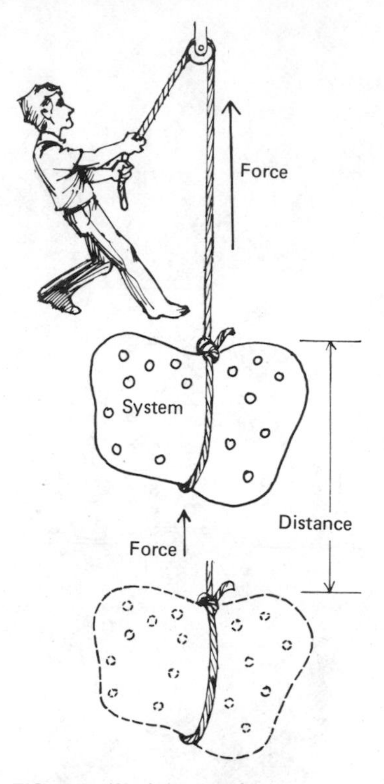

FIG. 3.9. Work interaction.

Solution

$$W = F(\Delta x)$$

$$= (700\ \text{N})(8\ \text{m}) = 5600\ \text{N}\cdot\text{m}$$

The basic unit of work in the SI system is the newton-meter. The newton-meter is also known as the joule.

$$1\ \text{N}\cdot\text{m} = 1\ \text{J}$$

EXAMPLE PROBLEM 3.2

Compute the work required to lift a weight of 150 lbf through a vertical distance of 25 ft.

Solution

$$W = F(\Delta x)$$

$$= 150\ \text{lbf}\ (25\ \text{ft}) = 3750\ \text{ft}\cdot\text{lbf}$$

The unit of work in the engineering system is the foot·lbf and has no alias. Notice also that we usually write ft·lbf rather than lbf·ft, which would seem to be more consistent with N·m.

The average *power* associated with a process is the quantity of work done divided by the time required for that quantity of work to be done. In other words, power is the time rate of doing work. We usually denote average power with the symbol $\dot{W}$. Thus

$$\dot{W} = \frac{W}{\Delta t}$$

where Δt is the change of time required to accomplish the amount of work W.

FIG. 3.10. Example Problem 3.3.

EXAMPLE PROBLEM 3.3

A certain person is assigned the task of lifting blocks weighing 200 N each from the floor onto a table 1.5 m above the floor (Figure 3.10). That person is able to lift 12 blocks in 1 minute. Compute the power output of the block lifter.

Solution

Using appropriate units we write:

$$\dot{W} = (12\ \text{blocks/min})(200\ \text{N/block})(1.5\ \text{m}) \times (1\ \text{min}/60\ \text{s})(1\ \text{J/N}\cdot\text{m}) = 60\ \text{J/s}$$

One joule per second is called a watt.

$$1\ \text{J/s} = 1\ \text{watt} = 1\ \text{W}$$

The power output of the block-lifting person is thus computed as 60 W.

EXAMPLE PROBLEM 3.4

If the person described in the previous example could maintain a power output of 60 W for a time of 2 hours, how much work would be done?

Solution

Noting that power is the time rate of doing work,

$$\dot{W} = \frac{W}{\Delta t}$$

it should be obvious that

$$W = \dot{W}(\Delta t)$$

Therefore

$$W = (60\ \text{J/s})(2\ \text{h})(3600\ \text{s/h})$$

$$W = 432\,000\ \text{J} = 432\ \text{kJ}$$

It is also customary to express work in units of watt-hours or kilowatt-hours. Thus

$$W = 60\ \text{W}(2\ \text{h})$$

$$= 120\ \text{W}\cdot\text{h} = 0.12\ \text{kW}\cdot\text{h}$$

EXAMPLE PROBLEM 3.5

A certain person is assigned the task of lifting blocks weighing 45 lbf each from the floor onto a table 4.92 ft above the floor. That person is able to lift 12 blocks from the floor to the table in one minute. Compute the power output of the block lifter.

Solution

$$\dot{W} = (12\ \text{blocks/60 s})(45\ \text{lbf/block})(4.92\ \text{ft}) = 44.3\ \text{ft}\cdot\text{lbf/s}$$

A power output of 550 ft·lbf/s is equivalent to 1 horsepower.

$$550\ \text{ft}\cdot\text{lbf/s} = 1\ \text{hp}$$

Therefore the power output of this block-lifting person is

$$\dot{W} = (44.3\ \text{ft}\cdot\text{lbf/s})\left(\frac{1\ \text{hp}}{550\ \text{ft}\cdot\text{lbf/s}}\right)$$

$$= 0.08\ \text{hp}$$

EXAMPLE PROBLEM 3.6

Suppose the person described in the previous example could maintain that same rate of work output for two hours. How much work would be done during the two-hour period?

Solution

$$W = \dot{W}(\Delta t)$$
$$= 44.3(\text{ft}\cdot\text{lbf/s})(2\ \text{h})(3600\ \text{s/h})$$
$$= 318\,960\ \text{ft}\cdot\text{lbf}$$

or

$$W = (0.08\ \text{hp})(2\ \text{h}) = 0.16\ \text{hp}\cdot\text{h}$$

In the engineering system of units, work is often expressed in ft·lbf or hp·h (horsepower hours).

PROBLEMS

3.1. The five basic energy resource categories mentioned include (a) solar energy used directly and (b) solar energy used indirectly. In which of these categories would each of the following energy conversion systems fit?

(a) Windmill
(b) Photovoltaic cell
(c) Flat-plate solar collector
(d) Alcohol from sugarbeets
(e) Alcohol from natural gas
(f) Passive solar home heating system
(g) Coal combustion
(h) Device to harness ocean currents
(i) A hydropower plant
(j) Natural-gas heating system

3.2. The earth's tidal power is estimated at 3 TW (terawatts). Using the information given in the text, estimate the rate at which solar energy strikes the earth. Express your answer in TW. What is the ratio of the rate at which solar energy strikes the earth to the rate of energy released from burning fossil fuels and operating nuclear reactors ($\sim$10 TW)?

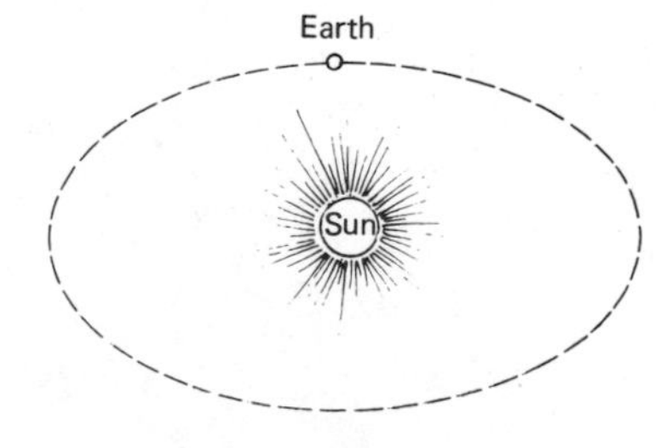

3.3. Noting that the earth's orbit about the sun has a radius of approximately 149×10^6 km and the earth's diameter is about 1.288×10^4 km, show that roughly two billion photons leave the sun for each photon that is intercepted by the earth. Assume that the sun radiates uniformly in all directions. The radius of the sun is 0.696×10^6 km.

3.4. How would you define the terms (a) system, (b) closed system, (c) open system? What is the basic difference between heat and work?

3.5. Discuss the difference between work and power? Which of the following are units of work and which are units of power?

(a) ft·lbf
(b) kW
(c) kW·h
(d) hp
(e) hp·h
(f) kJ
(g) kJ/s
(h) ft·lbf/s
(i) N·m

3.6. Calculate the work required to lift a loaded aircraft weighing 1 000 000 N to an altitude of 10 000 m.

3.7. Suppose your automobile stalls and you are required to push it. You find that by exerting a steady force of 200 lbf against the rear bumper, the car will roll along a level street. How much work is required to push the automobile 500 ft along a level street to a service station?

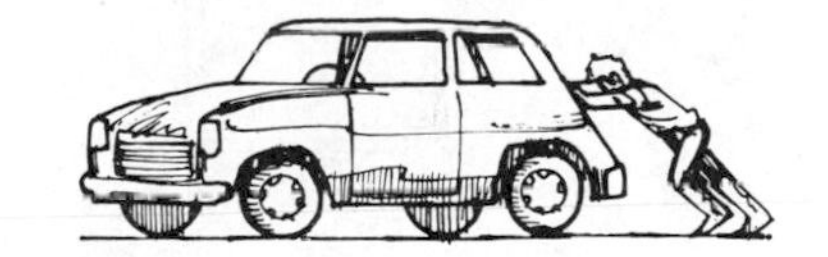

3.8. If you are able to push the car described in Problem 3.7 a distance of 500 ft in 5 min, calculate your power output. Express your answer in horsepower.

3.9. Show that 1 hp = 0.746 kW.

3.10. An aircraft weighing 10^6 N is expected to require only 200 s following lift off to reach an altitude of 10 000 m. What power is required to accomplish this? Express your answer in kilowatts and also in horsepower.

3.11. Perhaps you observed from working several of the previous problems that in some circumstances power may be computed as

$$\dot{W} = \text{(force)(distance/time)}$$

Notice that distance/time may be interpreted as velocity. Therefore

$$\dot{W} = \text{force (velocity)}$$

Based on these observations, calculate the power required to lift a loaded elevator weighing 20 000 N vertically upward at a uniform velocity of 1 m/s. Also calculate the power required to lift

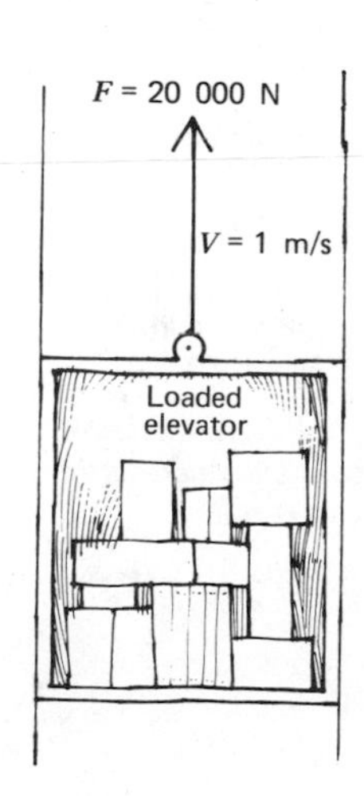

the same elevator vertically upward at a uniform rate of 10 m/s. Express your answer in kilowatts and in horsepower. Explain why it requires more power to lift things fast?

3.12. In order to estimate the total wind resistance, tire rolling resistance, and friction resistance in an automobile, the power output of the engine is carefully measured while driving along a perfectly level highway. It is found that the engine delivered 60 kW while traveling at 90 km/h. Calculate the total resistive force against the automobile. Express your answer in newtons. (Hint: see Problem 3.11.)

3.4/The Heat Interaction

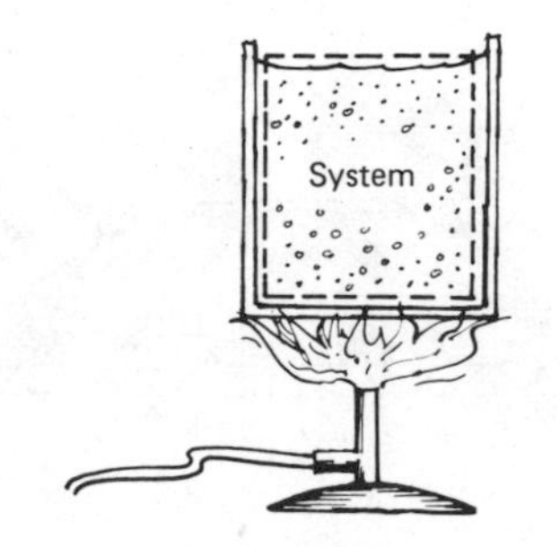

FIG. 3.11. Heat applied to a system increases the temperature.

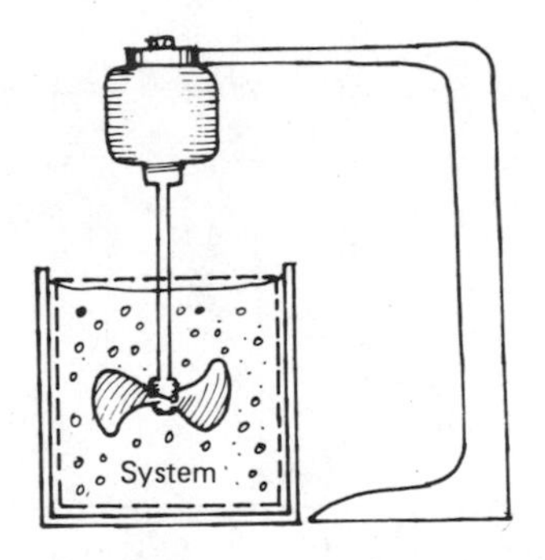

FIG. 3.12. Work interaction may increase the temperature of the water within the system.

Consider an experiment in which one kilogram of water in a container is placed over a flame, as suggested by Figure 3.11. We will treat the water as the system. Obviously this process involves a heat interaction between the system (water) and the surroundings (flame). This heat interaction causes the temperature of the system to increase. Now consider a similar container of 1 kg of water equipped with a stirring device as indicated in Figure 3.12. This process involves a work interaction between the system and the surroundings, but we would observe the temperature of the water increasing just as if the water were over a flame. Now suppose we carefully measured the amount of work which was done by the surroundings (turning shaft) on the system (water) and carefully measured the accompanying increase in temperature of the water. We would find that about 4.187 kJ of work would be required to increase the temperature of 1 kg of water by 1°C. The work and heat interactions caused a similar effect on the temperature of the water and thus it is not unusual to express heat in the same units as work. In the SI system, heat is measured in joules or kilojoules, just as work is. In the engineering system the customary unit of heat is the British thermal unit (Btu). The Btu is defined as the amount of work or heat required to increase the temperature of 1 lbm of water by 1 F. In a heating-stirring experiment, similar to that described in Figure 3.12, we would find that 778.16 ft·lbf of work would be equivalent to 1 Btu.

Another common unit is the kilocalorie. One kilocalorie is defined as the work or heat required to increase the temperature of one kilogram of water by 1°C. The calorie used in a food and nutrition context is generally equivalent to the kilocalorie defined here. Although based on the metric system, the kilocalorie is not an acceptable unit in the SI system.

EXAMPLE PROBLEM 3.7

We want to increase the temperature of 1 kg of water from 20 to 30°C. What would be the magnitude of the heat interaction required to accomplish this?

Solution

We have observed that 4.187 kJ would increase the temperature of 1 kg of water by 1°C. Therefore, we conclude that approximately 2(4.187 kJ) would be required to increase the temperature of 1 kg of water by 2°C. Similarly, a temperature rise of 3°C would require 3(4.187 kJ), 4°C would require 4(4.187 kJ), and so forth. Based on this logic, the heat required to increase the temperature of 1 kg of water from 20°C to 60°C would be

$$Q = (40°\text{C})(4.187\ \text{kJ/kg}\cdot°\text{C}) = 167.5\ \text{kJ/kg}$$

where we have used the symbol Q to represent the heat interaction required in this process.

EXAMPLE PROBLEM 3.8

It becomes necessary to increase the temperature of 400 kg of water from 20 to 60°C. How much heat is requried?

Solution

If 40(4.187 kJ) will increase the temperature of 1 kg of water by 40°C, then 400(40)(4.187 kJ) should increase the temperature of 400 kg of water by 40°C. Thus

$$Q = 400\ \text{kg}(40°\text{C})(4.187\ \text{kJ/kg}\cdot°\text{C}) = 6.7 \times 10^4\ \text{kJ}$$

The amount of work or heat required to increase the temperature of a unit mass of a particular substance by one degree is sometimes called the *specific heat* of that substance. Typical values are shown in Table 3.1.

The symbol C is often used to denote the specific heat of a substance. For a substance of mass M, the heat interaction Q, the temperature change ΔT, and specific heat C are related by the equation

$$Q = MC\Delta T$$

EXAMPLE PROBLEM 3.9

Compare the heat required to increase the temperature of 1 ton (2000 lbm) of water from 70 to 100 F with the heat required to increase the temperature of 1 ton of sandstone from 70 to 100 F.

TABLE 3.1. Approximate Values of Specific Heats of Various Substances

Substance	Specific Heat	
	kJ/kg°C	*Btu/lbm·F*
Aluminum	0.9	0.2
Brick	0.9	0.2
Copper	3.8	0.9
Concrete	0.8	0.2
Earth (dry)	1.3	0.3
Ethylene glycol (50% solution)	3.3	0.8
Glass	0.8	0.2
Gold	0.13	0.03
Ice (0°C)	2.0	0.5
Iron	0.5	0.11
Marble	0.9	0.21
Nickel	0.5	0.11
Plastic	1.7	0.4
Platinum	0.13	0.03
Sandstone	0.9	0.22
Sawdust	0.9	0.21
Silver	0.25	0.06
Water	4.2	1.0
Wood (oak)	2.1	0.5
Wood (pine)	2.5	0.6

Solution

$$Q = MC\Delta T$$

For the water

$$Q = 2000 \text{ lbm} \left(\frac{1.0 \text{ Btu}}{\text{lbm} \cdot \text{F}} \right) (30 \text{ F}) = 60\,000 \text{ Btu}$$

For the sandstone

$$Q = 2000 \text{ lbm} \left(\frac{0.22 \text{ Btu}}{\text{lbm} \cdot \text{F}} \right) (30 \text{ F}) = 13\,200 \text{ Btu}$$

The relative values of specific heat of various substances have important implications in many phases of engineering design.

3.5/Energy—A Property of Matter

You probably became acquainted with the word energy at a very early age. With the advent of the so-called *energy* crisis, it is difficult to listen to a news broadcast or read a newspaper or magazine without encountering some mention of energy or energy resources. So we all know what the word energy means. Or do we?

Engineers, in communicating with other engineers, must be very exact in the meaning of words. It is amazing, but true, that we don't know how to adequately define energy (see Figure 3.13). Some would say that energy is the capacity to do work (exert a force through a distance), but it is easy to point to systems that obviously have energy, yet have zero capacity to do work. So for the engineer who must communicate with precision, that simple definition will not do.

There are several things which we can say about energy. Energy seems to be a *property* of matter. That is, all systems possess energy. Furthermore, energy is an *extensive property* of matter. That means the energy of the whole system is the sum of the energy of various parts of the system. Pressure and temperature are examples of properties of matter that are *intensive* rather than *extensive*. The temperature of a system is not the sum of the temperature of various parts of the system nor is the pressure. The mass of a system, however, is equal to the sum of the mass of the various parts of the system (Figure 3.14). *Energy* thus seems to be associated with mass. In the sections that follow, we will describe some other characteristics of energy.

Energy apparently enters or leaves a system through heat and/or work interactions with the surroundings (Figure 3.15). Consequently, energy is expressed in the same units as work or heat (i.e., ft·lbf, Btu, or kJ). The rate at which a system uses or delivers energy is often expressed in power units such as Btu/h, hp, or kW.

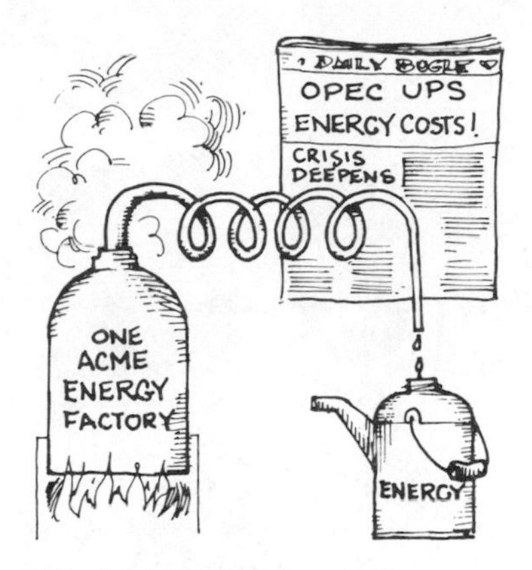

FIG. 3.13. Energy is a common term but hard to define.

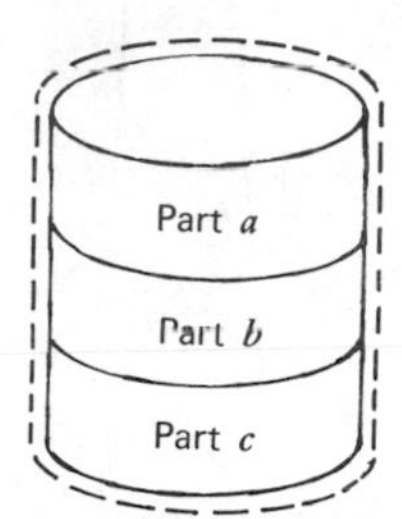

System energy
$E = E_a + E_b + E_c$

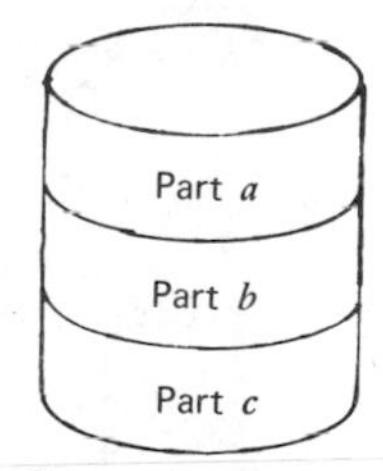

System temperature
$T \neq T_a + T_b + T_c$

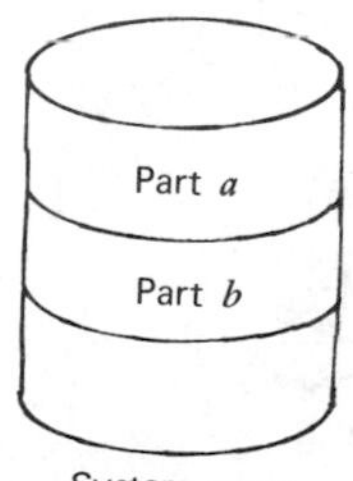

System mass
$M = M_a + M_b + M_c$

FIG. 3.14. Energy and mass are extensive properties, but temperature is not.

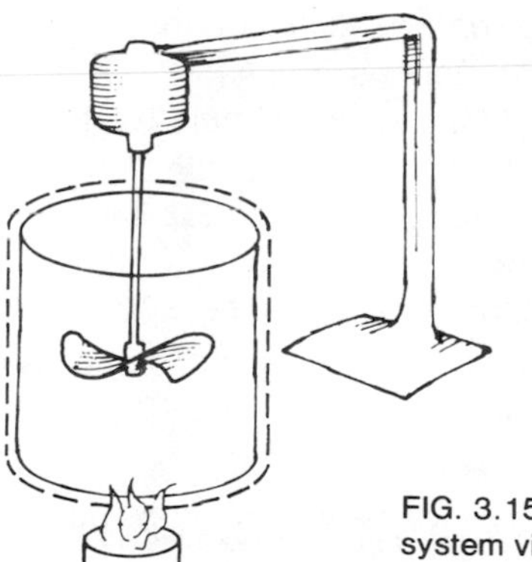

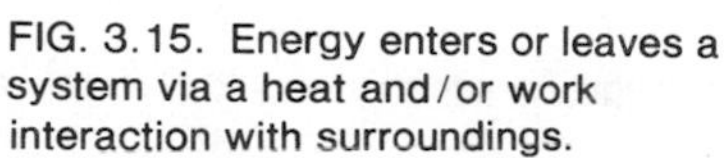
FIG. 3.15. Energy enters or leaves a system via a heat and/or work interaction with surroundings.

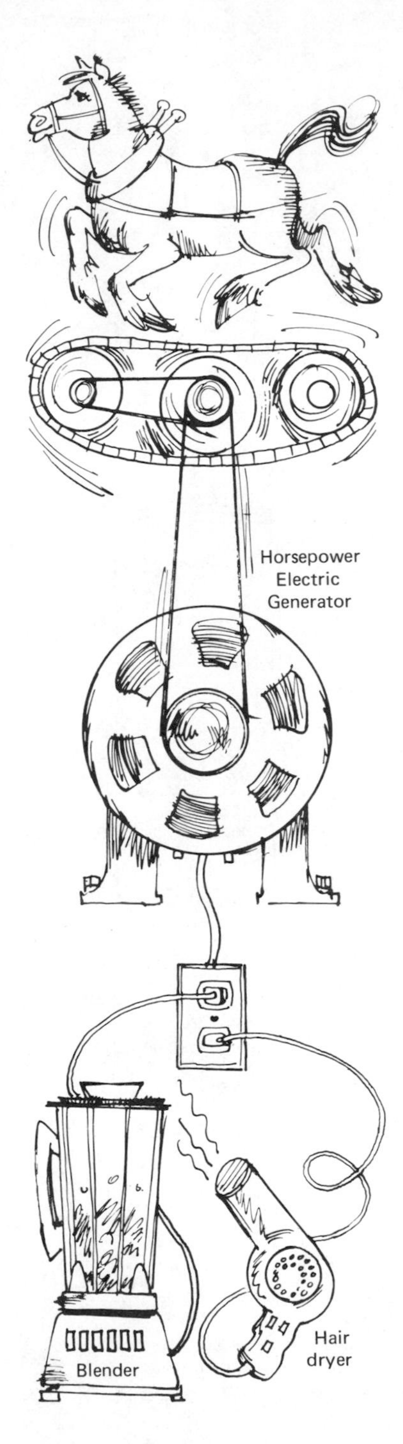

TABLE 3.2. Summary of Some Useful Energy and Power Equivalents

1 J = 1 N·m = 0.738 ft·lbf
1 Btu = 778.16 ft·lbf = 1.055 kJ
1 kcal = 4.187 kJ = 3.969 Btu
1 kW·h = 3600 kJ = 3413 Btu
1 hp·h = 0.746 kW·h = 2545 Btu
1 langley = 1 cal/cm^2 = 3.69 Btu/ft^2 = 41.9 kJ/m^2
1 hp = 550 ft·lbf/s = 0.746 kW = 0.707 Btu/s
1 kW = 1 kJ/s = 738 ft·lbf/s = 1.340 hp

TABLE 3.3. Typical Power Requirements of Some Household Appliances

Appliance	*Average Wattage*
Color television	240
Blender	300
Hair dryer	381
Vacuum cleaner	630
Dishwasher	1200
Toaster	1300
Water heater	4500
Electric range w/oven	12000

We again emphasize the difference between energy and power. Power is the time rate of energy flow. One kilowatt is the rate of energy flow, not a quantity of energy. Remember, 1 kW = 1 kJ/s. One horsepower is roughly the rate at which a strong, healthy horse can work on a continuous basis. We observe from Table 3.2 that 1 hp is equivalent to about $\frac{3}{4}$ kW. Table 3.3 suggests that if we harnessed the typical horse to an electrical generator, the horse could probably work fast enough to run a blender plus a hair dryer (Figure 3.16), but would have difficulty running a toaster on a continuous basis. We see that about 16 horses working steadily could generate enough electricity to operate an electric range with an oven. You might think of some familiar examples to get a feeling for horsepower. Your automobile probably produces somewhere from 20 to 80 hp. The engine on your lawnmower may deliver about 2 or 3 hp. A 100-W light bulb uses power at the rate of roughly $\frac{1}{7}$ hp. Large jet engines may produce 25 000 hp. A large rocket engine at launch may deliver as much as 20 000 000 hp for short periods of time.

The horsepower hour (hp·h) and the kilowatt hour (kW·h) are units of energy. One horsepower hour is the quantity of energy used in 1 h if the rate of use is 1 hp or 550 ft·lbf/s. Similarly, the 1 kW·h is the quantity of energy used in 1 h if the rate of use is 1 kW or kJ/s. Thus 1 kW·h = 3600 kJ since there are 3600 s in 1 h. Figures 3.17 and 3.18 offer an interesting comparison of different energy and power levels.

FIG. 3.16. A single horsepower would probably run a blender plus a hair dryer.

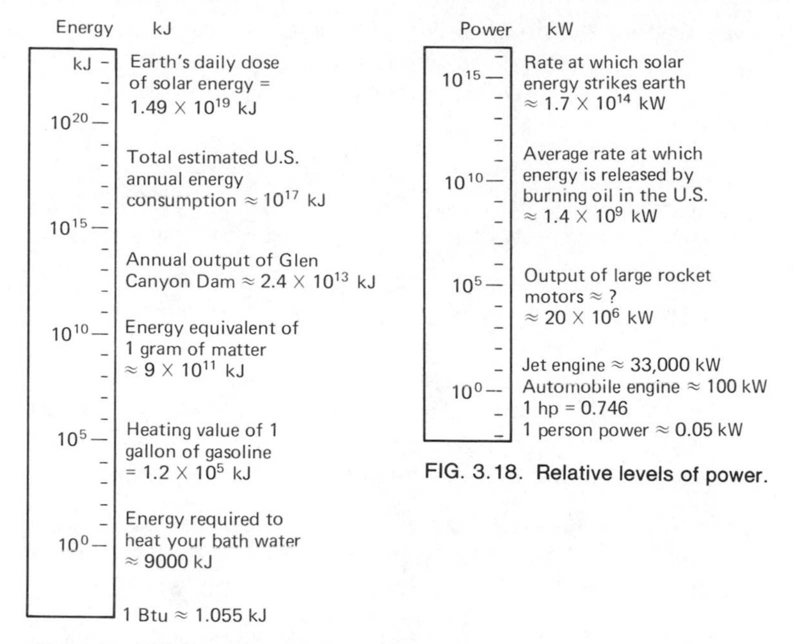

FIG. 3.18. **Relative levels of power.**

FIG. 3.17. **Relative amounts of energy.**

PROBLEMS

3.13. Calculate the magnitude of the heat interaction required to increase the temperature of 10 kg of water from 30 to 40°C. Express your answer in kilojoules.

3.14. At what rate must a heat interaction take place in order to increase 10 kg/s of water from 30 to 40°C? Express your answer in kilowatts.

3.15. Suppose you are responsible for designing a solar home-heating system. It is necessary to store energy collected on clear sunny days for use on cloudy days. It is estimated that 300 000 kJ is the daily requirement to keep the home comfortable. Assume that the energy is stored in water by increasing the water temperature from 25 up to 45°C. Estimate the mass of water required to provide storage for four days. Express your answer in kilograms.

3.16. Repeat Problem 3.15 assuming that the storage medium is sandstone rather than water.

3.17. It has been proposed that the Colorado River be used to cool a power plant by allowing the river to flow through the power plant condenser. The minimum flow rate of the Colorado

River is about 283 m^3/s. Each cubic meter contains about 1000 kg of water. It is desired to limit the temperature increase of the water to 10°C. What would be the maximum permissible heating rate?

3.18. (a) Write three true statements about energy.

(b) Explain the meaning of extensive property. Give three examples of extensive properties and three examples of intensive properties.

3.19. (a) Which is the larger unit of power, the kilowatt or the horsepower? The rated capacity of Hoover Dam on the Colorado River is 1344.8 MW. Express this capacity in horsepower.

(b) If the generators at Hoover Dam operated at 90 percent capacity for one year, how much energy would be produced? Express your answer in kilowatt hours and horsepower hours.

3.20. Suppose a TV set in your home is operated for 50 h per week, a toaster for 30 min per week, and a vacuum cleaner for 5 h per week. Which uses the most energy per week?

3.21. A strong individual can produce about $\frac{1}{20}$ hp continuously. Assume that he or she works for an 8 h day at the rate of $12 per hour.

(a) Determine the total energy produced by the person in one day in terms of kilowatt hours.

(b) Determine the manpower cost per kilowatt hour and compare with the cost of electricity that may be assumed to be 5¢/kw·h.

(c) At these manpower prices, what would be the cost per hour to burn a 100-W bulb?

3.22. The energy cost of producing unbleached kraft paper is approximately 11 800 kwh/ton. Paper towels used in rest-room dispensers weigh approximately 3 g/towel. An electrical hand dryer utilizes approximately 19.16 W·h/cycle. Determine how many towels are equivalent in terms of energy to one cycle of electric drying.

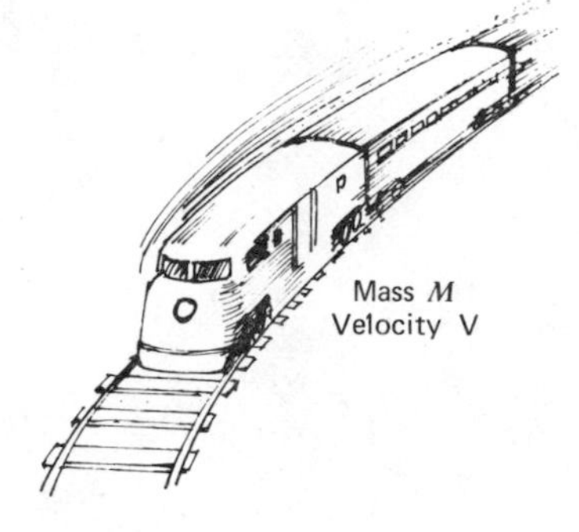

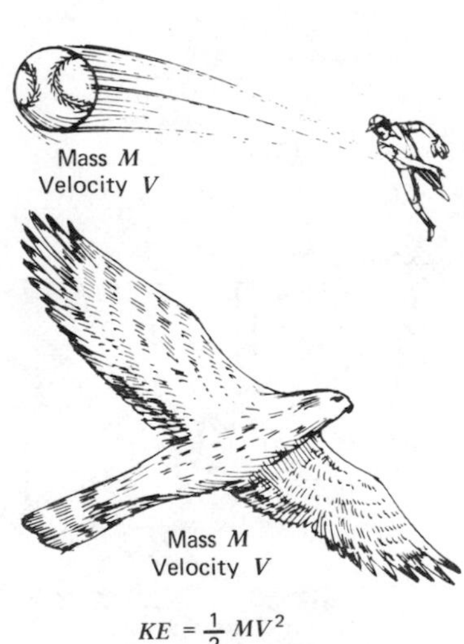

$$KE = \frac{1}{2}MV^2$$

FIG. 3.19. Objects in motion have kinetic energy.

3.6/Kinetic Energy

A moving object is said to have *kinetic energy.* A spinning wheel, a bird gliding through the air, a pitched baseball, flowing water, and a moving train all have kinetic energy. This energy is associated with both the translational and/or rotational motion of the bodies. The kinetic energy of a translating mass M moving with a velocity V is defined as (see Figure 3.19):

$$KE = \tfrac{1}{2}MV^2$$

EXAMPLE PROBLEM 3.10

A space laboratory having a mass of 100 000 kg is to be placed in orbit at a velocity of 30 000 km/h. Calculate the kinetic energy of the laboratory in orbit. Assume negligible rotation. (See Figure 3.20.)

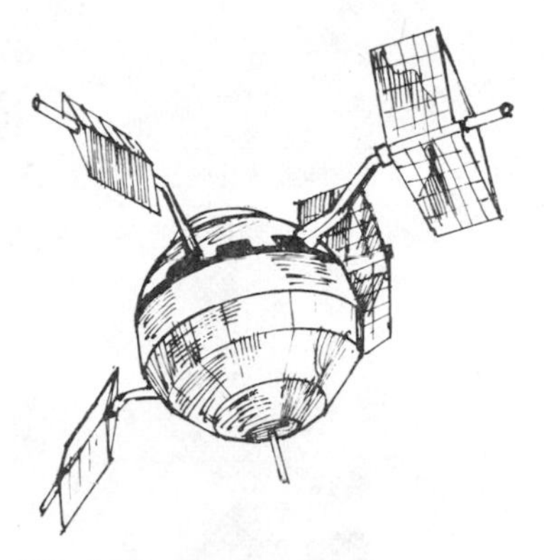

FIG. 3.20. Example Problem 3.10.

Solution

$$KE = \frac{MV^2}{2}$$

$$KE = \frac{(\frac{1}{2})(100\,000\text{ kg})(30\,000\,000\text{ m/h})^2}{(3600\text{ s/h})^2}$$

$$= 3.47 \times 10^{12}\text{ kg}\cdot\text{m}^2/\text{s}^2$$

Recalling that 1 kg·m/s² is equivalent to 1 N we write

$$KE = 3.47 \times 10^{12}\text{ N}\cdot\text{m} = 3.47 \times 10^{12}\text{ J}$$

EXAMPLE PROBLEM 3.11

Calculate the kinetic energy of a 4000-lbm automobile traveling at 80 miles/h.

Solution

$$KE = \frac{MV^2}{2}$$

$$V = (80\text{ miles/h})(5280\text{ ft/mile})(1\text{ h}/3600\text{ s})$$

$$= 117.33\text{ ft/s}$$

$$KE = \tfrac{1}{2}(4000\text{ lbm})(117.33\text{ ft/s})^2$$

$$= 2.75 \times 10^7\text{ ft}^2\cdot\text{lbm/s}^2$$

Recalling that 1 lbf is equivalent to

$$32.174\text{ ft}\cdot\text{lbm/s}^2$$

we write

$$KE = \left(\frac{2.75 \times 10^7}{32.174}\right)$$

$$= 8.56 \times 10^5\text{ ft}\cdot\text{lbf}$$

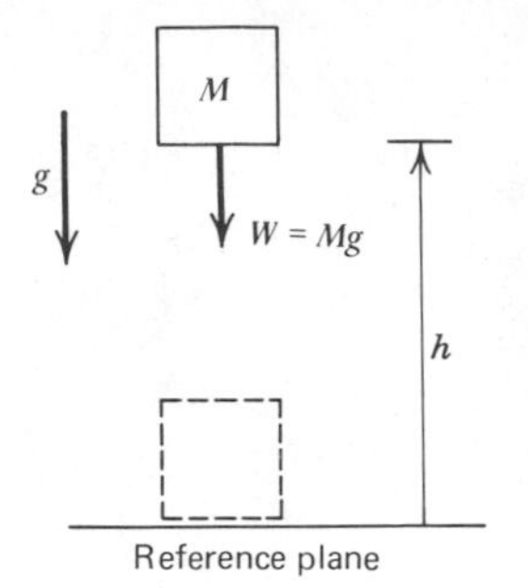

FIG. 3.21. An elevated mass has potential energy.

3.7/Potential Energy

Consider a body of mass M that is elevated a distance h above a reference plane as shown in Figure 3.21. The acceleration due to gravity is g and directed downward. The weight of the body of mass M is calculated as weight $= Mg$. The *potential energy* of the mass M in the elevated position is measured by the work that could be done by the body as it is lowered to the reference plane and is defined as

$$PE = Mgh$$

EXAMPLE PROBLEM 3.12

A space laboratory having a mass of 100 000 kg is to be placed into orbit at a height above the earth of 300 km. Calculate the potential energy of the laboratory in orbit (Figure 3.22).

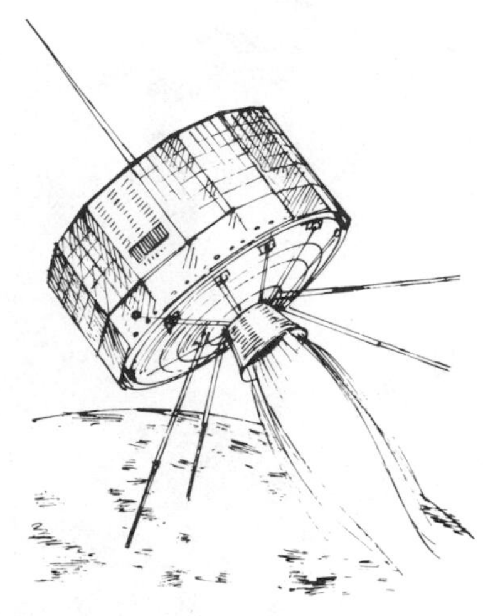

FIG. 3.22. Example Problem 3.12.

Solution

The weight of the laboratory may change slightly as it ascends because the force of gravity decreases as altitude increases. We will assume an average value for the acceleration due to gravity of 9.7 m/s^2 and compute the weight of the laboratory as:

$$\text{Weight} = Mg$$
$$= 100\,000 \text{ kg } (9.7 \text{ m/s}^2)$$
$$\text{Weight} = 9.7 \times 10^5 \text{ N}$$

The potential energy is thus calculated as

$$PE = (\text{weight})(h)$$
$$= (9.7 \times 10^5 \text{ N})(300\,000 \text{ m})$$
$$= 2.91 \times 10^{11} \text{ N}\cdot\text{m}$$
$$= 2.91 \times 10^{11} \text{ J} = 2.91 \times 10^8 \text{ kJ}$$

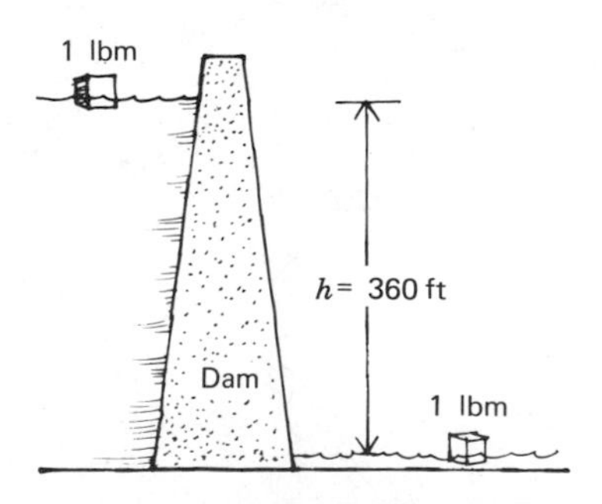

FIG. 3.23. Example Problem 3.13.

EXAMPLE PROBLEM 3.13

Calculate the potential energy of 1 lbm of water located at the surface of a reservoir 360 ft above the river below the dam (Figure 3.23).

Solution

A mass of 1 lbm weighs approximately 1 lbf at most locations on the earth. Therefore,

PE = weight (h)

= 1 lbf (360)ft = 360 ft·lbf

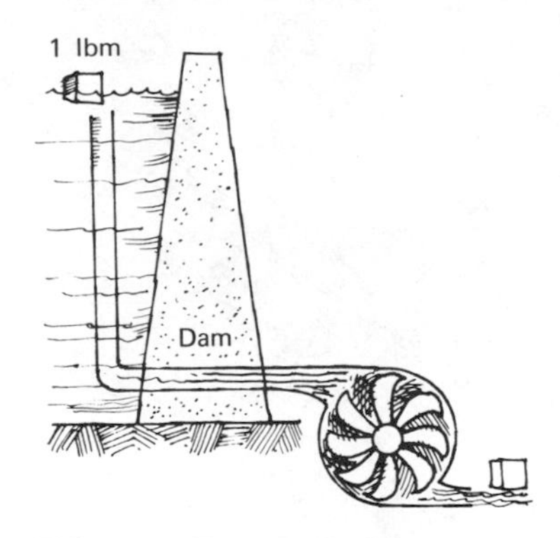

FIG. 3.24. Example Problem 3.14.

EXAMPLE PROBLEM 3.14

Water from the reservoir described in Example 3.13 flows through turbines at the rate of 100 000 lbm/s. The turbines are located at the level of the river, which is 360 ft below the surface of the reservoir. What would be the maximum power expected to be generated by the water flowing through the turbines (Figure 3.24)?

Solution

We calculated in Example 3.13 that the potential energy of each pound mass of water was 360 ft·lbf greater at the surface of the reservoir than at the surface of the river where the water escapes from the turbines. We assume that this potential energy leaves the turbines as work in the form of a turning shaft. Because there are 100 000 lbm of water per second flowing through the turbines and a maximum of 360 ft·lbf of energy may be extracted from each pound mass, the maximum power generated would be

$\dot{W}$ = (360 ft·lbf/lbm)(100 000 lbm/s)

= 3.6×10^7 ft·lbf/s

Using

1 hp = 550 ft·lbf/s

we find that

$\dot{W}$ = 65 455 hp

PROBLEMS

3.23. Calculate the kinetic energy of a 1500-kg automobile traveling at 90 km/h. Express your answer in kilojoules.

3.24. Oil pipelines are often hundreds of kilometers long and thus contain huge quantities of oil. When all that oil gets moving, the kinetic energy involved is rather awesome. Calculate the kinetic energy of a column of oil having a mass of 10^9 kg and moving at 3 m/s. Express answer in kilojoules.

3.25. Determine the kinetic energy of a 10000-lbm satellite moving at 18600 miles/h.

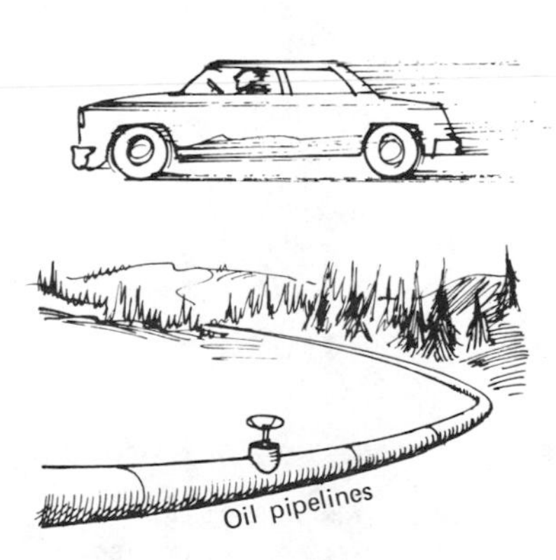

FIG. 3.25. Macrosopic kinetic and potential energy are associated with the system as a whole.

3.26. With what velocity must a 150-lbm person move in order to have the same kinetic energy as a 180-grain bullet moving at 2000 ft/s (1 lbm = 7000 grains).

3.27. Estimate the translational kinetic energy of the earth as it moves in its orbit about the sun. Assume that the mass of the earth is 5.98×10^{24} kg and the radius of its orbit 149.5×10^6 km.

3.28. Ten people board an elevator on the ground floor of a tall building. The elevator carries the people to the one-hundredth floor, 300 m above the ground floor. Calculate the increase in potential energy of these 10 people due to their elevator ride? Assume that the average mass of each person is 65 kg. Express your answer in kilojoules.

3.29. Calculate the potential energy of 1 kg of water vapor trapped in a cloud 1500 m above the earth. Express your answer in kilojoules per kilogram.

3.30. Rain begins to fall from clouds 1500 m above the earth. The rain continues until a depth of 2.5 cm has fallen over an area of 10 000 km^2. Estimate the decrease in potential energy of the water as a result of the rainstorm. Recall that 1 m^3 of water contains 1000 kg. What happens to this energy?

3.31. Calculate the power released by the rainstorm described in Problem 3.30 assuming that the water falls in a 2-hour period of time.

3.8/Internal Energy

Basically there are only two kinds of energy associated with matter. One kind is the energy associated with motion that we have called kinetic energy. The other kind, associated with position or proximity to other matter, is called potential energy (Figure 3.25). We have seen that it is not difficult to calculate the simple translational kinetic energy of individual bodies of mass *M*. It is also not difficult to compute the change in potential energy of a body of mass *M* as it moves in a uniform gravitational field. Consider a piece of coal thrown up into the air. Obviously it has some kinetic and potential energy that you could now calculate if you were given the mass, velocity, and distance above the ground (reference plane) of the piece of coal. We call those the *macroscopic* kinetic and potential energy modes. But if we could peer into the *microscopic* world of individual atoms, the basic particles of which the chunk of coal consists, we would observe these atoms vibrating back and forth. Because the individual atoms have motion, they have microscopic kinetic energy modes (Figure 3.26). In spite of the motion, most of the atoms remain locked in their relative position within the chunk of coal by the forces exerted on them by their neighboring atoms. Some of these forces are called chemical bonds.

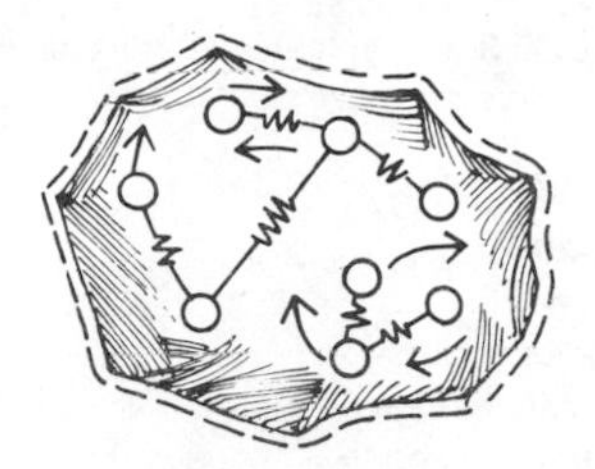

FIG. 3.26. Microscopic kinetic and potential energy are associated with the individual atomic particles which make up the system.

The work required to position an atom in the force field existing among its neighbors gives us a measure of the microscopic potential energy of the atom associated with these chemical bonds (Figure 3.27). If the temperature of the chunk of coal is increased, the motion and thus the kinetic energy of the individual atoms tends to increase. If the temperature becomes great enough, as in a combustion process, the atoms may break the bonds holding them to their neighbors, thus releasing some of the microscopic potential energy stored in the chemical bonds. It would be a complex task to account for the kinetic and potential energies of all the individual particles of the piece of coal. Therefore, we usually refer to the sum of those microscopic kinetic and potential energy modes as the *internal energy* of the piece of coal. Using the symbol U for internal energy we write

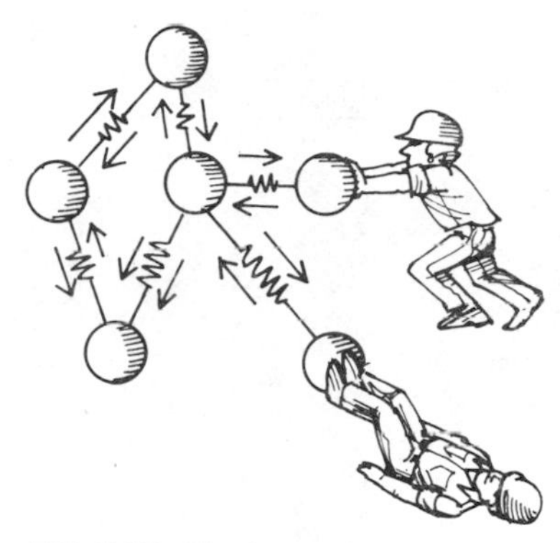

FIG. 3.27. The internal energy may be influenced by interactions at the boundary of the system.

$$U = KE_{\text{micro}} + PE_{\text{micro}}$$

The total energy of our piece of coal is thus the sum of the internal energy and the macroscopic kinetic and potential energies of the lump as a whole. Using the symbol E to represent total energy of our system, thus:

$$E = U + KE + PE$$

If the macroscopic kinetic and potential energies, KE and PE, are zero or negligible, then the total energy E is equal to the internal energy U.

Consider a 1-kg lump of coal that possesses internal energy by virtue of the microscopic potential and kinetic energies of its atoms. Suppose that a 1-kg lump of coal is burned in a calorimeter where the energy is released into a certain quantity of water as indicated in Figure 3.28. If the increase in temperature of the water is carefully measured, we would find that about 30 000 kJ of energy would be transferred to the water through a heat interaction. We conclude that the internal energy of the coal decreased by 30 000 kJ on combustion. Shown in Table 3.4 are similar quantities for a number of fuels. The values shown in the table represent the heating value of the fuel, which is the internal energy released on combustion.

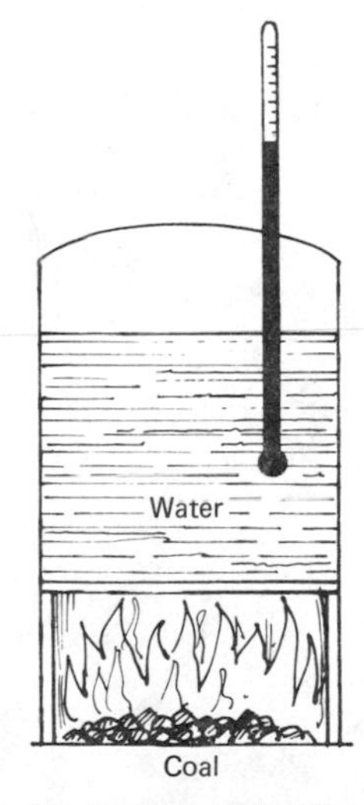

FIG. 3.28. During combustion, the internal energy of the coal is transferred to the water.

TABLE 3.4. Typical Heating Values of Various Fuels

	kJ/kg		Btu/lbm	
Wood	20 470		8 800	
Peat	20 930		9 000	
Lignite coal	25 590		11 000	
Subbituminous coal	29 070		12 500	
Bituminous coal	34 890		15 000	
Anthracite coal	30 240		13 000	
Methane	50 010	(33 750 kJ/m³)	21 500	(906 Btu/ft³)
Propane	46 350		19 930	
Gasoline	44 800	(34 870 kJ/l)	19 260	(125 000 Btu/gal)
Fuel oil	45 520	(39 300 kJ/l)	19 570	(141 000 Btu/gal)
Ethanol	26 520	(21 180 kJ/l)	11 400	(76 000 Btu/gal)
Methanol	19 890	(15 890 kJ/l)	8 550	(57 000 Btu/gal)

EXAMPLE PROBLEM 3.15

The United States burns approximately six billion barrels of oil each year. Calculate the energy that is released by this process.

Solution

One barrel is equivalent to 42 gallons. We assume that the heating value of crude oil is approximately that of fuel oil as given in Table 3.4. The heating value of one barrel of oil is, therefore,

$$(42 \text{ gal/bbl})(141\,000 \text{ Btu/gal})$$

$$= 5.9 \times 10^6 \text{ Btu/bbl}$$

Thus, the total energy released by burning oil and products made from oil in the United States is approximately

$$(6 \times 10^9 \text{ bbl/year})(5.9 \times 10^6 \text{ Btu/bbl})$$

$$\simeq 3.6 \times 10^{16} \text{ Btu/year}$$

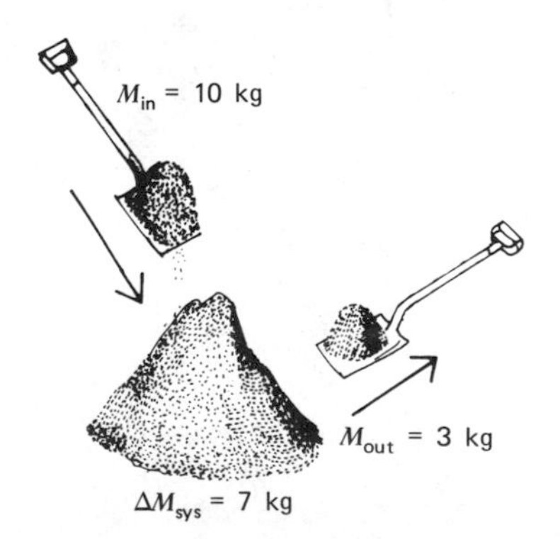

FIG. 3.29. Conservation of mass.

3.9/Mass and Energy Are Conserved

The notion that mass is conserved is a familiar one. For example, if 10 kg of sand were added to a sandpile while 3 kg of sand were removed from the pile, we would expect the mass of the sandpile to increase by 7 kg during this process (Figure 3.29). Mathematically we could state this as

$$M_{in} = M_{out} + \Delta M_{sys}$$

where

M_{in} = mass that enters the system

M_{out} = mass that leaves the system

ΔM_{sys} = change in mass of the system during the process in question

The conservation of energy notion, often referred to as the first law of thermodynamics, similarly states that all of the energy that enters a system during a certain process is equal to the energy that leaves the system during that process plus the change in the quantity of energy stored within the system. This is expressed mathematically as

$$E_{in} = E_{out} + \Delta E_{sys}$$

where

E_{in} = energy that enters the system during the process in question

E_{out} = energy that leaves the system during the process in question

ΔE_{sys} = change in the amount of energy stored within the system

Consider a system in which 10 kJ of energy enter as work, while 3 kJ are leaving as heat (Figure 3.30). We conclude that during this process, the energy stored within the system increased by 7 kJ. The concept of conservation of energy is basic to a large segment of engineering science, particularly thermodynamics. Thermodynamics is the study of energy and its transformation from one form to another. As an engineer you will most likely utilize the first law of thermodynamics in the solution of many engineering problems.

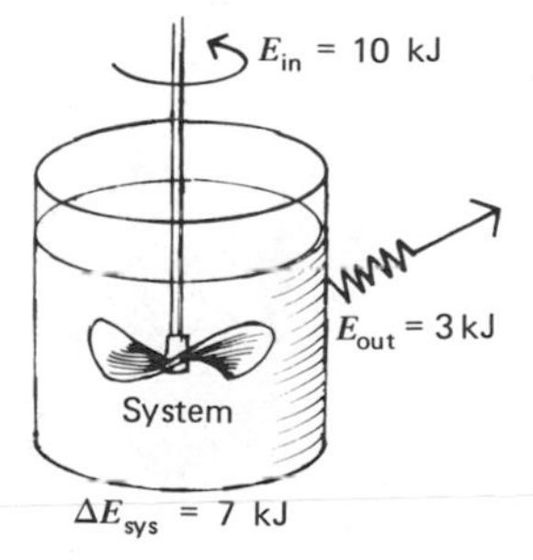

FIG. 3.30. Conservation of energy.

EXAMPLE PROBLEM 3.16

Suppose the energy content of your lunch is 2000 kJ. During the time you eat your lunch your body loses about 100 kJ of energy through a heat interaction with the surroundings. Estimate the increase in internal energy of your body during the lunch-eating process.

Solution

We treat your body as the system and apply the first law of thermodynamics:

$$E_{in} = E_{out} + \Delta E_{sys}$$

$$2000 \text{ kJ} = 100 \text{ kJ} + \Delta E_{sys}$$

$$\Delta E_{svs} = 1900 \text{ kJ}$$

EXAMPLE PROBLEM 3.17

After lunch you return to your job of stacking blocks. During a four-hour period your body loses about 1800 kJ of energy through heat interaction with the surroundings and you lift a total of 1000 bricks weighing 200 N each through a distance of 1.5 m. You find time to eat a candy bar containing 400 kJ of energy. Estimate the change in internal energy of your body during this process.

Solution

Again we treat your body as the system. The work done by you in lifting the blocks is

$$W = 1000 \text{ bricks } (200 \text{ N/brick})(1.5 \text{ m})$$

$$= 300\,000 \text{ N}\cdot\text{m} = 300 \text{ kJ}$$

therefore,

$$E_{in} = E_{out} + \Delta E_{sys}$$

$$400 \text{ kJ} = 1800 \text{ kJ} + 300 \text{ kJ} + \Delta E_{sys}$$

$$\text{Candy bar} = \text{heat loss} + \text{work done} + \text{change in energy stored}$$

$$\Delta E_{sys} = -1700 \text{ kJ}$$

Thus the internal energy of your body *decreases* by 1700 kJ during the process in question.

It is often convenient to express the conservation of energy concept as a *rate equation.* We thus say that at any instant of time the rate at which energy enters a system is equal to the rate at which energy leaves the system plus the rate at which the energy stored within the system is changing. Mathematically this is expressed as follows:

$$\dot{E}_{in} = \dot{E}_{out} + \frac{\Delta E_{sys}}{\Delta t}$$

where

$\dot{E}_{in}$ = rate at which energy enters the system

$\dot{E}_{out}$ = rate at which energy leaves the system

$\frac{\Delta E_{sys}}{\Delta t}$ = rate at which energy of the system changes

The corresponding rate equation for conservation of mass would be

$$\dot{M}_{in} = \dot{M}_{out} + \frac{\Delta M_{sys}}{\Delta t}$$

where the meaning of the symbols should be obvious.

EXAMPLE PROBLEM 3.18

Referring to Example Problem 3.16, suppose the time required to eat your lunch is 15 min. At what rate is the internal energy of your body changing during this time?

Solution

$$\dot{E}_{in} = \dot{E}_{out} + \frac{\Delta E_{sys}}{\Delta t}$$

$$\frac{2000 \text{ kJ}}{0.25 \text{ h}} = \frac{100 \text{ kJ}}{0.25 \text{ h}} + \frac{\Delta E_{sys}}{\Delta t}$$

$$\frac{\Delta E_{sys}}{\Delta t} = 7600 \text{ kJ/h}$$

An important class of engineering problems deals with steady-state systems in which neither the mass of the system nor the energy stored within the system changes with time. An example of such a system could be a steadily operating automobile engine. As suggested by Figure 3.31, we would expect the mass flow rate of the fuel and air that enter the system to exactly equal the mass flow rate of the exhaust gases that leave the engine. We would not expect the mass of the engine to change significantly during that hour of operation. For this case

$$\dot{M}_{in} = \dot{M}_{out}$$

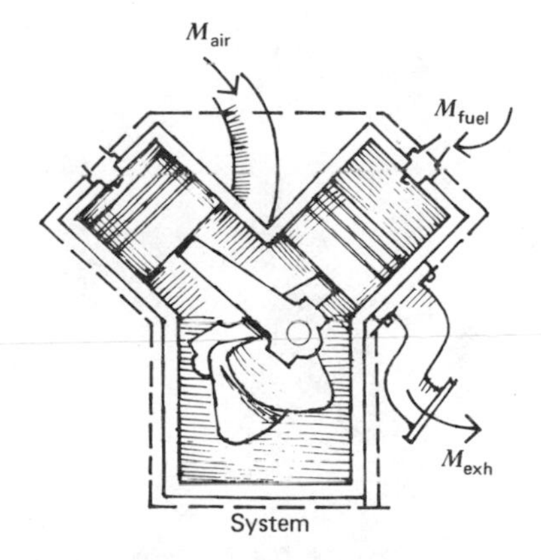

FIG. 3.31. In steady state systems, the mass flow rate in equals the mass flow rate out.

Similarly we would expect that as the engine operates steadily, the rate at which energy enters the engine is equal to the rate at which energy leaves the engine. The amount of energy stored within the engine would not change. That is,

$$\dot{E}_{in} = \dot{E}_{out}$$

Energy enters the engine mainly as the internal energy (heating value) of the fuel and leaves as work delivered by the turning shaft and heat lost to the surroundings. The *thermal efficiency* of an engine is defined as

$$\eta = \frac{\text{shaft work delivered}}{\text{heating value of fuel}}$$

Typical values of thermal efficiency of automobile engines lie in the 12 to 20 percent range. That is, approximately 12 to 20 percent of the heating value of the fuel is converted to shaft work. The balance is lost as heat.

EXAMPLE PROBLEM 3.19

An automobile engine steadily produces 56 kW of power while burning 32 kg of gasoline per hour. At what rate does energy leave the engine via the exhaust gases and as heat to the surroundings?

Solution

$$\dot{E}_{\text{in}} = \dot{E}_{\text{out}}$$

We calculate

$$\dot{E}_{\text{in}} = (32\ \text{kg/h})(44\ 800\ \text{kJ/kg})$$

$$= 1.43 \times 10^6\ \text{kJ/h}$$

$$\dot{E}_{\text{out}} = \dot{W} + \dot{E}_{\text{lost}} = 1.43 \times 10^6\ \text{kJ/h}$$

$$\dot{W} = (56\ \text{kJ/s})(3600\ \text{s/h})$$

$$= 0.20 \times 10^6\ \text{kJ/h}$$

therefore,

$$\dot{E}_{\text{lost}} = (1.43 - 0.20) \times 10^6\ \text{kJ/h}$$

$$= 1.23 \times 10^6\ \text{kJ/h}$$

The thermal efficiency of this engine is

$$\frac{\dot{W}}{\dot{E}_{\text{in}}} = \frac{0.20 \times 10^6}{1.43 \times 10^6} = 0.14$$

According to the theory of relativity, mass and energy are related by the well-known equation

$$E = Mc^2$$

where

E = energy of the system

M = mass of the system

c = velocity of light = 2.9979×10^8 m/s

This suggests that if the energy of a system changes so does its mass. If that is true, then both energy and mass are not really conserved as we have assumed. It is appropriate to examine the magnitude of the error involved in typical engineering calculations by assuming the conservation laws are valid.

EXAMPLE PROBLEM 3.20

We note from Table 3.4 that the heating value of 1 kg of gasoline is about 44 800 kJ. Complete combustion of 1 kg of gasoline requires that 15 kg of air be provided (Figure 3.32). Suppose a closed system consisting of 1 kg of gasoline and 15 kg of air is ignited. Following combustion, 44 800 kJ of energy are transferred from the system, thus returning the temperature to its initial value and decreasing the energy of the system by 44 800 kJ. We wish to estimate the change in mass of the closed system associated with this decrease of energy.

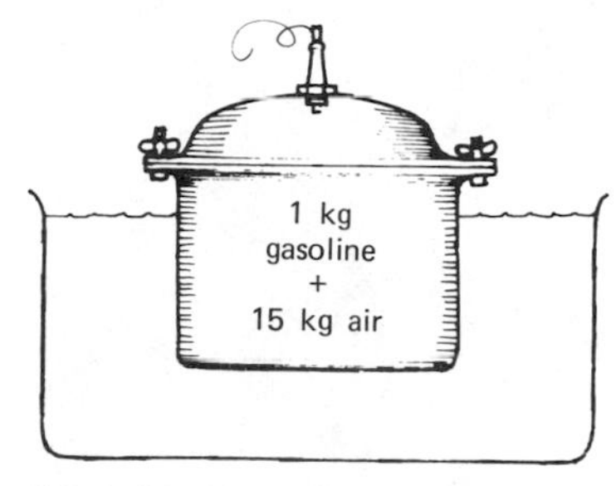

FIG. 3.32. Example Problem 3.20.

Solution

$$\Delta M = \frac{\Delta E}{c^2}$$

$$44\,800\,000 \text{ N}\cdot\text{m}/(3 \times 10^8 \text{ m/s})^2 =$$

$$4.98 \times 10^{-10} \text{ kg}$$

On a percentage basis the mass decreased by

$$\frac{100 \times 4.98 \times 10^{-10}}{16} = 0.000\,000\,003\,1 \text{ percent}$$

A fractional change of this magnitude is practically impossible to detect by even the most sensitive instruments. We conclude that for ordinary systems of interest in energy conversion calculations, the relativistic effects may be neglected and the conservation of mass and energy concept remains valid.

PROBLEMS

3.32. Calculate the number of hours a 100-W light bulb could be operated for an expenditure of energy equal to the heating value of 1 gal of gasoline.

3.33. Assuming that wood costs $40/ton (where 1 ton =

2000 lbm) and natural gas costs \$2.50/1000 ft^3, which is the better energy bargain? Justify your answer. Assume that natural gas has the same heating value as methane.

3.34. Suppose a certain person eats food having an energy content of 2000 kcal/day. Assuming that the body converts 20 percent of that energy to work done by the muscles and 70 percent to heat (thus utilizing 90 percent of the food energy) compute the average rate of heat loss by the body. Express your answer in watts. Compare this average rate of heat dissipation from the body with that from a 100-W light bulb. (A kilocalorie is the energy required to increase the temperature of 1 kg of water by 1°C and is equivalent to 4.186 kJ.)

3.35. A typical home in the northern part of the United States requires about 500 000 kJ/day of energy for heating purposes in January. It may be assumed that upon combustion, approximately 60 percent of the heating value of a particular fuel is released into the home and about 40 percent of the heating value is lost up the chimney. Calculate the amount of (a) wood, (b) bituminous coal, (c) methane, (d) propane, and (e) fuel oil that must be burned each day to heat a typical home. Express your answers in kilograms per day.

3.36. When gasoline is burned in an automobile engine, about 14 percent of the heating value is typically converted to mechanical energy delivered to the drive shaft. The remainder is transferred via a heat interaction to the surroundings. Calculate the rate of fuel consumption of a gasoline engine producing 100 hp as mechanical power output. Express your answer as liters per hour and gallons per hour.

3.37. Suppose you are able to drive your automobile 7 km while it burns 1 liter of gasoline. How much energy will be expended as fuel during a typical year in which you drive 15 000 km? Express answer in liters and kilojoules.

3.38. Suppose the rate of heat transfer from each of 20 000 people seated in a large auditorium to the air is about 100 W. (a) At what rate must fuel oil be burned to provide the same rate of heating to the air? Express your answer in liters per hour. (b) At what rate must air conditioning be provided in the auditorium to remove the heat transferred from the audience to the air? Express your answer in kilojoules per hour.

3.39. Fifty kilograms per hour of diesel having a heating value of 45 000 kJ/kg, flow steadily into an engine producing work at the rate of 125 kW. Determine the rate of heat loss by the engine. Express your answer in kilojoules per hour. Also, determine the thermal efficiency of the engine.

3.40. A thermal electrical generating plant produces 3000 kW of power. The thermal efficiency of the plant is estimated as 36 percent. Estimate the rates at which heat must be added to and rejected from the cycle.

3.41. Suppose the power plant of the previous problem is cooled by evaporating water in a cooling tower. Approximately 2430 kJ of energy are required to evaporate 1 kg of water. Estimate the required flow of cooling water. Express your answer in kilograms per second.

3.42. You throw a ball weighing 0.1 lbf vertically upward to a height of 60 ft. Calculate the potential energy of the ball just as it ceases moving upward. Express your answer in ft·lbf. Where did this energy come from? What happens to this energy as the ball returns to the ground?

3.43. The water level in a certain large man-made lake is 140 m above the level of the river below the dam. (a) Compute the reduction in potential energy of each kg of water as it is lowered from the surface of the lake to the river surface below the dam. Express your answer in kilojoules per kilogram. (b) Given that the average flow of water through the turbines is 280 m^3/s, estimate the power output of the dam. Assume that each cubic meter of water contains 1000 kg and that 90 percent of the potential energy removed from each kilogram of water is converted to electricity.

3.44. Make a realistic estimate of the energy required to place a space laboratory having a mass of 70 000 kg into an orbit 480 km above the earth at a velocity of 32 000 km/h. Express your answer in kilojoules and in gallons of gasoline equivalent. (Hint: Calculate the kinetic and potential energy of the orbiting satellite and apply the first law of thermodynamics.

3.45. (a) Write a statement of the first law of thermodynamics both in words and by equation.

(b) Under what conditions must relativistic effects be considered when applying the principle of conservation of energy and mass?

3.10/Limitations on Energy Conversion Efficiency

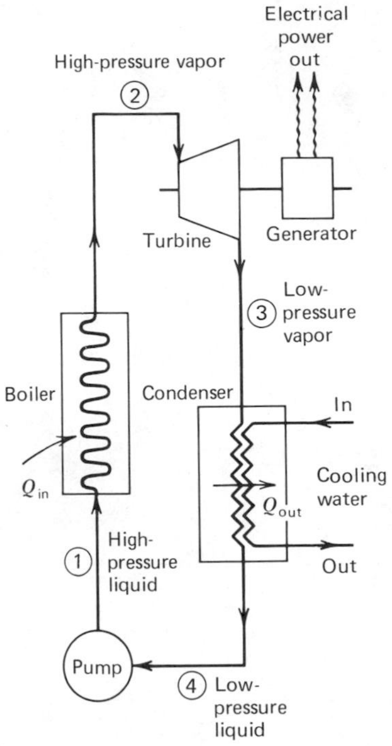

About 96 percent of the electrical power used in the United States is generated by the Rankine (steam) power cycle, shown in Figure 3.33. A high temperature is created in the boiler by burning fuel, a nuclear reaction, concentrating solar energy, or tapping a geothermal well. High-pressure liquid enters the boiler at point one to emerge as a high pressure high temperature vapor at point two. The high-pressure steam then flows through a turbine causing the turbine shaft to spin. The turbine shaft in turn drives an electrical generator. The steam leaves the turbine at a very low pressure (point three) and enters a condenser where it is converted back to liquid (point four). The condensed liquid is then pumped back up to a high pressure enabling it to enter the boiler and begin the cycle again. Energy must be removed as heat from the steam entering the condenser in order to effect the required phase change from vapor to liquid. This energy is carried away by cooling water, which also flows through the condenser. Thus, energy enters the cycle as heat in the boiler and leaves the plant as electricity (work) and also as heat transferred to the cooling water and the environment.

Electrical power out
High-pressure vapor
2
Turbine
Generator
3 Low-pressure vapor
Boiler
Condenser
In
Cooling water
Q_{in}
Q_{out}
1 High-pressure liquid
Out
Pump
4 Low-pressure liquid

FIG. 3.33. Rankine power cycle.

The thermal efficiency of such a power plant is defined as the net power delivered in the form of electricity divided by the rate at which energy is added as heat to the boiler

$$\eta = \frac{\text{net work}}{\text{heat added}} = \frac{\dot{W}_{net}}{\dot{Q}_{in}}$$

EXAMPLE PROBLEM 3.21

A certain thermal electrical generating plant produces 3000 MW of electrical power. The thermal efficiency of the power plant is 40 percent. At what rate must heat be added, and at what rate must heat be rejected (Figure 3.34)?

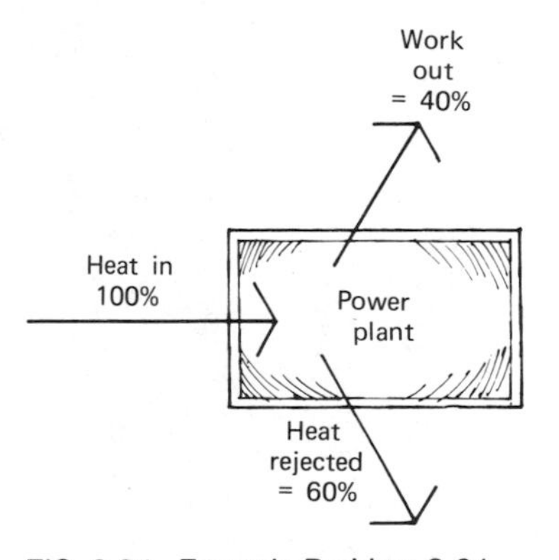

FIG. 3.34. Example Problem 3.21.

Solution

If the plant operates steadily, the rate at which energy enters the plant in the form of heat must leave the plant as electrical power and as heat to the cooling water.

$$\dot{E}_{in} = \dot{E}_{out}$$

$$\dot{Q}_{in} = \dot{W} + \dot{Q}_{out}$$

where

$\dot{W}$ = electrical power produced

$\dot{Q}_{out}$ = cooling rate

also

$$\dot{W} = 0.40\,(\dot{Q}_{in})$$

therefore

$$\dot{Q}_{in} = \frac{\dot{W}}{0.40} = \frac{3000\ \text{MW}}{0.40}$$

$$\dot{Q}_{in} = 75\,000\ \text{MW}$$

$$\dot{Q}_{out} = 4500\ \text{MW}$$

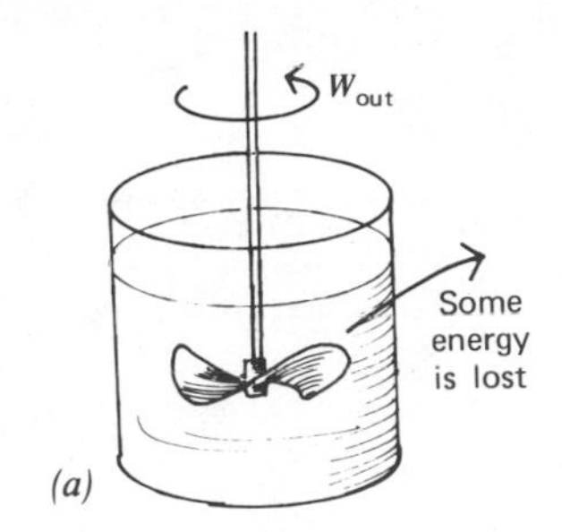

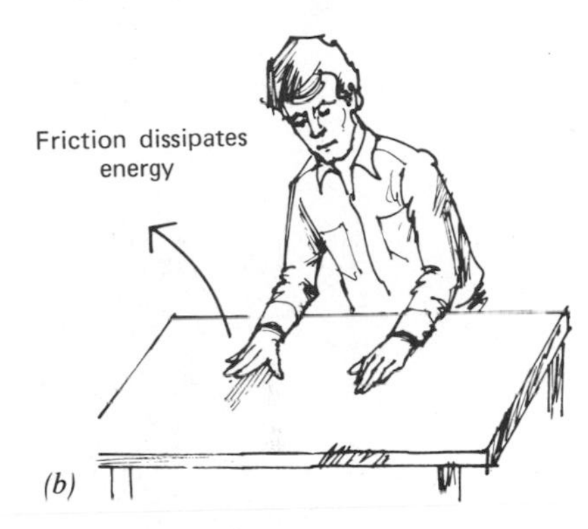

FIG. 3.35. It is impossible to convert all disorganized (internal) energy into organized energy (work).

It may seem rather unfair that we are able to convert only 40 percent of the energy input to the cycle into electrical power. Attempts to convert energy from one form to another are often frustrated by realities imposed by a principle known as the second law of thermodynamics. It was observed previously that stirring a quantity of liquid could have the same effect on temperature as heating that liquid (Figure 3.35*a*). You might vigorously rub the surface of your desk with the palm of your hand and observe a similar phenomenon (Figure 3.35*b*). These are examples of converting organized macroscopic energy available for good uses such as lifting a weight (in other words, a work interaction) entirely into internal (microscopic) energy, which is highly disorganized. However, if we expect the warm molecules of a quantity of water in a container to organize themselves and cause the shaft to spin, we will always be disappointed. Neither should we expect the randomly moving molecules of the desk to become organized and throw our hand off the surface. The second law of thermodynamics suggests that it is impossible to convert disorganized microscopic (internal) energy entirely into organized macroscopic energy (work). We observed from the Rankine cycle described in Example Problem 3.21 that only part of the heat released in the boiler was converted into organized energy leaving the turbine via the spinning shaft.

In Figure 3.36, $\dot{Q}_H$ represents the rate of energy transfer as heat from the energy source to the steam at a high temperature T_H. $\dot{Q}_L$ represents the rate of energy transfer as heat from the steam to the cooling water at a low temperature T_L, and $\dot{W}$ represents the net power delivered by the turbine via the spinning shaft. The thermal efficiency of the cycle would thus be

$$\eta = \frac{\dot{W}}{\dot{Q}_H}$$

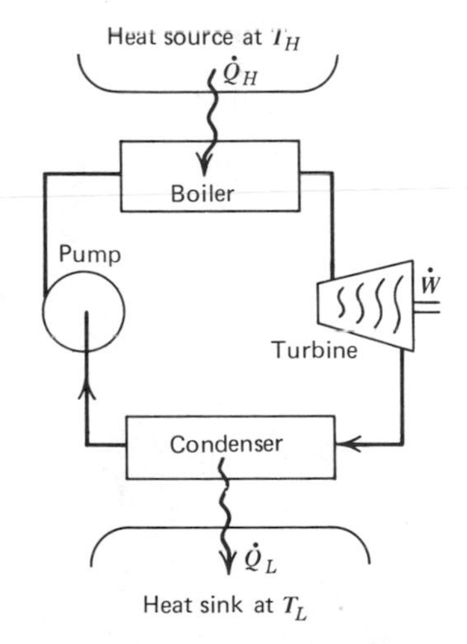

FIG. 3.36. A typical heat engine.

Engineers strive to build power cycles with the highest possible thermal efficiency. The second law of thermodynamics tells us that it is impossible to construct a power cycle that operates with 100 percent efficiency. It also tells us what maximum theoretical efficiency we can expect. The maximum thermal efficiency of the ultimate perfect power cycle is given by

$$\eta_{max} = 1 - \frac{T_L}{T_H}$$

where

T_H = maximum absolute temperature of the energy source (where $\dot{Q}_H$ comes from)

T_L = minimum absolute temperature of the energy sink (where $\dot{Q}_L$ goes)

We thus observe that the thermal efficiency of power cycles might be improved by increasing T_H, the temperature of the heat source, or decreasing T_L, the temperature of the heat sink.

EXAMPLE PROBLEM 3.22

The maximum temperature attainable in thermal power plants is limited by the material properties of the equipment. For instance, if the steam temperature is too high, the turbine blades may become overly flexible. Present materials limit the maximum allowable steam temperature to about 650°C. Assuming that a cool river at 25°C is available as a heat sink, calculate the maximum theoretical efficiency of any power cycle that operates between these temperatures.

Solution

$$\eta_{max} = 1 - \frac{T_L}{T_H}$$

$$= 1 - \frac{298}{923}$$

$$= 0.677$$

These kinds of calculations tell us that even a perfect power plant operating between a heat source at 650°C and a heat sink at 25°C, could not convert more than 68 percent of the input heat into electricity.

As materials and systems are developed that can tolerate higher temperatures, then we can expect the conversion efficiency of heat into work to improve. That is the name of the game for engineers dealing in energy conversion.

The second law of thermodynamics verifies that energy in a highly organized form, available for doing work, is generally more useful and valuable than an equal quantity of disorganized energy.

The relative value to society of various forms of energy is indicated in Table 3.5. Notice that internal energy in the form of coal or natural gas is very inexpensive compared to the more highly organized electrical or mechanical energy. Energy in the form of human labor is extremely valuable in our society compared to other forms of energy. In some societies, human labor may be relatively inexpensive and forms of energy such as electricity may not be available at any price.

Important lessons to be learned from Table 3.5 are as follows: (1) energy comes in many different forms, (2) the value of energy to society depends on the mode by which the energy is transferred, and (3) energy transferred in more highly organized forms (work) is usually more valuable than energy transferred in less organized forms (heat).

Engineers who are concerned with energy conversion are

TABLE 3.5. An Indication of the Relative Value of Various Forms of Energy

Form of Energy	*Assumptions*	*Value of 10^6 kJ (rounded to nearest dollar)*
Fossil Fuels		
Natural gas	36.4 MJ/m^3, \$0.07/m^3	\$2.00
Coal	30.3 kJ/kg, \$0.06/kg	2.00
Diesel	40.8 MJ/l, \$0.25/l	6.00
Gasoline	34.7 MJ/l, \$0.25/l	7.00
Electricity		
Industrial	3.0¢/kWH	8.00
Residential	6.0¢/kWH	16.00
Mechanical Energy		
Farm tractor	\$15 000/6 000 h, 7.6 l/h of diesel, 20% eff.	71.00
Automobile	18¢/km, 6.4 km/l, 15% eff.	221.00
Human Labor		
Manual labor	29.3 kJ/min, 20% eff., \$5.00/h	\$ 14 220.00
Skilled labor	12.6 kJ/min, 20% eff., \$10.00/h	\$ 66 360.00
Professional	8.4 kJ/min, 20% eff., \$15.00/h	\$149 308.00

FIG. 3.37. Brick walls tend to become less organized if left by themselves.

familiar with a property of matter, which is a measure of disorganization. That property is called *entropy.* Systems, if left to themselves, always seem to go from an organized condition to a disorganized condition which means an increase in entropy of the system (Figure 3.37). The second law of thermodynamics may even be stated as:

The entropy of things left by themselves always increases.

You may wonder about living things. As you look at your own body, you realize that the molecules that make up your skin, eyes, hair, lips, fingers, nose, and toes were very likely once scattered about the earth in air, grass, corn, wheat, cows, pigs, chickens, and perhaps even another human being such as Julius Caesar. If you happen to be an environmentally conscious person, it may be comforting to know that your body is organized from mostly recycled material! Does the fact that these molecules have been assembled into that marvelously organized thing you call your body violate the second law of thermodynamics? The answer is *no,* because you have not been left to yourself. The organization accomplished in assembling your body (and the structure of every other living thing) took place at the expense of a great amount of disorganization elsewhere in the universe. As we pointed out at the beginning of this chapter, vast amounts of energy flow from the sun where it is localized, concentrated, and somwhat organized. Most of this energy goes directly to outer space where it apparently becomes totally disorganized and unavailable for good uses. A tiny fraction is captured on the earth through photosynthesis into organized or organic structures. Your highly organized body is now in a state of reduced entropy, but the entropy of the universe as a whole has increased tremendously as a result of the process required to assemble that body. The second law of thermodynamics also suggests that if you were locked in a room all by yourself for about 100 years or so, that you would become considerably less organized (in a state of higher entropy) than you are now. Such truths appear rather self-evident!

The second law of thermodynamics raises interesting philosophical questions about the origin, nature, and final destiny of the universe. Considerable controversy exists regarding the seeming contradiction of the theory of evolution by the principle of increase of entropy. As a student of energy conversion, you will learn more about the second law and this fascinating property of matter called entropy as your engineering education progresses.

PROBLEMS

3.46. Write a one-page essay describing the operation of a Rankine power cycle. Include a discussion of why it is necessary that energy be rejected from the cycle as heat. In other words, why must the thermal efficiency of power cycles be less than 100 percent? Comment also on how the temperatures of the heat source and heat sink influence the thermal efficiency of the power cycle.

3.47. (a) In a certain nuclear power plant reactor the maximum temperature that can be tolerated by the reactor core is 700°C. Calculate the maximum possible thermal efficiency of this power plant. Assume that a heat sink (the atmosphere) is available at 25°C.

(b) A magnetohydrodynamic power generator has temperatures near 1650°C. Compare the maximum possible thermal efficiency of this system with that of the nuclear power plant described in part (a).

3.48. In certain societies, manual tasks such as lifting ore from a mine or water from a well are accomplished by human muscle energy. This suggests that in those societies mechanical energy is more expensive than human muscle energy.

(a) Estimate the annual wages of a person producing work at the rate of 30 kJ/min, but paid at the value of mechanical energy produced by a farm tractor listed in Table 3.3.

(b) Write a brief essay discussing how the "quality of life" or "standard of living" is influenced by the relative value of different energy forms and their availability.

3.49. Write a brief essay summarizing the second law of thermodynamics.

3.50. We have learned that both energy and entropy are properties of matter. Energy is conserved in most energy conversion processes. Is entropy also conserved? Discuss your answer.

PART B/SELECTED ENERGY TOPICS

3.11/Organic Fuels

Fossil fuels act as long-term storage systems for solar energy. A small fraction of long-ago incident solar energy, trapped as microscopic potential energy in the organized structure of organic sys-

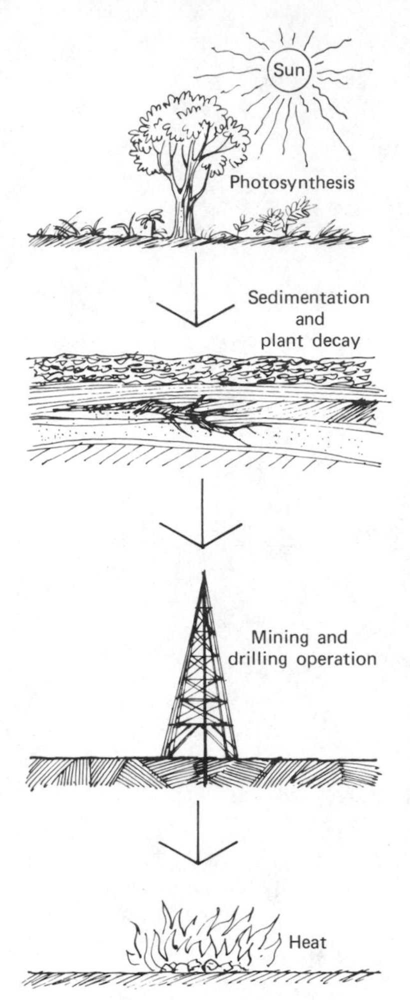

FIG. 3.38. Fossil fuels store solar energy.

tems became, over eons of time, those marvelously concentrated energy forms—natural gas, oil, and coal—on which our way of life has come to depend. (See Figure 3.38.)

3.11.1/Natural Gas and Oil

Consider people so fortunate as to live in a society where a miracle fuel, piped directly into their homes and places of employment, is automatically burned by sophisticated equipment in just the right manner to warm or even cool them as desired. This fuel burns so cleanly that almost no pollutants are released into the air on combustion. For most people, an entire year's supply of this miracle fuel could be purchased for about three day's wages. Such has been the case with natural gas.

The cost of delivering natural gas must increase as supplies in convenient locations near the surface of the earth dwindle. Those people who make it their business to find and deliver this precious substance must continually hunt further and drill deeper to find enough to keep the pipelines full and the innumerable fires at the end of those pipelines burning.

There now seems to be less new gas discovered each year in the United States than is burned in that same time period. It might be possible to temporarily reverse that trend for a time, but inevitably we must find a substitute for the approximately 23 billion ft^3 of natural gas burned in the United States each year to provide about one-fourth of our total energy needs.

Contemplate the impact the availability of immense quantities of low-cost oil has had on our society. Oil stores so much energy in such a small volume that only a cupful burned in an engine could lift you nearly 10 000 ft into the sky. A 20-gal tank of gasoline (refined from oil) will transport an entire family and their automobile several hundred miles across the landscape. The fuel costs for such a trip have usually amounted to only two or three hours wages. Oil, and fuels derived from oil, provided the ordinary person with instant mobility. Use of mass-transit systems declined and immense freeway systems were developed in response to the clamor for more automobiles.

As a result, some 130 billion gal of gasoline and diesel are converted into heat, water vapor and pollution on U.S. highways and byways each year. Even so, transportation uses only about half the oil consumed. Roughly 25 billion gallons flow into the fires of electrical generating plants, about 25 billion gal into furnaces to heat homes and commercial enterprises, some 35 billion gal feed the fires of industry, and nearly 40 billion gal go for nonenergy uses where it is not burned directly but converted into essential materials. All together, we require the stupendous quantity of 250 billion gal of oil (some 6 billion bbl) annually to keep our society running (Figure 3.39).

No one knows for sure how much oil and natural gas remain in that part of the earth's crust controlled by the United States, but estimates of our undiscovered resources are impressive. The U.S. Geological Survey estimates onshore recoverable deposits in the United States at 200 to 350 billion bbl of oil and 900 to 1600 trillion ft^3 of natural gas. Adding the estimated potential of 65 to 130 billion bbl of oil and 400 to 800 trillion ft^3 of gas thought to exist under water along the U.S. outer continental shelves, the potential supply is truly remarkable.

In a country so lavishly endowed with these miracle fuels, why are we not able to produce all that we need? Perhaps the answer is that we are spoiled. For several years, oil and gas were produced in abundance from relatively shallow wells and the cost was low. As wells had to be drilled deeper, costs naturally increased. Drilling a 20 000-ft well involves much more than simply using more pipe. Much larger rigs are required and all costs increase tremendously as the bits go deeper (Figure 3.40). The problems and costs associated with drilling underwater are enormous.

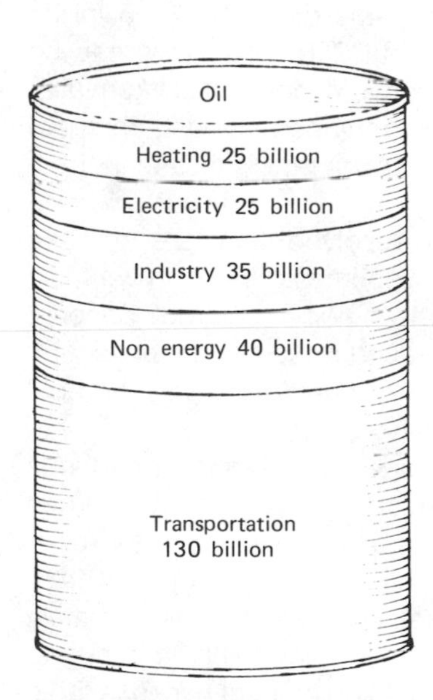

FIG. 3.39. The use of oil in the United States.

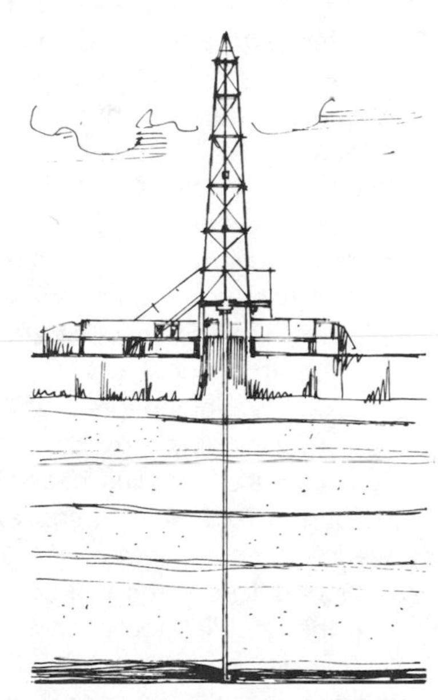

FIG. 3.40. Larger oil rigs and deeper wells are required.

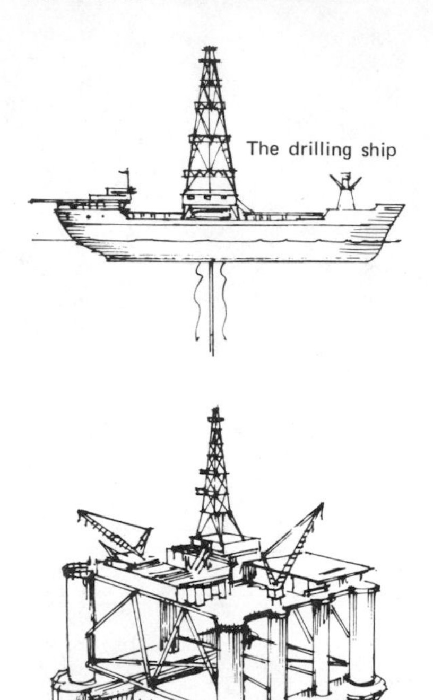

FIG. 3.41. Several types of oil rigs are now being used for offshore drilling.

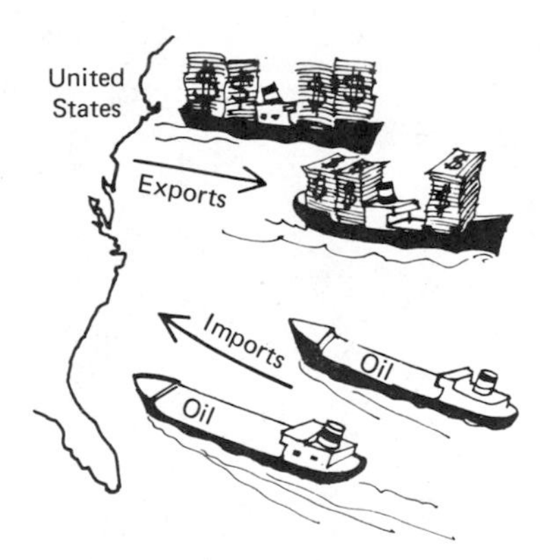

FIG. 3.42. The United States depends on foreign sources for nearly half of their oil.

The United States has become dependent on foreign sources for nearly half of the oil it uses. The oil-producing and exporting countries (OPEC) have raised their prices in response to the decreasing value of the dollar. In 1959, a barrel of oil cost $2.50 and gold was $35 an ounce on the world market. In other words, an ounce of gold would buy 14 bbl of oil. Twenty years later a barrel of oil was priced at $40 and gold at $600 an ounce. An ounce of gold still bought 14 bbl of oil, but the value of the dollar had slipped precipitously.

In summary, we depend on oil and natural gas for more than three quarters of our total energy needs. While the low-cost, easily recoverable supplies in the United States seem to be nearing an end, there are probably large reserves at deeper levels and in the outer continental shelf. We face some major challenges. We must curb the fantastic rate at which we are burning these precious fuels. We must locate and develop additional domestic reserves in an economically competitive and environmentally acceptable manner (Figure 3.41). We must vigorously begin the awesome task of finding substitutes for naturally occurring oil and gas in our total mix of energy resources (Figure 3.42).

3.11.2/Coal

The United States seems to have a lot of coal. A generally accepted estimate of known recoverable deposits is 1.5 trillion tons. Another estimate suggests about 200 billion tons can be recovered under present local economic conditions using available technology. These large numbers may be placed in perspective by noting that present U.S. production rate is less than 1 billion tons per year.

Three major questions must be resolved if we are to make effective use of our abundant coal resources. First, can we mine the coal at the tremendous rates that would be required? Second, can we burn the coal at vastly increased rates without unacceptable environmental degradation? Third, can we convert the coal to other forms of energy at competitive cost?

Where the coal lies near the surface, strip-mining methods may be employed. Giant shovels first strip away the overburden covering the coal, then scoop up the coal and load it into trucks or other devices that transport it to the point of use (Figure 3.43).

If the coal lies too deep to be strip mined then people must go underground after it. In shaft mining, a vertical hole is dug to provide access to the coal (Figure 3.44). The coal mined is lifted vertically to the surface. In drift mining a horizontal tunnel is made

FIG. 3.43. An open coal mine. (Photos courtesy of Bridger Coal Company, Rock Springs, Wyoming)

into a hillside to reach the coal and in slope mines a sloping tunnel reaches down at some angle. Coal mining has traditionally been dirty, dangerous work.

Think of the challenges involved in increasing coal production by 1 billion tons per year. An immense increase in people, mining machines, trains, and other equipment would be needed.

It is difficult to burn coal in vast quantities without environmental risk. Coal often contains pollutants such as sulfur, which on combustion, is released into the air. Most of the sulfur dioxide that goes up the stack of a power plant is converted into sulfuric acid and is brought back to earth. There is evidence that this acid rain may be damaging to the certain parts of the ecosystem. Burning an additional billion tons of coal would release nearly 4 billion tons of carbon dioxide into the air. Some have expressed concern regarding the possible impact on climatic conditions. Increased CO_2 levels in the atmosphere tend to retard the radiation heat transfer between earth and outer space leading possibly to a slight increase in average temperature of the earth's surface. This, in turn, could precipitate melting of polar ice, raise the level of the oceans, and bring disaster to coastal cities.

Coal can be converted into more convenient energy forms. In coal gasification plants, coal is converted into a pipeline quality gas capable of performing all the functions of natural gas. Roughly 16 500 ft^3 of gas can be produced from 1 ton (2 000 lb) of coal. If we were to look to gas manufactured from coal to

FIG. 3.44. Coal is plentiful, but there are problems associated with the mining and burning of coal.

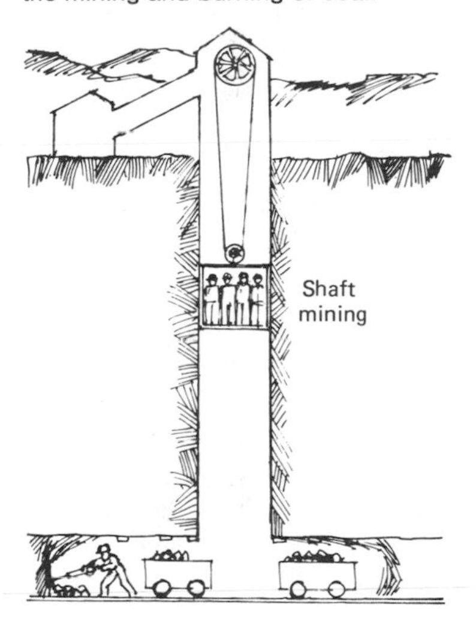

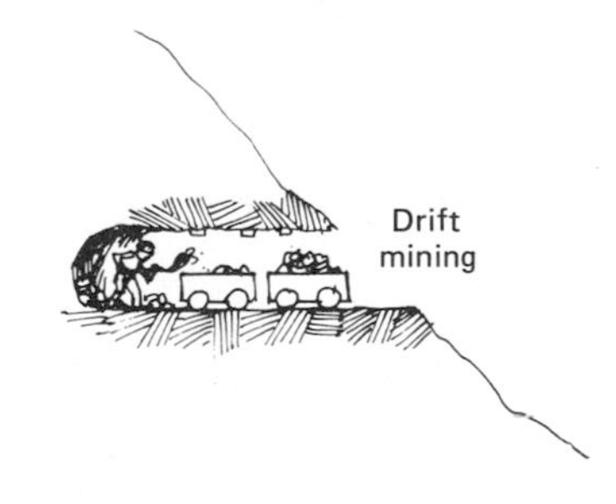

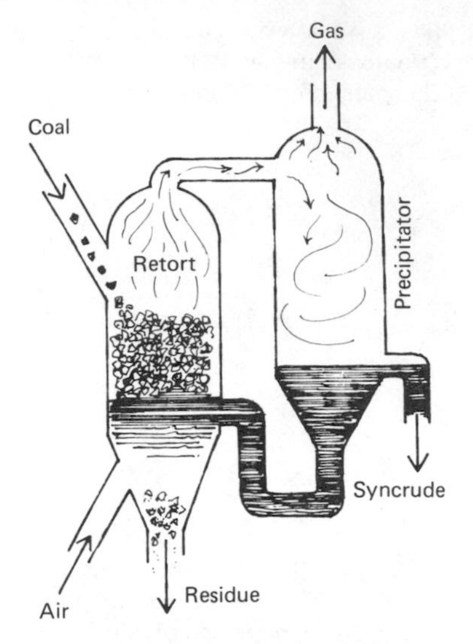

FIG. 3.45. Gas and oil may be made from coal.

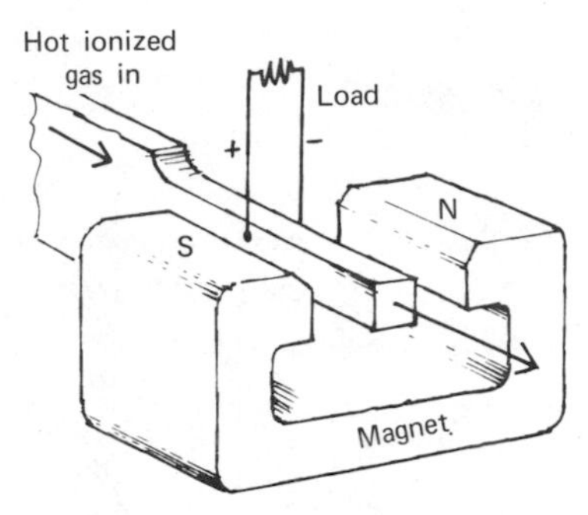

FIG. 3.46. Magnetohydrodynamic (MHD) power generation.

replace half of the 23 trillion ft^3 of natural gas this country uses each year, about three quarters of a billion tons of coal would have to be mined and processed annually for just this purpose (Figure 3.45).

A synthetic oil can be made from coal at the rate of about 3 to 3.5 bbl of oil per ton of coal. If we were to obtain half of the 6.5 billion bbl of oil presently used in this country from coal liquification processes, about 1 billion tons of coal would have to be mined and processed for just this purpose.

Of course, coal can be converted into electricity. A great engineering effort is being devoted to learn how to burn coal more cleanly and make the conversion process more efficient. We learned earlier that the efficiency of power cycles can be improved by increasing the temperature of the heat source. One exciting technique for converting coal into electricity is magnetohydrodynamics (MHD).

It is possible to raise the coal combustion temperature so high that electrons are knocked free from some of the energetically colliding gas molecules. Such a mixture of free electrons, ions, and neutral particles is called an ionized gas. Referring to Figure 3.46, if an ionized gas is passed through a duct located between the poles of a superconducting magnet, the charged particles interact with the lines of force in such a way that the negatively charged electrons are deflected upward while the positively charged ions are deflected downward thereby creating a charge separation and, hence, an electric field. If the upper and lower walls of the duct are made of a suitable electrode material, and connected through an external load, the electric field causes a d.c. current to flow through the external circuit. This device for extracting electric power from ionized gases without any moving parts is known as magnetohydrodynamic (MHD) power generation (Figure 3.47).

Since the gas that emerges from an MHD generator is still very energetic, additional power can be extracted by allowing it to drive a conventional Rankine cycle power generator.

The combined MHD/Rankine cycle generator is of considerable current interest since, for each 100 kJ of input fuel energy, 15 kJ can be extracted from the MHD generator and 40 kJ from the Rankine cycle generator. Thus, the combined thermal efficiency is an attractive 55 percent.

The improvement of efficiency of the combined MHD-Rankine cycle over the Rankine cycle alone primarily results from the much higher temperature allowable in the MHD channel than can be tolerated in the turbine. There are no moving parts in the MHD channel while the rapidly rotating turbine blades are highly stressed.

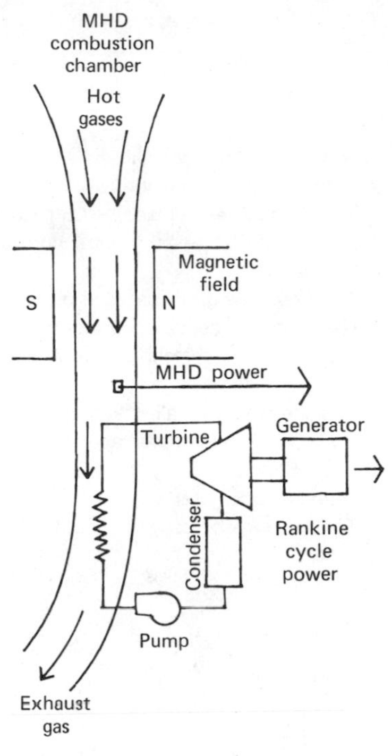

FIG. 3.47. The combined MHD/Rankine cycle generator.

EXAMPLE PROBLEM 3.23

Suppose 20 million tons of coal having a heating value of 12 000 Btu/lbm are burned daily in coal-fired electrical generating plants with an overall thermal efficiency of 36 percent. (a) Calculate the amount of electrical energy generated daily. (b) If the overall efficiency of conversion could be increased to 55 percent, how much coal could be saved while producing the same quantity of electricity?

Solution

(a) Power generated $= \dot{W}$

$$\dot{W} = \frac{(0.36)(20 \times 10^6)(2000)(12\,000)(\text{Btu/day})}{(3413\ \text{Btu/kW}\cdot\text{h})}$$

$$= 5.06 \times 10^{10}\ \text{kW}\cdot\text{h/day}$$

(b) If efficiency increased to 55 percent, the coal burned would be

$$(20 \times 10^6)\left(\frac{0.36}{0.55}\right) = 13.09 \times 10^6\ \text{tons/day}$$

Coal saved would be 6.9×10^6 tons/day. It becomes obvious why there is so much interest in improving energy conversion efficiency.

In summary, the United States has immense coal supplies that potentially can be converted into electricity, and both gaseous and liquid fuels. However, the mining and the conversion of coal at the stupendous rates necessary and in an environmentally acceptable manner present many engineering challenges.

3.11.3/Oil Shale and Tar Sands

Oil shale appears to be a promising source of synthetic liquid fuel. The Green River Formation underlying large areas of Utah, Colorado, and Wyoming, is the largest known oil shale deposit in the world. Estimates of the oil potential of the Green River Formation run into trillions of barrels as suggested by Table 3.6.

The only known method of releasing the kerogen from the rock involves heating to about 900 F. The two ways of doing this being pursued are surface retorting and in-place (in-situ) conversion. Surface retorting requires that the rock be mined, hauled to

TABLE 3.6. Reserves of Green River Oil Shale Coal and Oil Sands

		Equivalent Btu Contents
Green River Formation Oil Shale		
Total resources in place, +10 g/tons grade	4×10^{12} bbl oil	24×10^{18} Btu
Total resources in place, +25 g/tons grade	1×10^{12} bbl oil	6×10^{18} Btu
Prime reserves, +25 g/tons, current technology	0.3×10^{12} bbl oil	1.8×10^{18} Btu
U.S. Coals		
Total remaining resources in place to 6,000	3.2×10^{12} tons	64×10^{18} Btu
Recoverable reserves, current technology	553×10^{9} tons	11×10^{18} Btu
Prime, surface-mineable reserves	360×10^{9} tons	7.2×10^{18} Btu
Alberta Oil Sands		
In place reserves, all stratigraphic units	895×10^{9} bbl oil	5.4×10^{18} Btu
Surface-mineable reserves	74×10^{9} bbl oil	0.44×10^{18} Btu
U.S. Oil Sands		
Utah in-place reserves	28×10^{9} bbl oil	0.17×10^{18} Btu
Total U.S. reserves	30.3×10^{9} bbl oil	0.18×10^{18} Btu

Taken from "Synthetic Fuels," a report published by Cameron Engineers, Inc., 1315 South Clarkson St., Denver, Colorado.

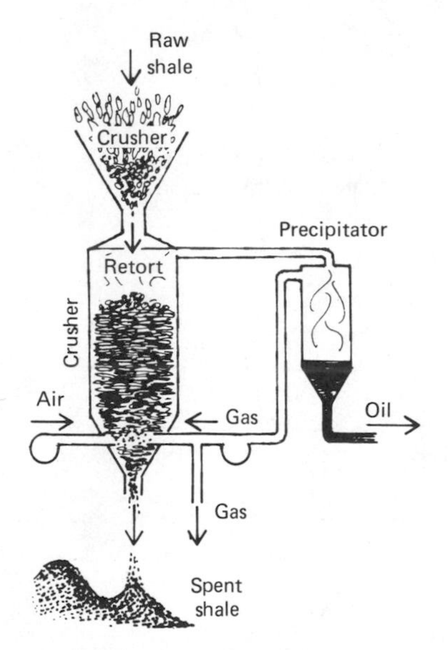

FIG. 3.48. Synthetic liquid fuels may be produced from oil shale and tar sands.

the surface, crushed, and dumped into a large vessel to have most of the oil cooked from it. The spent shale, occupying a volume about 13 percent larger than the raw shale, must then be removed from the retort and disposed of (Figure 3.48).

Techniques for in-situ recovery of kerogen from oil shale generally involve (1) drilling a number of wells from the surface into the oil shale, (2) fracturing the shale by explosives, high pressure, or even high-voltage electricity, (3) pumping air down one or more injection wells, (4) igniting the shale at the base of these wells, (5) recovering the oil generated through recovery wells (Figure 3.49). This has the advantage of eliminating the mining and disposal problems, but a much smaller fraction of the kerogen is recovered.

A combination of mining and in-situ processing shows promise. About 25 percent of the shale is first removed from the lower part of the shale formation, hauled to the surface, and processed. This provides space for the shale above the mined area to settle into when fractured, thus greatly increasing its porosity and improving the conditions for in-situ recovery.

EXAMPLE PROBLEM 3.24

A surface retorting plant is to be constructed with a capacity of 50 000 bbl per day of syncrude from oil shale containing 25 gal of kerogen per ton of shale. Estimate the amount of oil shale that must be mined annually to supply the plant.

Solution

We recall that 1 bbl is equivalent to 42 gal. We might expect to recover $\frac{25}{42}$ = 0.60 bbl of oil per ton. Thus we must mine

$$\frac{50\,000 \text{ bbl/day}}{0.6 \text{ bbl/ton}} = 83\,333 \text{ ton/day} = 30.4 \text{ million tons/yr}$$

This is truly a stupendous mining operation when one considers that the largest underground coal mine in the United States currently produces about 4 million tons per year.

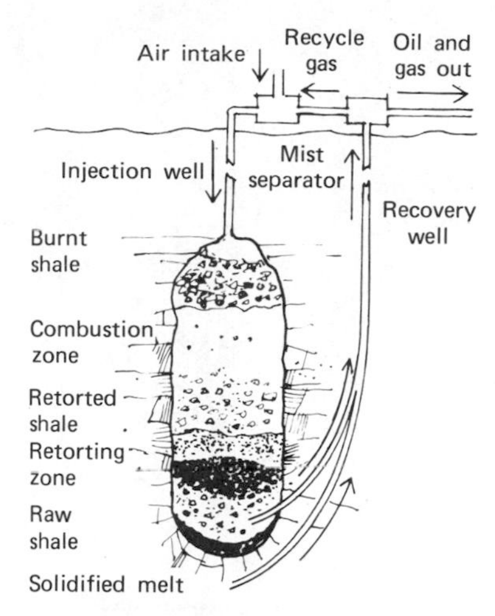

FIG. 3.49. In situ recovery of kerogen from oil shale.

The United States has an estimated 30 billion bbl of oil contained in tar sand. About 85 percent of this total is in Utah. As the name implies, the oil is more like tar than oil and will not flow easily unless heated. It has been suggested that one possible technique for heating and recovering oil in tar sands involves using radio waves. Radio wave electrodes would be placed in horizontal rows in the tar sand deposits. Using the same principle as a microwave oven, the radio waves would gradually heat the tar sand allowing the oil to be displaced by a fluid such as water (Figure 3.50).

The method is indicative of the exciting kinds of engineering research that will be involved in tapping such energy resources as oil shale and tar sands.

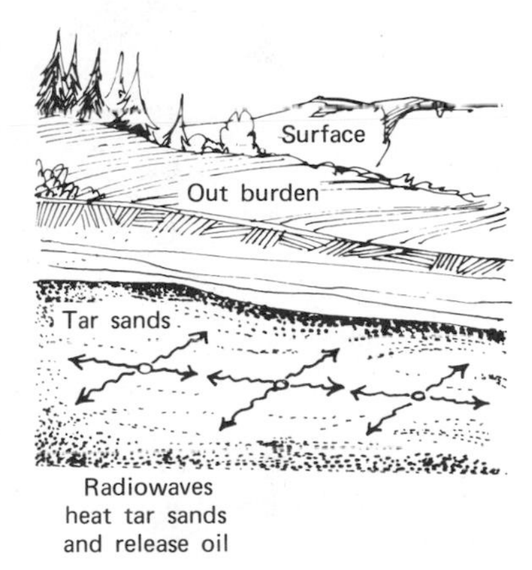

FIG. 3.50. One scheme for heating tar sands using radio waves.

3.11.4/Energy Via Bioconversion

Biomass is a general term referring to organic material such as grass, trees, leaves, wheat, corn, alfalfa, algae, seaweed, or any of an infinite variety of growing things. Bioconversion seems to refer not only to the photosynthesis by which solar energy is stored in the organized structure of these materials but also to additional processes by which the plant material is converted to other forms such as oil, gas, or alcohol. Bioconversion is becoming an area of intense interest for many engineers.

Only a small fraction of the solar energy striking the earth can be potentially trapped by growing things (Figure 3.51). Of the sunlight entering the atmosphere only about 53 percent survives the scattering, reflecting, and absorbing processes to reach the ground. Of that which does reach the ground only about 25 percent is of the proper wavelength to stimulate photosynthesis and the plant can utilize only a fraction of that. It would be unreasonable to expect more than about 1 or 2 percent of the total incident sunlight to be captured by a crop! In spite of these limitations the quantities of dry matter that can be produced are impressive.

FIG. 3.51. Plant material may be converted to fuels such as oil, gas, and alcohol.

TABLE 3.7. Yield of Selected Crops

Crop	*Location*	*Yield Tons/ Acre—Year*
Sorghum	Puerto Rico	30
Sorghum	Mississippi	8
Corn	Georgia	7
Kenaf	Forida	20
Water hyacinth	Florida	16
Sugar cane	Mississippi	20
Sudan grass	California	15
Napier grass	Puerto Rico	40
Alfalfa	New Mexico	8
Sycamore	Georgia	4
Red alder	Washington	10
Eucalyptus	California	24
Kelp	Ocean	25
Algae	Ponds	30

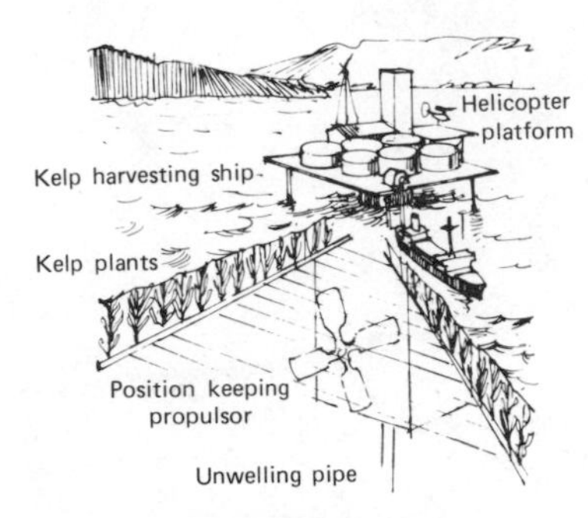

FIG. 3.52. Methane gas may be converted from kelp through ocean farming.

Table 3.7 indicates quantities of biomass production which might be expected.

Generally the heating value of dry biomass is about 7000 to 8000 Btu/lb. The fuel potential from such crops may be estimated as roughly one to two barrels of oil or 4000 to 8000 ft^3 of methane per ton depending on the kind of crop.

There are those who argue that we will need our arable land to produce food and timber rather than fuel and thus it would be impractical to plan on devoting huge areas to the production of energy crops. The notion of ocean farming in which kelp (seaweed) is produced, harvested, and converted to methane in an integrated system is depicted in Figure 3.52. Such a system would not utilize valuable farm or forest lands.

EXAMPLE PROBLEM 3.25

The U.S. uses about 23 trillion ft^3 of natural gas per year. Estimate the area of ocean required to produce that quantity of methane from kelp.

Solution

We will use the value of 25 tons per acre per year from Table 3.7 and assume that 6000 cubic feet of methane may be produced from a ton of dry kelp. Thus one acre would produce

$$(25 \text{ tons/acre-yr})\,(6000 \text{ ft}^3\text{/ton}) = 150\,000 \text{ ft}^3\text{/acre-yr}$$

The area required to produce 23 trillion ft^3/yr would then be

$$(23 \times 10^{12}\text{ft}^3/\text{yr})/(1.5(10)^5/\text{acre-yr}) = 1.5 \times 10^8 \text{ acres}$$
$$= 2.3 \times 10^5 \text{ miles}^2$$

This would be an area about 500 miles on each side.

Alcohol production from biomass is an ancient art. Now serious attention is being devoted to large scale production of alcohol for fuel. Gasohol (a blend of alcohol and gasoline) is finding considerable demand as a fuel. Table 3.8 indicates how much alcohol one might expect from several crops.

TABLE 3.8. Typical Alcohol Yield from Selected Crops

Crop	*Typical Crop Yield*	*Alcohol Yield*
sugar beets	20 ton/acre	23 gal/ton
wheat	40 bu/acre	2. gal/bushel
corn	90 bu/acre	2.7 gal/bushel
potatoes	15 ton/acre	28 gal/ton

EXAMPLE PROBLEM 3.26

The U.S. burns approximately 100 billion gal of gasoline per year for highway travel. Suppose an effort were launched to produce 10 billion gal of alcohol per year as part of a national gasohol program. The alcohol is to be produced principally from surplus corn and wheat. If produced from wheat, estimate the number of acres required.

Solution

The wheat required would be 10×10^9 gal/(2.6 gal/bu) = 3.8×10^9 bushels. The farming area required would be roughly 3.8×10^9 bu/(40 bu/acre) = 95 million acres.

There is little question that solar energy can be converted to useful forms via bioconversion. Again, the engineering challenge is to accomplish that conversion in an economically competitive and environmentally acceptable fashion.

PROBLEMS

3.51. If you were the president of the United States, what specific steps would you take to (a) reduce oil consumption and (b) increase domestic supply of oil?

3.52. Estimate the number of barrels of oil that have been burned in automobiles transporting you about thus far in your life. Estimate how many barrels of oil you will use in automobile travel for the rest of your life. Clearly state all the assumptions you make.

3.53. Compare the weight that an oil-drilling rig must be capable of lifting with (a) 25 000-ft drill string and (b) with the weight of a 5000-ft drill string. Assume drill pipe weighs 200 lb/ft.

3.54. Discuss why natural gas is such a popular heating fuel as compared with coal.

3.55. Estimate the weight that one gallon of gasoline burned in a 20 percent efficient engine will lift to a height of two miles.

3.56. Discuss the engineering problems associated with recovering oil from beneath the continental shelves.

3.57. What are the major engineering problems associated with mining and burning coal?

3.58. An idea for transporting coal is to mix it with water (50 percent coal - 50 percent water by weight), fine grind the coal and pump the resulting slurry through pipelines. One of the objections to coal slurry pipelines is the exporting of water from the water-short regions where western coal is often found. Estimate the amount of water required to operate a coal slurry pipeline transporting 20 million tons of coal per year. Express your answer in acre feet per year and cubic feet per second.

3.59. Suppose you are project engineer for a coal-liquifaction plant producing 50 000 bbl of syncrude per day from coal. Estimate the rate at which coal must be supplied to your plant. Express your answer in pounds per hour, tons per day and tons per year. Assume that 3.5 bbl of syncrude can be obtained from one ton of coal.

3.60. A 3000-MW, coal-fired power plant is planned. Conventional technology would yield a coal to electricity conversion efficiency of 36 percent. By incorporating an MHD cycle in the power plant it is hoped to increase the efficiency to 55 percent. Estimate the amount of coal saved over the expected 40-year life of the plant by incorporating the MHD technology.

3.61. Explain how an MHD power generator works. Use a sketch to help clarify your explanation. Why is there such great interest in developing MHD systems?

3.62. Suppose you are a project engineer for a plant to produce 1 000 000 bbl of syncrude per day from high-grade oil shale. Estimate the rate at which oil shale must be supplied to your plant. Express your answer in pounds per hour, tons per day and tons per year. Assume that 25 gal of syncrude can be extracted from one ton of oil shale.

3.63. Devise a method for in-situ recovery of oil from oil shale and tar sands. Use sketches as required to effectively convey your ideas. What volume of in-place shale would be required to yield 1 000 000 bbl of syncrude per day? Assume the shale weighs 2700 lb/yd^3, has an oil content of 25 gal/ton, and that the in-situ process extracts 70 percent of the oil content.

3.64. About 20 gal of synthetic fuel can be obtained from 1 ton (2000 lbm) of high-grade western oil shale. How many tons of oil shale must be mined hourly to provide 3 billion bbl of fuel per year. Note that 3 billion bbl represents about 50 percent of the U.S. demand in 1980. How does your calculated rate of mining shale compare with the total rate of mining coal in the United States? The estimated total amount of coal mined in the United States in 1980 is approximately 800 million tons.

3.65. Assuming that 10 gal of fuel oil can be produced from a ton of forest products and that the heating value of the forest products is 7500 Btu/lb, estimate the efficiency of the conversion process.

3.66. Approximately 1 ft^3 of methane may be produced from 1 lbm of steer manure. A large steer produces about 7 lb of solid wastes per day. One medium-size coal gasification plant is designed to produce 250 million ft^3 of methane per day. The wastes from how many steers would be required to produce an equivalent quantity of gas? What do you conclude from your answer?

3.67. Suppose you have a home in the country on a 2-acre lot. In a typical winter you burn 1000 gal of fuel oil to heat your home. You decide to grow your own heating fuel and plant a rapidly growing variety of sycamore trees with the idea of burning the wood in a stove or furnace. Estimate the quantity of wood required and whether or not you could expect to produce that much on your 2-acre lot.

3.68. Prepare a table comparing the energy content of the synthetic fuel obtained from 1 ton of high-grade western oil shale with the energy content of (a) 1 ton of bituminous coal, (b) methane produced from 1 ton of dry waste from dairy cows or beef cattle (0.05 lbm of methane per lbm of dry waste, (c) methane produced from 1 ton of dry alfalfa hay (0.20 lbm of methane per lbm of dry hay), (d) ethanol produced from 1 ton of wheat (0.28 lbm of ethanol per lbm of wheat), and (e) methanol from 1 ton of forest wastes (0.04 lbm of methanol per lbm of forest wastes). What do you conclude?

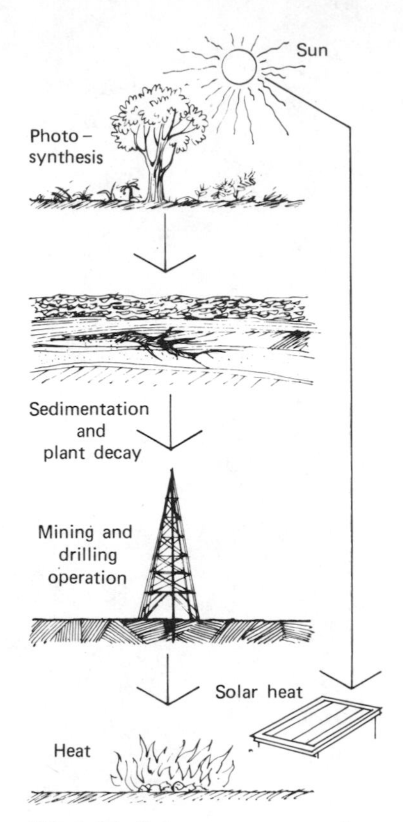

FIG. 3.53. Solar energy conversion devices enable the direct use of sunshine and bypass photosynthesis.

3.12/Solar Energy

Unlike coal, oil, natural gas or uranium, sunshine has been widely regarded as being free. If we view the photons or packets of sunshine in their natural form just as they arrive on our back on a warm, summer day, this philosophy is true. However, if we desire to convert this free energy into a more useful organized form, we must often employ sophisticated mechanical devices that are themselves products of fossil fuel using industries. In terms of dollars or energy units, the conversion system required at a solar installation is no more without cost than are the machines and devices at a hydropower dam or a nuclear reactor generating station. In many instances useful energy from sunshine is much more costly than energy from conventional sources. There is a great deal of effort being devoted to design and construct less expensive methods of utilizing solar energy (Figure 3.53).

3.12.1/Solar Collection Surfaces

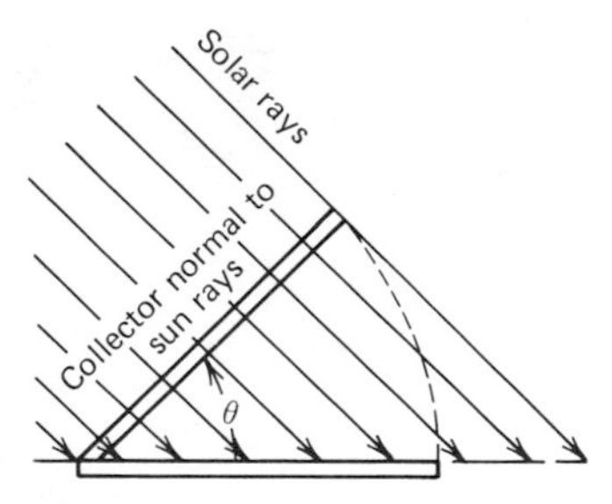

FIG. 3.54. A collector oriented normal to the sun's rays intercepts more energy than a horizontal collector.

The average rate at which sunshine falls on a collector surface differs widely from location to location and season to season as suggested by the average amount of solar energy incident on a flat horizontal surface in any one day at the location and month indicated in Table 3.9.

A little thought suggests that more solar energy can be intercepted by a collector oriented normal to the sun's rays than by a horizontal collector (Figure 3.54). Simple trigonometry indicates that energy incident on a collector normal to the sun's rays = (energy incident on horizontal collector)/cos θ.

An important question is, at what angle will the collector be normal to the sun's rays and thus intercept the most solar energy? To answer that, we must consider the relationship between the sun and the earth. The earth's axis is tilted about 23.45° from the axis of revolution of the earth about the sun so that winter occurs

TABLE 3.9. Mean Daily Solar Radiation (langleys)[a]

Location	Aprox. Lat°	Jan.	Feb.	Mar.	Apr.	May	Jun.	Jul.	Aug.	Sep.	Oct.	Nov.	Dec.	Anl. Ave.
Alaska Fairbanks	65	16	71	213	376	471	504	434	317	180	82	26	6	224
Arizona Tucson	32	315	391	540	655	729	699	626	588	570	442	356	305	518
California China Lake	36	306	412	462	683	772	819	772	729	635	467	363	300	568
Colorado Grand Junction	59	227	324	434	546	615	709	676	595	514	373	260	212	465
Florida Miami	26	349	415	489	540	553	532	532	505	440	384	353	316	451
Hawaii Honolulu	21	363	422	416	559	617	615	615	612	573	507	426	371	516
Idaho Boise	44	138	236	342	485	585	636	670	576	460	301	182	124	395
Kansas Dodge City	38	255	316	418	528	568	650	642	592	493	380	285	234	447
Louisiana Lake Charles	30	245	306	397	481	555	591	526	511	449	402	300	250	418
Maine Caribou	47	133	231	364	400	476	470	508	448	336	212	111	107	316
Massachusetts Boston	42	129	194	290	350	445	483	486	411	334	235	136	115	301
Michigan East Lansing	43	121	210	309	359	483	547	540	466	373	255	136	108	311
Montana Great Falls	48	140	232	366	434	528	583	639	532	407	264	154	112	366
Nevada Las Vegas	36	277	384	519	621	702	748	675	627	551	429	318	258	509
New Mexico Albuquerque	35	303	386	511	618	686	726	683	626	554	438	334	276	512
New York Central Park	41	130	199	290	369	432	470	459	389	331	242	147	115	298
Oregon Astoria	40	90	162	270	375	492	469	539	461	354	209	111	79	301
Utah Salt Lake City	41	163	256	354	479	570	621	620	551	446	316	204	146	394
Washington Spokane	48	119	204	321	474	563	596	665	445	404	225	131	75	361
Wyoming Laramie	41	216	295	424	508	554	643	606	536	438	324	229	186	408

[a]One langley = 41.9 kJ/m^2 = 3.69 Btu/ft^2.

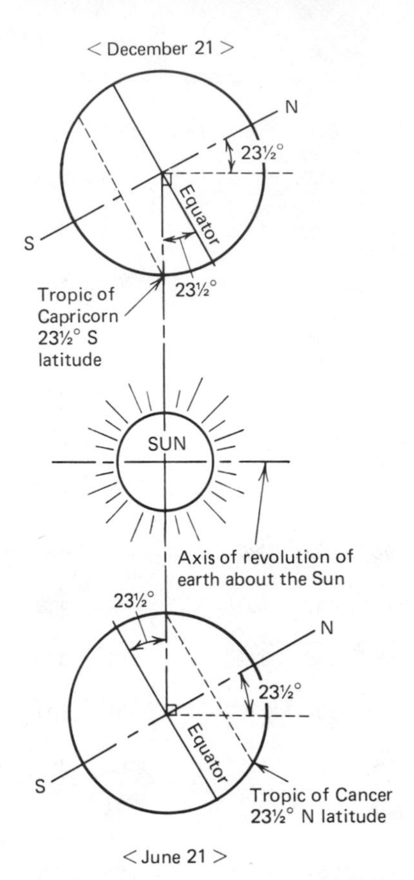

FIG. 3.55. The earth's axis is tilted. The northern hemisphere has winter while the southern hemisphere has summer.

in the northern hemisphere while summer occurs in the southern hemisphere and vice versa as suggested by Figure 3.55. The summer solstice, that point at which the sun appears furthest north from the equator is called the Tropic of Cancer. The sun reaches its summer solstice on about June 21, or the 172nd day of the year. The winter solstice, that point at which the sun appears furthest south from the equator is called the Tropic of Capricorn and occurs on about December 21, the 355th day of the year.

The declination, D, of the sun is the angular position of the sun at solar noon with respect to the plane of the equator and can be found from the approximate equation

$$D = 23.45 \sin\left(360 \frac{284 + n}{365}\right)$$

where n is the day of the year beginning January 1 (see Table 3.10). Notice that on June 21, $D = 23.45°$ while on December 21, $D = -23.45°$. In the Northern hemisphere, D is negative between September 21 and March 21.

The latitude, L, is the angular position on the surface of the earth with respect to the plane of the equator, as shown in Figure 3.56. The angle between the latitude vector and the sun's ray is, therefore, $L + (-D)$. The angle N between the horizontal surface (normal to the latitude vector) and the collector surface (normal to the sun's rays at solar noon) is also given by $N = L + (-D)$.

EXAMPLE PROBLEM 3.27

At what angle with the horizontal should a solar collector located in Salt Lake City, Utah, be placed so as to be normal to the rays from the midday sun on (a) October 2, (b) December 21, and (c) March 15.

Solution

(a) From Table 3.9, we observe that the latitude of Salt Lake City, Utah, is about 41°. From Table 3.10, we observe that October 2 is the 274th day of the year. The declination is:

$$D = 23.45 \sin\left(360 \frac{284 + 274}{365}\right) = -4.2°$$

Consequently, the appropriate collector angle for October 2 would be

$$N = L + (-D)$$

$$N = 41° + (+4.2) = 45.2°$$

TABLE 3.10. The Number of Each Day of the Year

Day of Mo.	*Jan.*	*Feb.*	*Mar.*	*Apr.*	*May*	*Jun.*	*Jul.*	*Aug.*	*Sep.*	*Oct.*	*Nov.*	*Dec.*	*Day of Mo.*
1	1	32	60	91	121	152	182	213	244	274	305	335	1
2	2	33	61	92	122	153	183	214	245	275	306	336	2
3	3	34	62	93	123	154	184	215	246	276	307	337	3
4	4	35	63	94	124	155	185	216	247	277	308	338	4
5	5	36	64	95	125	156	186	217	248	278	309	339	5
6	6	37	65	96	126	157	187	218	249	279	310	340	6
7	7	38	66	97	127	158	188	219	250	280	311	341	7
8	8	39	67	98	128	159	189	220	251	281	312	342	8
9	9	40	68	99	129	160	190	221	252	282	313	343	9
10	10	41	69	100	130	161	191	222	253	283	314	344	10
11	11	42	70	101	131	162	192	223	254	284	315	345	11
12	12	43	71	102	132	163	193	224	255	285	316	346	12
13	13	44	72	103	133	164	194	225	256	286	317	347	13
14	14	45	73	104	134	165	195	226	257	287	318	348	14
15	15	46	74	105	135	166	196	227	258	288	319	349	15
16	16	47	75	106	136	167	197	228	259	289	320	350	16
17	17	48	76	107	137	168	198	229	260	290	321	351	17
18	18	49	77	108	138	169	199	230	261	291	322	352	18
19	19	50	78	109	139	170	200	231	262	292	323	353	19
20	20	51	79	110	140	171	201	232	263	293	324	354	20
21	21	52	80	111	141	172	202	233	264	294	325	355	21
22	22	53	81	112	142	173	203	234	265	295	326	356	22
23	23	54	82	113	143	174	204	235	266	296	327	357	23
24	24	55	83	114	144	175	205	236	267	297	328	358	24
25	25	56	84	115	145	176	206	237	268	298	329	359	25
26	26	57	85	116	146	177	207	238	269	299	330	360	26
27	27	58	86	117	147	178	208	239	270	300	331	361	27
28	28	59	87	118	148	179	209	240	271	301	332	362	28
29	29	[a]	88	119	149	180	210	241	272	302	333	363	29
30	30		89	120	150	181	211	242	273	303	334	364	30
31	31		90		151		212	243		304		365	31

[a]In leap years, after February 28, add 1 to the tabulated number.

(b) Similarly for December 21,

$$D = 23.45 \sin\left(360 \frac{284 + 355}{365}\right) = -23.45^\circ$$

and the appropriate collector angle would be

$$N = 41^\circ + (23.45) = 64.45^\circ$$

(c) Also, for March 15,

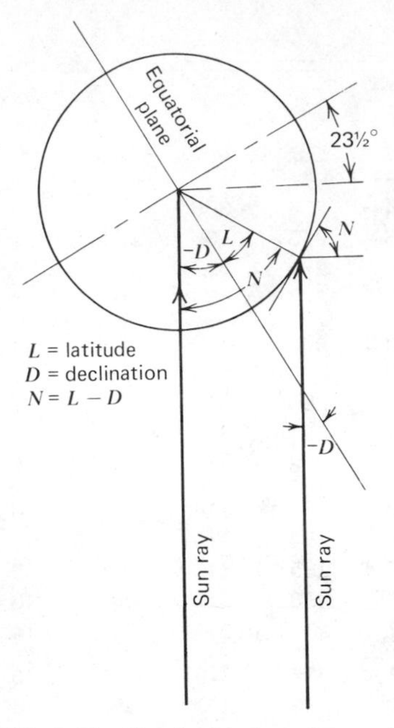

FIG. 3.56. The declination of the sun.

$$D = 23.45 \sin\left(360\,\frac{284 + 75}{365}\right) = -2.4°$$

$$N = 41° + (+2.4°) = 43.4°$$

The heating season in the Northern hemisphere generally extends between late September and late March. So, an angle equal to the latitude plus 10 to 15° for solar collectors is often recommended. This would be slightly steeper than optimum in the early and late parts of the heating season and slightly flatter than optimum during the middle of the heating season but acceptable over the entire period.

Anyone who has used a prism or magnifying glass lens to *focus* or *concentrate* the sun's rays from a larger to a smaller area knows that very high temperatures can be obtained in this manner. Theoretically a temperature equal to the temperature of the surface of the sun can be achieved by a concentrating collector. A major engineering effort is underway to develop systems that collect and concentrate sunshine for useful purposes including the generation of electricity. In the power tower concept, shown in Figure 3.57, solar energy is collected by a large area of

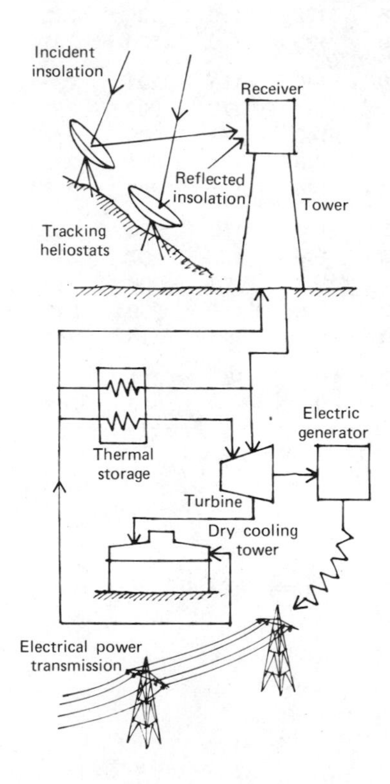

FIG. 3.57. The power tower concept may be used to convert solar energy to electricity.

reflective devices called heliostats, which automatically follow the sun and reflect the incident energy onto a steam boiler located at the top of a tall tower. The steam is then used to drive a turbine in a conventional Rankine power cycle.

These kinds of concentrating collectors depend not only on maintaining a highly reflective surface, but a surface where the reflection is highly *specular.* Specular reflection means that the reflected light remains coherent so that the angle of reflection (θ_r in Figure 3.58) is equal to the angle of incidence (θ_i in Figure 3.58). A convenient and much-used configuration for collecting and concentrating sunshine is the *parabolic trough.* The parabola is shaped such that all parallel rays striking the surface are reflected to the focal point, as suggested by Figure 3.59. Thus, most of the solar energy entering the wide aperture of the parabola can be concentrated onto a pipe or other absorbing device running along the focus. Fluid circulating in this pipe can thus be vaporized at rather high temperatures and used to drive the conventional Rankine power cycle. Usually the surface at the focus is black so the energy will be *absorbed* rather than reflected. The relective surface and/or the focus must be properly oriented with the sun's rays to operate effectively. Most concentrating collectors are therefore equipped with a device that tracks the sun and automatically positions the collector components. It is also necessary to surround the absorbing surface with a *transparent* material having a high *transmissivity,* such as glass, which will allow the energy to pass through via incoming short wave length *radiation,* but will help prevent the loss of energy from the hot surface back to the surrounding air via *convection* and long-wave radiation.

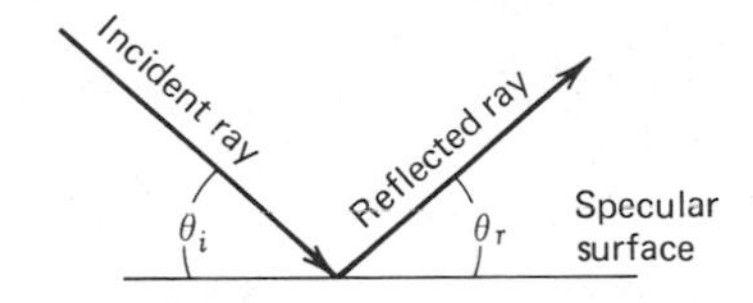

FIG. 3.58. Specular reflection.

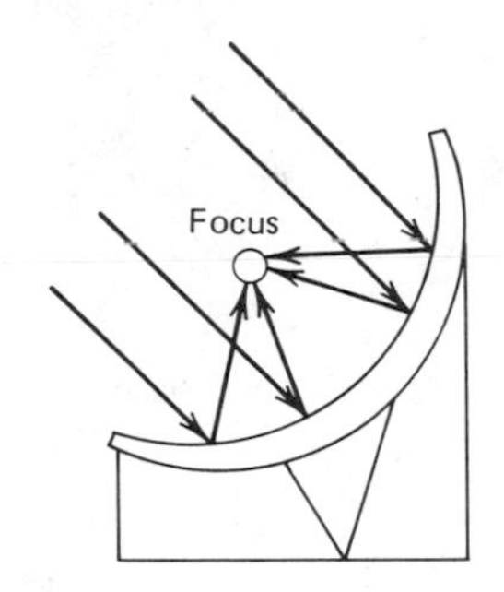

FIG. 3.59. A parabolic collector concentrates solar energy at the focal point.

The technical problems associated with constructing concentrating solar collectors of the kinds mentioned can be solved, but it remains to be demonstrated when and where the highly sophisticated technology and materials required are economically competitive with alternative approaches. One of the problems associated with any type of concentrating collector that reflects the sun's rays is maintaining surface *reflectivity* and *specularity* (Figure 3.60). Dust, blowing sand, oxygen, rain, ice, resting birds, and the sunlight itself can all have a devastating effect on the reflecting surface. Another problem is that the need for energy is often greatest during cold seasons when sunshine is least available. Pumping irrigation water for agriculture might prove to be a very attractive use of solar energy because the need for water is greatest during warm clear summer days when sunshine is most available.

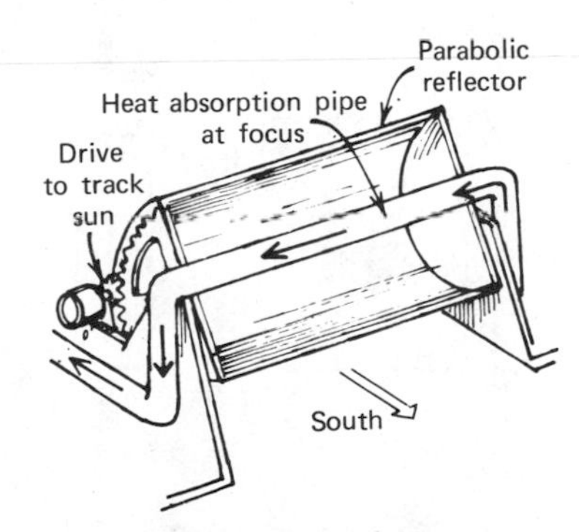

FIG. 3.60. A tracking parabolic collector.

For applications where high temperatures are not required,

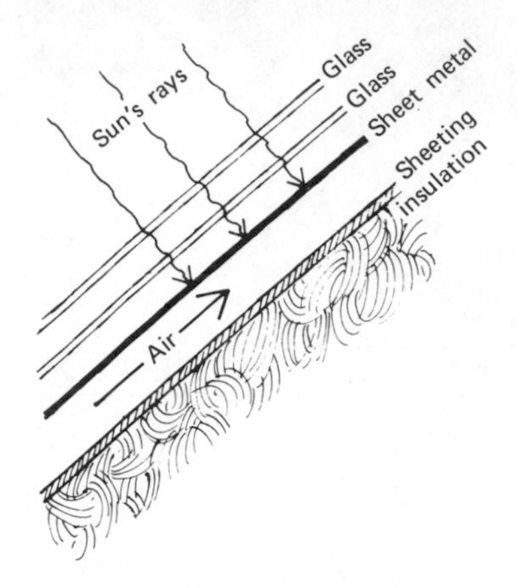

FIG. 3.61. A flat plate collecter with air as a collecting fluid.

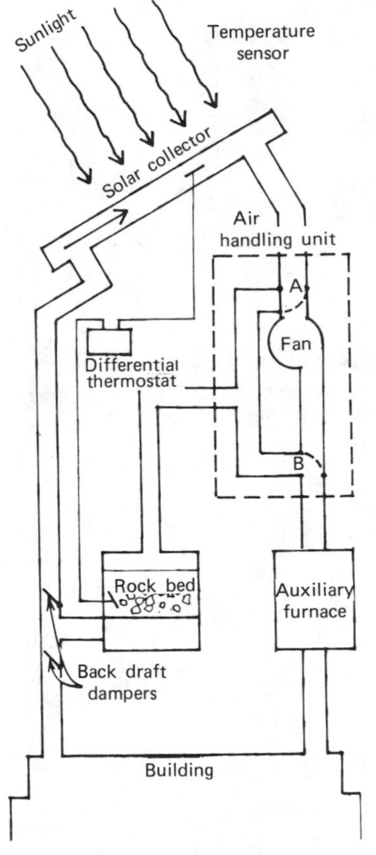

FIG. 3.62. A possible solar collection and storage system.

such as water or space heating, the flat-plate solar collector configuration is generally less costly than the concentrating collector. The flat-plate collector is so named because it usually consists of a black flat plate covered by one or more layers of a transparent material such as glass.

The incoming sunlight passes through the transparent material to be absorbed on the black plate. A fluid such as water or air is then brought in contact with the warm surface to carry the heat to the point of use or to storage. The cross section of a typical collector with air as the collector fluid is shown in Figure 3.61. Notice that the backside of the collector is well insulated to minimize heat loss to the surroundings.

A disadvantage of solar heating systems lies in the unfortunate fact that sunshine is often least available during those times when it is needed most. How does one utilize solar energy on cloudy days and at night? Obviously the energy must be collected when it is available and stored for use when it is not. Figure 3.62 shows a possible configuration for collecting, and storing solar energy. Air passes through the collector where it is heated. If heat is presently needed in the building, dampers A and B are positioned to allow the air to pass through the fan and the auxiliary furnace and into the building. If heat is not required in the building at a time when the sun is shining, the dampers are so positioned as to divert the heated air through the rock bed warming up the large mass of rock, and thus storing the energy for later use. It is left as an excercise for the student to describe the operation of the system when the controls call for heat to be delivered from the rock-bed storage to the building.

A typical liquid-based solar collection and storage system is shown in Figure 3.63. The collector loop fluid generally contains an antifreeze compound to prevent freezing in the collector during cold nights. It is rather expensive to add antifreeze to the large quantity of water in the storage loop. Consequently, the collector fluid and storage fluid are separated as shown. It is an interesting exercise to trace through the operation of this system.

EXAMPLE PROBLEM 3.28

We want to construct a solar heated home in Salt Lake City, Utah. The average heating requirement for the home in January is 400 000 kJ/day. Assuming that 60 percent of the solar energy striking the collector is delivered to the inside of the home, estimate the required collector area facing due south and oriented at 60° from the horizontal.

Solution

We find from Table 3.9 that the average solar insolation at Salt Lake City in January on a horizontal surface is 163 langleys/day. The solar energy striking the surface sloped 60° from the frame would be something less than

$$\frac{163 \text{ langleys/day}}{\cos 60°} = 326 \text{ langleys/day}$$

since the morning and afternoon sun's rays are not normal to the collector. We will estimate the average insolation as about 300 langleys/day. Therefore, 0.60(300 langleys/day)(41.9 kJ/m^2·langley)(area)m^2 = 400 000 kJ/day.

$$\text{area} = \frac{400\,000}{0.60(300)(41.9)} = 53 \text{ m}^2$$

EXAMPLE PROBLEM 3.29

It is desired to construct a storage system with the collector described in the previous example. Estimate the mass and volume of a rock-storage system capable of storing energy equivalent to four days heating requirements in January. Repeat the calculations for water storage and compare the results.

Solution

We observe from Table 3.4 that the specific heat of stone is about 0.9 kJ/kg·°C. This means that 0.9 kJ of heat are stored in each kilogram of stone if its temperature increases by 1°C. It seems reasonable to assume that if the storage temperature drops below about 20°C (86 F), the stored energy would not be very useful for space heating. Also, we know from experience that storage temperatures may not realistically be expected to exceed about 55°C (131 F).

From the previous example, we learn that the heating requirements for four days would be

$$4 \text{ days } (400\,000 \text{ kJ/day}) = 1\,600\,000 \text{ kJ}$$

This energy must be stored in M kg of rock changing temperature by 25°C. Thus we write

$$(0.9 \text{ kJ/kg}\cdot°\text{C})(25°\text{C})(M \text{ kg}) = 1\,600\,000 \text{ kJ}$$

$$\text{or } M_{\text{rock}} = \frac{1\,600\,000}{(0.9)(25)} = 71\,111 \text{ kg}$$

The density of rock is about 2670 kg/m^3. One could expect a void fraction of about 0.40 for the loose rock placed in the stor-

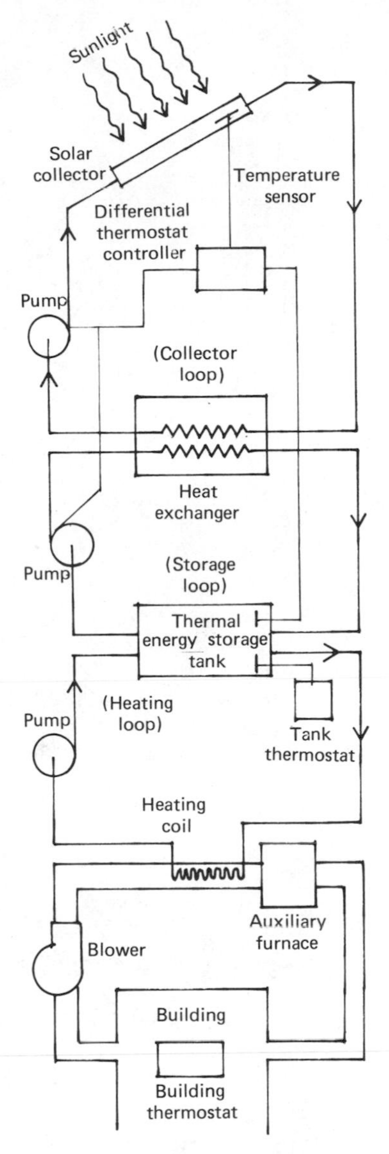

FIG. 3.63. A liquid based solar collection and storage system.

age bin. The required volume of the storage bin would thus be

$$\text{Volume of rock} = \frac{71\,111 \text{ kg}/(2670 \text{ kg/m}^3)}{(0.60)} = 44 \text{ m}^3$$

Assuming that water has a specific heat of 4.2 kJ/kg·°C and a density of 1000 kg/m^3, we calculate the required mass as

$$M_{\text{water}} = \frac{1\,600\,000}{(4.2)(25)} = 15\,238 \text{ kg}$$

and the volume as

$$\text{Volume of water} = \frac{15\,238}{1000} = 15.24 \text{ m}^3$$

We thus see that the required mass of water is about 21 percent of the mass of rock and the volume of water required is 35 percent of the volume of rock required to provide the needed energy storage.

3.12.2/The Solar Pond

An interesting approach to solar energy collection and storage is provided by the solar pond. Water heated by the sun ordinarily rises to the pond surface where it is cooled by evaporation and wind. But consider a shallow pond so constructed that less saline water (insulating layer) floats on top of heavy salty brine (storage layer). As shown in Figure 3.64 some of the solar energy goes through the lighter surface water to be trapped in the brines below. This brine is so heavy due to the salts in solution that it remains on the bottom, where it continues to absorb energy. In a carefully constructed solar pond, temperatures near the boiling point of water may be attained. Energy collected in the summertime can be stored and used months later.

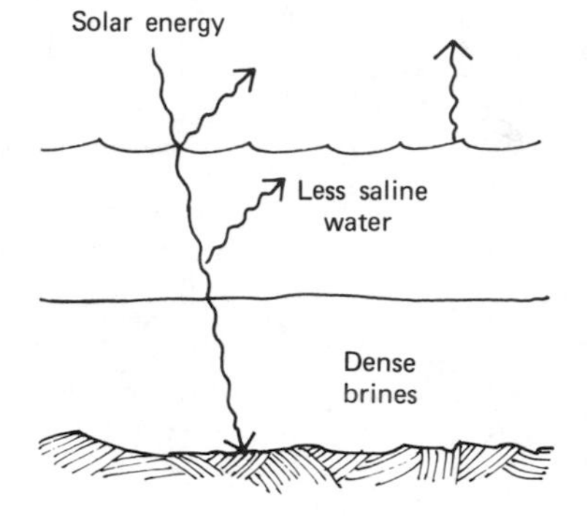

FIG. 3.64. A solar pond.

EXAMPLE PROBLEM 3.30

It is proposed to utilize a solar pond to provide heat for a group of condominium type homes. Assume that (a) each apartment (home) requires 200 000 Btu/day, (b) the average daily solar insolation is 250 langleys, and (c) 20 percent of the incident solar radiation incident on the pond can be delivered to the homes. Estimate the number of homes which could be heated by a solar pond covering 1 acre (43 560 ft^2) (Figure 3.65).

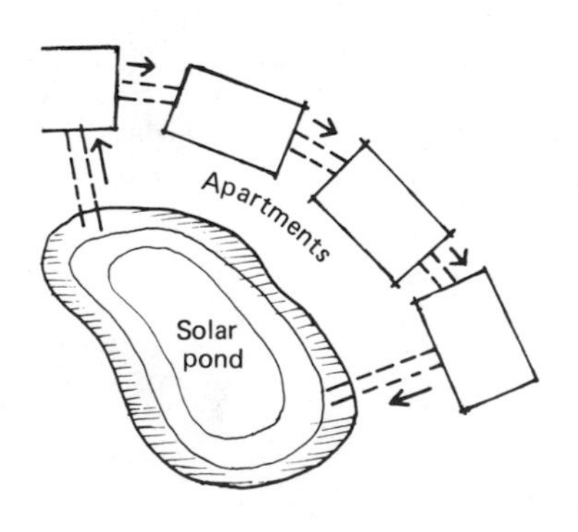

FIG. 3.65. Example problem 3.30.

Solution

$$0.20\ (250 \times 3.69\ \text{Btu/ft}^2\ \text{day})\ \frac{43\ 560\ \text{ft}^2}{200\ 000\ \text{Btu/day}}$$

$= 40$ homes

3.12.3/Ocean Thermal Energy Converters

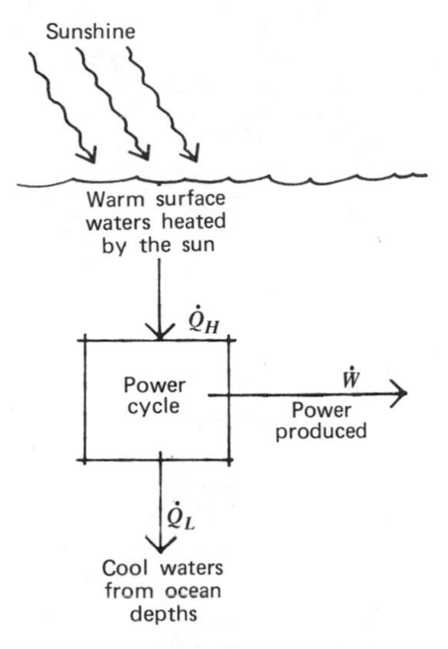

FIG. 3.66. Ocean thermal energy converters could produce power.

A particularly intriguing energy conversion notion is to use the oceans as solar collectors. Covering more than 70 percent of the earth's surface, the oceans absorb tremendous quantities of solar energy. It is estimated that the Gulf Stream alone carries enough energy northward to generate over 75 times the total electric power production of the United States. To produce electricity from the sea is not easy, however. We learned previously that in order to generate power, a relatively high temperature heat source and a relatively low temperature heat sink are required. There are many places in tropical waters such as the Caribbean Sea and the Gulf Stream where the surface waters are warmed by the sun to about 30°C while the water at a depth of 100 m is a cool 5°C. Chilled in the Arctic and Antarctic regions, this water has settled to the depths and flowed along the bottom of the oceans to the tropics (Figure 3.66).

A French engineer named D'Arsonval suggested as early as 1882 that power could be generated by utilizing this temperature difference found in the oceans. A friend of D'Arsonval, Georges Claude, directed the construction of a land-based unit in Matanzas Bay, Cuba, in 1926. The plant failed to produce its design output of 40 kW due to a multitude of technical difficulties such as corrosiveness of the sea water, low turbine efficiency, excessive heat flow into the long (2 km) lines carrying cold water from the depths, and difficulty in maintaining large leak-proof connections. The French do not give up easily, however, and in the 1950s a corporation (Energie Electrique de la Cote d'Ivoire) planned a 7000-kW plant in Abidjan in the Ivory Coast. A large pipe reaching 1 km into the depths was carefully positioned, but full production was never realized due to technical difficulties. Much was learned from these preliminary efforts, and the United States is now seriously developing Ocean Thermal Energy Conversion (OTEC) systems.

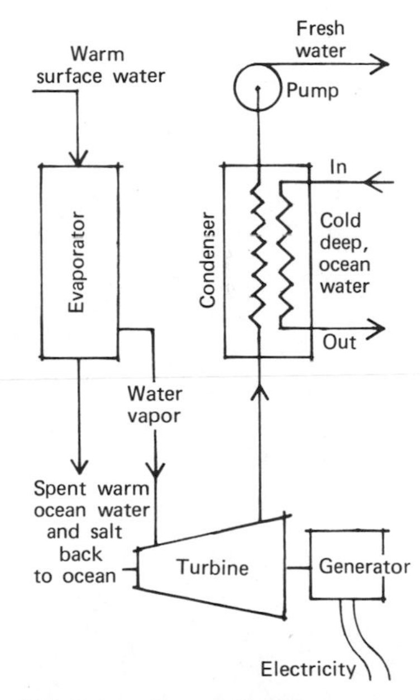

FIG. 3.67. A controlled flash evaporation system produces both electricity and fresh water.

The controlled flash-evaporation approach as shown in Figure 3.67 produces both electricity and fresh water. Warm sea water enters a special chamber where the pressure is reduced, converting part of the water to vapor. This vapor flows through a

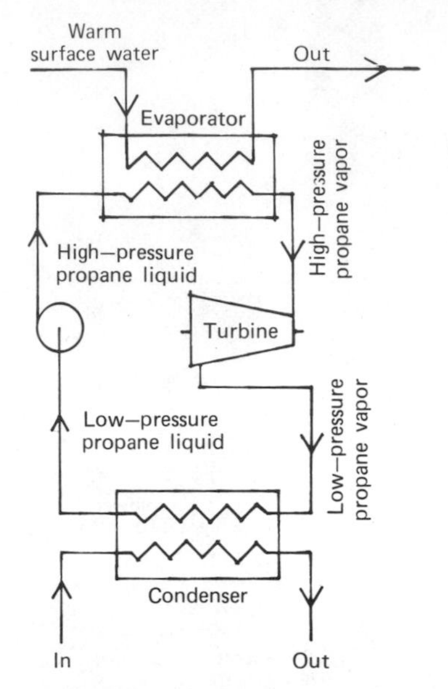

FIG. 3.68. An ocean thermal energy converter that uses a working fluid such as propane.

turbine, which drives an electrical generator. On leaving the turbine, the vapor enters a condenser to be chilled by the cold sea water from the depths and is converted into fresh pure water. This water is then pumped back up to a pressure greater than atmospheric to be used for any good purpose.

Another approach, shown in Figure 3.68, utilizes a more easily vaporized working fluid such as propane. Heat from the warm surface waters vaporizes the high-pressure propane liquid in the evaporator. The propane vapor then flows through the turbine, producing power. The low-pressure vapor leaving the turbine is condensed back to a liquid by the cold deep-sea waters. The liquid propane is then pumped to a high pressure to begin the cycle again. This scheme has a number of technical advantages, but fresh water is not produced. An important side benefit of both schemes is the mariculture (fish farming) industry, which could thrive in the nutrient-laden waters brought up from the depths.

If you become an engineer, it is not unlikely that you will help design an OTEC system.

EXAMPLE PROBLEM 3.31

Suppose you are selecting a site for an OTEC project. At location A, the surface waters are at 30°C, with deep water at 7°C being available. At site B, the surface waters are at only 27°C, but deep water at 5°C is available. At which site would an OTEC system be expected to have the greater thermal efficiency?

Solution

We learned in Section 3.11 that the efficiency of any thermal power plant depends on the absolute temperature of the heat source and the heat sink.

$$\eta_{\max} = 1 - \frac{T_L}{T_H}$$

$$\eta_{\max\ A} = 1 - \frac{280.15}{303.15} = 7.59 \text{ percent}$$

$$\eta_{\max\ B} = 1 - \frac{278.15}{300.15} = 7.33 \text{ percent}$$

We conclude that the maximum theoretical efficiency of a plant at site A is slightly greater than at site B. This may not be the overwhelming consideration, however. Other factors such as distance to market for the power produced or intensity of expected storms must be considered in the site selection.

3.12.4/Solar Cells

Imagine a device capable of converting sunlight directly into electricity without the necessity of the turbines, boiler, condensors, and rotating electrical generation equipment associated with the conventional power cycles. Furthermore, this device would be almost environmentally benign, having none of the pollution problems associated with combustion of fuel or nuclear reactors. Solar cells are part of the space-age electronic revolution brought about by the development of semiconductors. They represent perhaps the most sophisticated of solar conversion technologies offering at once the greatest promise and also the greatest challenges. A detailed discussion of solar cells is beyond the scope of this text.

EXAMPLE PROBLEM 3.32

We want to utilize solar cells to provide electricity for pumping irrigation water with the rationale that the sun shines most at the same time the water is most needed during the hot dry clear days of summer. At the proposed location, the average July solar insolation is 700 langleys/day. The solar cells are capable of converting about 10 percent of the incident solar energy into electricity. How large of an area is required to operate a 100-hp motor?

Solution

We will assume that 80 percent of the solar energy arrives during the eight-hour period nearest midday. The average incident solar energy rate during that period would therefore be

$$\frac{0.80(700 \text{ langleys}/8 \text{ h})(41.9 \text{ kJ/m}^2\cdot\text{langley})}{(3600 \text{ s/h})} = 0.815 \text{ kW/m}^2$$

At 10 percent efficiency 1 m^2 of solar cell would produce $\sim$ 0.08 kW. The required area would be approximately

$$A = \frac{(100 \text{ hp})(0.746 \text{ kW/hp})}{0.08 \text{ kW/m}^2} = 932 \text{ m}^2$$

PROBLEMS

3.69. Define latitude and declination. Show that, according to the approximate equation,

$$D = 23.45 \sin\left[360\,\frac{(284 + n)}{365}\right]$$

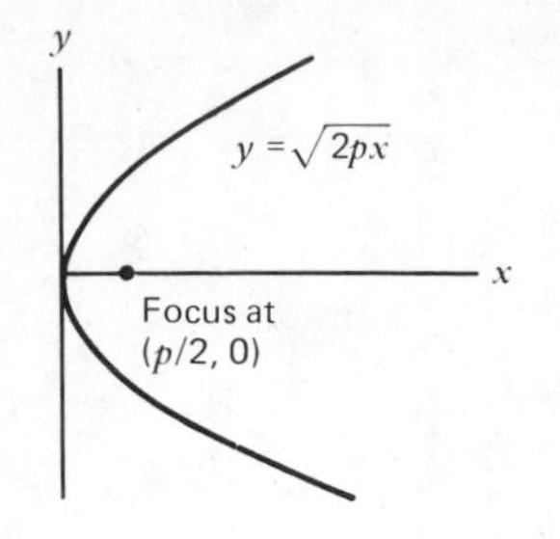

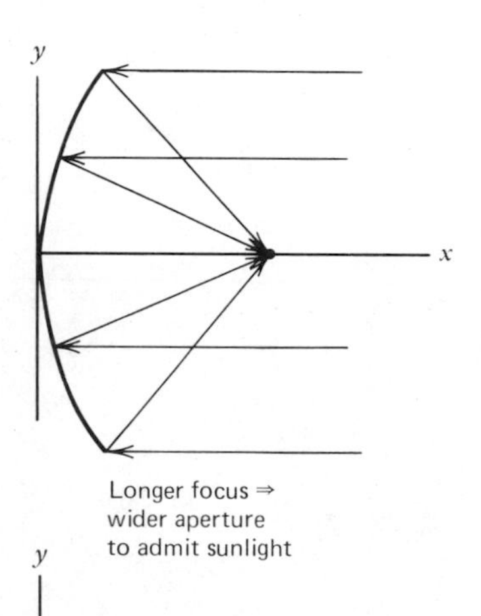

the declination is negative between September 21 and March 21 in the northern hemisphere.

3.70. We want to construct a flatplate solar collector in Caribou, Maine, oriented at such an angle as to be normal to the noon sun on December 21. At what angle with the horizontal should the collector be placed?

3.71. A reflective surface shaped in the form of a parabolic trough makes an effective concentrating solar collector. The equation for a parabola is given by

$$y^2 = 2px$$

where $p/2$ is the distance from the origin to the focus located on the x axis. Decreasing the value of p yields a parabola with a narrow aperture for a given collector area but one that is easily focused. Increasing the value of p gives a wider aperture for a given collector area, but increases the difficulty of focusing. Plot parabolas with (a) $p = 1$ unit, (b) $p = 2$ units, and (c) $p = 4$ units. Show graphically that any ray entering the parabola parallel to the x axis will reflect to the focus if the angle of reflection equals the angle of incidence.

3.72. Construct a small concentrating collector using aluminum foil or mirrored mylar shaped into a parabolic trough. One possible approach is to place a sheet of graph paper on the side of an empty cereal box and plot an appropriate parabola. Cut the box along the parabola and fit in the reflective surface with a piece of heavy paper as a backup. Water in a blackened test tube placed at the focus should boil in 5 or 10 min when the device is aimed at the sun.

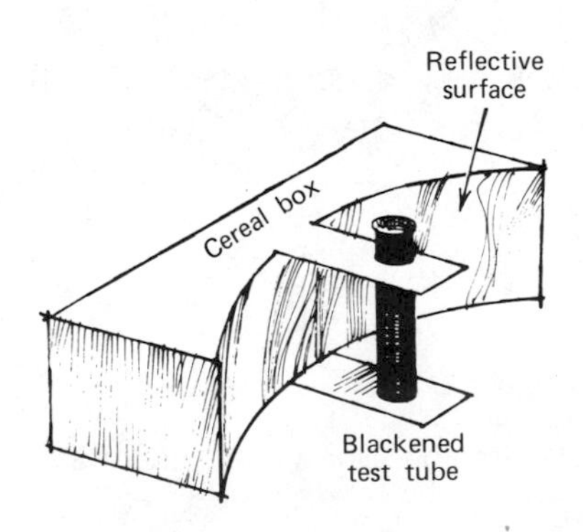

3.73. Explain why double layers of transparent material are required in the construction of flat plate solar collectors designed for use in cold climates.

3.74. Resketch the solar heating system shown in the diagram at right with the dampers oriented for the system to deliver heat from the rock storage to the building. Indicate the air flow path.

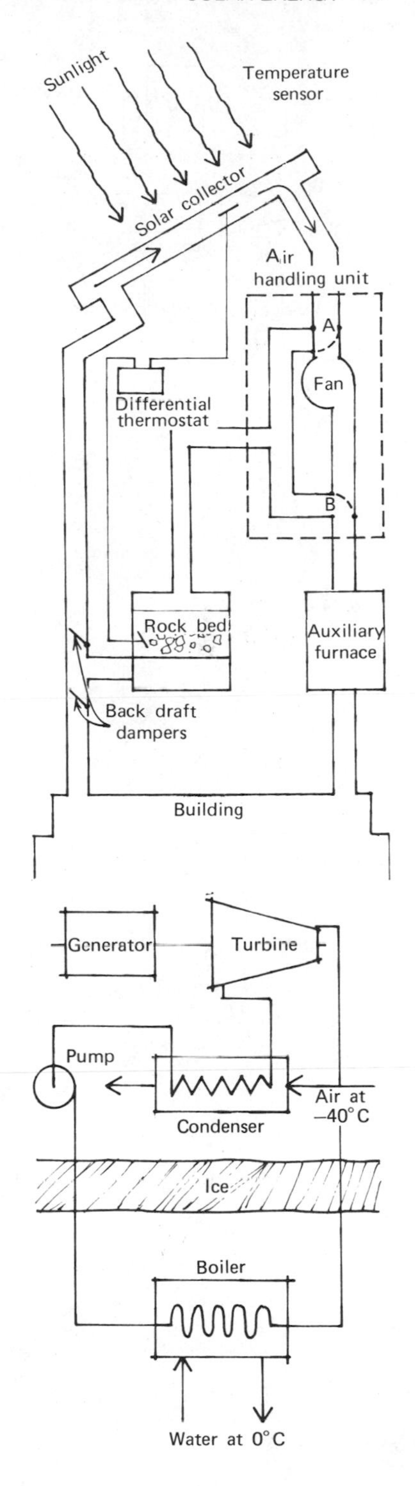

3.75. We want to construct a home in Caribou, Maine, heated with an active solar heating system. Assume the average heating requirement during January is 300 000 kJ/day. Select an appropriate collector angle and area. State your assumptions clearly.

3.76. Describe the principle of operation of a solar pond.

3.77. Calculate the mass of salt in a one-acre (43 560 ft^2) solar pond containing a layer of brine 4 ft deep, which is 25 percent salt by weight. Assume fresh water has a density of 62 lb/ft^3.

3.78. Estimate the area of solar pond required to provide an average heating requirement of 200 000 Btu/day over the heating season. Assume that the pond is located in Dodge City, Kansas and operates with an efficiency of 20 percent. Clearly state other assumptions.

3.79. Prepare a schematic diagram of a controlled flash evaporation OTEC system in which seawater becomes the working fluid. Why is this approach sometimes referred to as an *open-cycle* method? Also prepare a schematic diagram of a closed-cycle OTEC system using a working fluid such as propane. Can you think of any other fluids that might be suitable? Discuss the relative advantages and disadvantages of the open- and closed-cycle approaches.

3.80. An ocean thermal energy conversion (OTEC) system has been proposed at a location where the surface (heat source) temperatures average 30°C and the temperature of the depths (heat sink) is about 5°C. Assume that the temperature of the cooling water is increased from 5 to 7°C as it passes through the condenser. Estimate the minimum theoretical flow rate of cooling water in a power plant producing 1000 MW of electrical power. Express answer in kilograms per second.

3.81. It is proposed to operate a power cycle in the Antarctic utilizing the relatively warm water beneath the ice as a heat source and the colder air above the ice as a heat sink. Assuming

a water temperature of 0°C and an air temperature of −40°C, estimate the maximum possible thermal efficiency of such a cycle. Would this cycle work better on stormy cold days than on warm sunny days? Why?

3.82. Keeping in mind that approximately two-thirds of the earth's supply of fresh water is captured in the Antarctic ice pack, it has been proposed that atomic powered tugboats retrieve bergs from that southernly region and tow them thousands of miles to ports in water-short regions such as Southern California. Perhaps the icebergs could be covered with a plastic blanket to reduce the problem of melting during the long trip north. Once in port, the melting ice could serve not only as a pure water supply for thirsty thousands but as a heat sink for a power cycle. The power cycle could absorb heat from the ocean water at a temperature of 25°C and reject heat to the melting ice at 0°C. Given that 333 kJ of energy are required to melt 1 kg of ice at 0°C, estimate the minimum theoretical rate at which ice must be supplied for the power plant to produce 1000 MW of electricity. Express your answer in kilograms per hour and kilograms per year.

3.83. Do some outside reading in order to explain the basic principle of operation of a solar cell. Include in your explanation a description of bonding electrons, free or conducting electrons, "holes," *p*-type crystals, and *n*-type crystals.

3.84. Suppose you live in Laramie, Wyoming, and wish to operate a television set 10 h per day in January. The power is to come from 10 percent efficient solar cells coupled with a storage battery. What solar cell area would be required?

3.85. You are considering installing enough solar cells on your roof to provide 10 kW·h of electrical energy during a January day. Assume that you live in a location where the average daily solar insolation in January is 400 langleys. Estimate the area of solar cells required. Assume a 10 percent conversion efficiency.

3.86. The rate at which solar energy reaches the earth's outer atmosphere is about 442 Btu/h·ft^2 or 120 langleys/h or 1.4 kW/m^2 and is often called the solar constant. Estimate the area of 20 percent efficient solar cells that would have to be placed in synchronous orbit in order to provide 10 percent of the 1980 U.S. electrical energy requirement. The 1980 U.S. annual electrical energy requirement is estimated at 2.8 trillion kW·h.

3.13/Energy from the Wind

Winds are caused by unequal heating of land, water, and air by the sun. It is the temperature difference between the warm tropics and the cool polar regions that are primarily responsible for both gentle, pleasant spring breezes and violent destructive hurricanes. This circulation of air is an essential part of earth's intricate life-support systems. Atmospheric currents perform such tasks as transporting moisture from the ocean to inland areas and moving vast quantities of energy via heated or cooled air from one region to another thus dramatically influencing the climate.

Winds have been utilized in energy conversion devices for thousands of years. There is evidence that Egyptians used sails to drive their ships as early as 2800 B.C. Sailing ships continued to carry much of the world's cargo until largely replaced in the late eighteenth century by coal-fired steamships. The first recorded use of windmills seems to have been pumping irrigation water in Persia in about the seventh century A.D. Of course, the windmill played a key role in Western Europe, the Netherlands, and the settling of the American Great Plains. Perhaps there is no energy conversion device that has caught the attention of as many inventive minds as has the windmill. It is claimed that more patents for windmill designs have been applied for than for any other type of device. A few of the many types of wind machines are depicted in Figure 3.71. Almost certainly the optimum design is yet to be thought of.

There is ample reason for the long and continued interest in devices to convert wind energy to more useful forms. It is estimated that the surface winds over the continental United States,

Sun
Rain
Winds
Evaporation
Sea

FIG. 3.69. Atmospheric air transports moisture from the sea to inland areas.

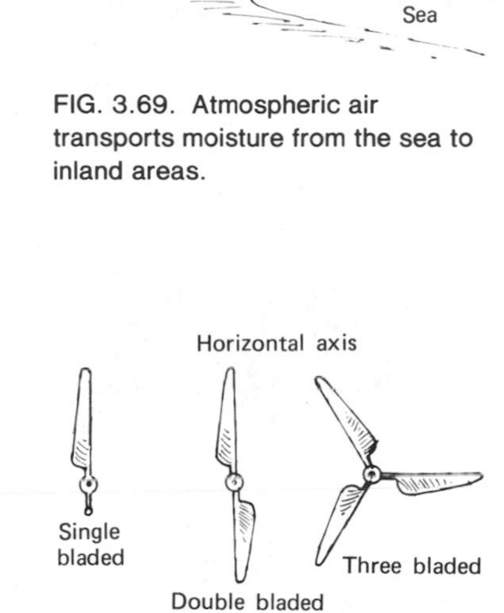

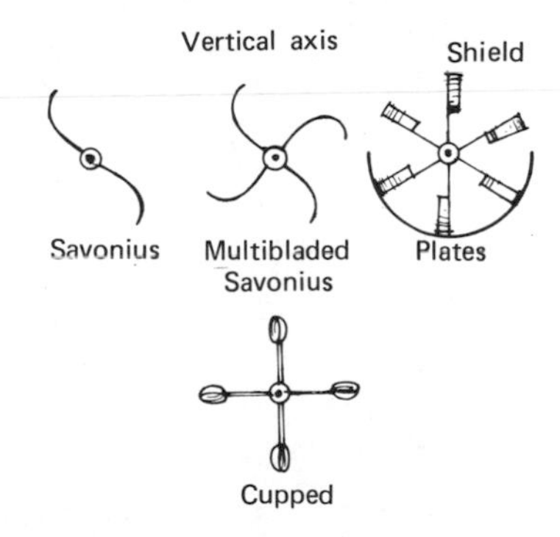

FIG. 3.71. Some typical wind machines.

FIG. 3.70. The energy in the wind has been converted through various devices for years.

the Aleutian Arc and the Eastern Seaboard could provide more than 100 times the total U.S. electrical generating capacity in 1980. In spite of this enormous potential, few major power utilities are enthusiastic about obtaining large quantities of power from the wind. Every single effort to date, both in this country and abroad, has failed to be economically competitive with conventional methods of producing electrical power.

In spite of the economic failures, wind-power potential is so tantalizing that almost everyone with an inventive bent to their nature allows the design of a better wind machine to challenge his or her ingenuity. A look at the theory of wind energy conversion will help us better appreciate the problems one might encounter in building a successful windmill.

$A = \pi r^2$

Air cylinder

$L = V\Delta t$

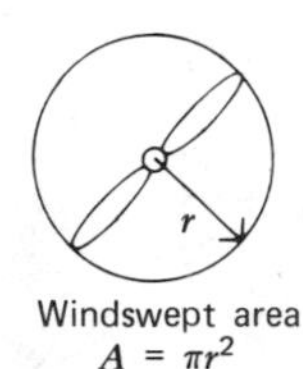

FIG. 3.72. Kinetic energy is extracted from the wind by a windmill.

A windmill is a device that converts the kinetic energy of moving air to mechanical energy. Consider the cylinder of air moving at velocity V, which in the next interval of time Δt will encounter a windmill, shown in Figure 3.72. If the windmill were able to bring the air completely to rest with no losses, the energy converted would be the total kinetic energy of the cylinder of air or

$$\text{Total energy} = \frac{MV^2}{2}$$

where M is the mass of air and V is its velocity. The air cylinder has length

$$L = V\Delta t$$

and cross-sectional area A equal to the area swept by the windmill so that the volume of the air cylinder is

$$\text{Volume} = V\Delta tA$$

Multiplying the cylinder volume by the density ρ (mass per unit volume) gives the mass of air that strikes the windmill in time interval Δt

$$\dot{M} = \frac{\rho V\Delta tA}{\Delta t} = \rho AV$$

The maximum rate at which kinetic energy would be converted to mechanical energy if the air were all completely stopped by the windmill then would be

$$\frac{1}{2}\dot{M}V^2 = \frac{1}{2}\rho AV^3$$

A windmill does not completely bring to rest all the air that passes through the swept area. A German engineer named Albert Betz,

TABLE 3.11. Air Density, ρ(kg/m³) as Function of Elevation and Temperature

Elev., m \ Temp., °C	−40	−30	−20	−10	−0	10	20	30	40
0	1.51	1.45	1.39	1.34	1.29	1.25	1.21	1.17	1.13
1000	1.35	1.29	1.24	1.19	1.14	1.11	1.07	1.04	1.00
2000	1.20	1.15	1.10	1.06	1.02	0.99	0.95	0.92	0.89
3000	1.06	1.02	0.98	0.94	0.91	0.87	0.84	0.82	0.79
4000	0.94	0.90	0.87	0.83	0.80	0.78	0.75	0.72	0.70
8000	0.56	0.54	0.52	0.50	0.48	0.46	0.45	0.43	0.42
10 000	0.43	0.42	0.40	0.39	0.37	0.36	0.35	0.33	0.32

Multiply by $1.940(10)^{-3}$ to obtain ρ in units of slugs/ft³.

was able to demonstrate in 1927, that even a perfect windmill could extract only $\frac{16}{27}$ths or about 59.3 percent of the winds total kinetic energy. But of course, no windmill is perfect. Even the best machines obtain only about 80 percent of Betz's theoretical maximum. We thus write

$$\text{Windmill power} = \eta\left(\frac{1}{2}\right)(\rho A V^3)$$

where η is the overall efficiency of the windmill. This overall efficiency may be expected to be less than 50 percent.

The density, ρ, of air depends on atmospheric pressure and temperature. Atmospheric pressure decreases as elevation increases. Table 3.11 shows how significantly air density can vary with elevation and temperature.

EXAMPLE PROBLEM 3.33

You purchase a horizontal-axis type windmill sweeping through a diameter of 4m to provide supplemental power for your home. Assuming that a wind is blowing at 25 km/h, what power could you expect to generate with this windmill at a location 1500 m above sea level and a temperature of 25°C?

Solution

$$A = \frac{\pi D^2}{4} = \pi\left(\frac{4V^2}{4}\right) = 16.6 \text{ m}^2$$

From Table 3.11 we estimate the air density to be very nearly 1.0 kg/m³

$$25 \text{ km/h} = 25\left(\frac{1000}{3600}\right) = 6.94 \text{ m/s}$$

$$\text{Power} = \frac{\eta\rho A V^3}{2} = \frac{0.50\,(1.0 \text{ kg/m}^3)(16.6)(6.94)^3}{2}$$

$$\text{Power} = 1387\ \text{kg}\cdot\text{m}^2/\text{s}^3 = 1387\ \text{W}$$

(Recalling that $1\ \text{kg}\cdot\text{m/s}^2 = 1\ \text{N}$,

$1\ \text{N}\cdot\text{m} = 1\ \text{J}$ and $1\ \text{J/s} = 1\ \text{W}$.)

It is interesting to compare this with the typical power consumption of some household appliances as shown in Table 3.3.

The ideal location for a windmill would be one where the wind blows steadily at a velocity that is neither too low nor too high. Hurricane force winds have high-energy densities, but are also capable of inflicting much damage. A structure must be strong enough to stand the stresses of the maximum gust of wind. Wind speeds five times as great as the mean design speed are not uncommon. Under such conditions the mechanical stresses that are proportional to the square of the wind speed would be 25 times the stress experienced at design speeds. The power available in the air, which is proportional to the velocity cubed, would be 125 times as great as the design power. It is apparent that special considerations must be made for the windmill to accommodate high speeds. To date, wind systems such as the one described in Example Problem 3.33. have not been economically competitive where distribution lines for electricity generated from large-scale, fossil-fuel-fired power plants or hydropower plants are available. Many thousands of windmills generating electricity and pumping water in remote areas across the country were largely replaced in the 1940s when the Rural Electrification Administration (REA) brought low-cost electricity to rural America.

It is often claimed that windmills are environmentally benign, a totally nonpolluting method of energy conversion. That position is not entirely defensible if one considers the materials required to manufacture wind machines. Windmills are usually made from various kinds of steel, steel alloys, aluminum, copper, and other metals. The device is often mounted on tall steel towers placed on concrete foundations and footings. These materials must be mined, transported, and processed, which requires energy inputs and almost inevitably pollutes the environment.

Shown in Table 3.12 is a comparison of the material and energy inputs to a hydropower system and a windpower system. Notice that under the stated assumptions the wind system must operate about 16 times as long as the hydrosystem must operate in order to return to society the energy invested in constructing the system. In addition it is interesting to note that approximately 600 000 windmills of 9.14 m diameter would have to be deployed in locations where the average wind speed is 16.09 km/h in order

TABLE 3.12

	Installed Mass or Quantity (Million kg)	Mean Replacement Schedule	Net Mat. Energy Investment TJ (10^{12}J)
Project: Glen Canyon Dam			
Installed capacity: 900 000. kW net energy investment is fossil fuel equiv.			
Ave. Output for 1973: 525 000. kW ave. output in electricity kW			
Evaluation period of 50 years			
Steel Rebar	12.998	0.0	544.084
Steel Carbon	46.392	0.0	2 718.761
Aluminum	0.950	1.1	376.133
Copper	0.250	1.3	62.777
Cement	969.834	0.0	12 179.175
Aggregate	9 298.408	0.0	428.152
Excavation and Fill	11 997.946	0.0	351.565
Turbines Hydro.	4.426	1.2	611.441
Generators	5.625	1.3	930.083
		Total energy investment	18 202.172
		Hours to break even point at average output	9 631
		Hours to break even point at total capacity	5 618
	Installed Mass or Quantity (1000 kg)	Mean Replacement Schedule	Net Mat. Energy Investment GJ (10^{9}J)
Project: horizontal axis windmill			
Installed capacity: 2 kW net energy investment is fossil fuel equiv.			
Ave. output for 1976: 2 kW ave. output in electricity kW			
Evaluation period of 50 years			
Steel rebar	2.041	0.0	85.428
Steel carbon	4.923	1.0	432.778
Cement	9.480	0.0	119.054
Aggregate	47.402	0.0	2.181
Generators	0.113	0.0	9.494
		Total energy investment	648.935
		Hours to break even point at average output	90 131
		Hours to break even point at total capacity	90 131

to have the same power capability as Glen Canyon Dam. These windmills would require 71 times as much steel and almost six times as much concrete as required to construct Glen Canyon Dam. From that point of view, windmills of the type analyzed in Table 3.12 do not seem to be the wisest method of deploying our resources. Efforts are being made to develop windmills that use less material per unit area swept by the rotors. Vertical axis machines of the type shown in Figure 3.73 have a mass of less than 10 kg/m^2 of swept area. As new high-strength, low-cost plastics and other synthetic materials are developed, we might

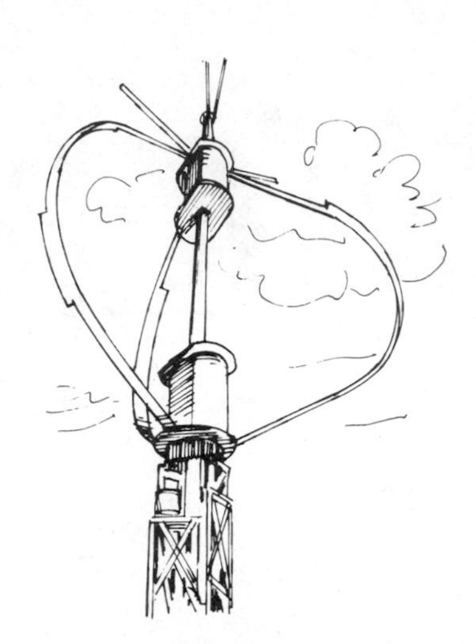

FIG. 3.73. Vertical axis windmill.

FIG. 3.74. Horizontal axis windmill.

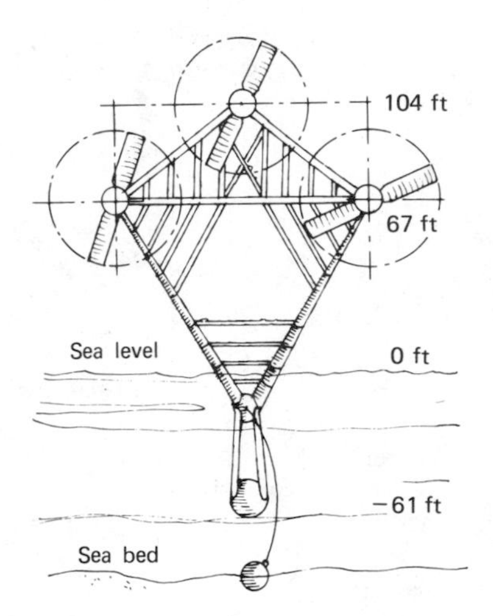

FIG. 3.75. Ocean-based windmill.

expect to see them used to effectively reduce the mass/area ratio even more.

A great many imaginative large-scale wind power projects have been proposed. As fossil fuel costs continue to escalate some of the more innovative of these projects will almost certainly be constructed and tested in the most suitable locations. A proposed wind farm is shown in Figure 3.74.

Unfortunately, when the wind slows or stops so does the windmill. One solution to this problem is to operate a windmill system in combination with a hydropower system. Under this concept the release rate from the reservoir would be reduced when the wind blows, thus conserving water.

A very ambitious plan for generating electricity on a large scale has been suggested to take advantage of the high winds existing off the New England coast. According to this scheme a complex of large ocean based semisubmersible wind power stations similar to that shown in Figure 3.75 would generate electricity. The electricity thus generated could be used to produce hydrogen from sea water through an electrolysis process. The hydrogen would be collected and carried by underground pipeline to shore to be used as a clean virtually nonpolluting fuel. If implemented as envisioned, this would be the largest ocean engineering project ever undertaken. Just imagine how exciting it would be to be part of such a vast effort.

Another suggestion is to tap energy from high-altitude jet-stream winds. The jet stream, according to its discoverer, "is a belt of great wind speeds at altitudes of approximately 10 km which surrounds the whole hemisphere in wavy meanders." Jet-stream velocities seem to be in the 100 to 700 km/h range and consequently have much greater energy densities than lower velocity surface winds. Air turbines and generators would be constructed as part of a large glider tethered by cable in a relatively fixed position in the jet stream. The cable would also conduct the electricity generated to the ground (Figure 3.76). The practical engineering problems that must be solved in order to implement such a scheme are immense.

Windmills may yet prove to be an important contributor to the total "energy mix." Certainly they present a great engineering challenge and the answers are not all in yet. They may assume new and different forms that make them more efficient users of material and hence more competitive with other energy converters (Figure 3.77), or they may become a part of other structures. Perhaps you will one day be part of an engineering team that solves some of these challenging problems.

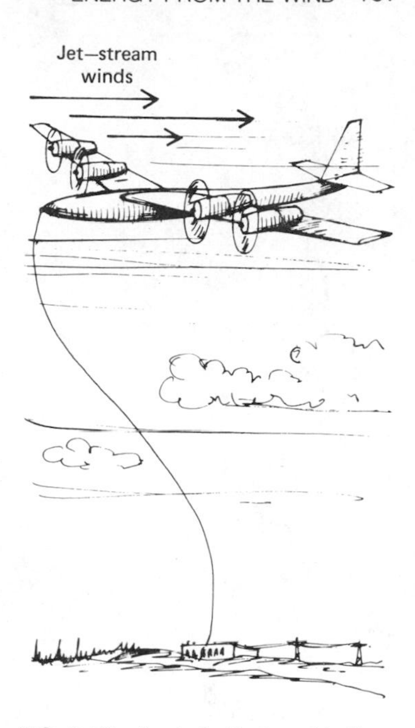

FIG. 3.76. A windmill placed in the jet stream.

PROBLEMS

3.87. The windmills of Western Europe were often rather large structures able to convert about 25 percent of the kinetic energy of the wind intercepted into mechanical energy. Compute the power output of such a windmill having a rotor diameter of 30 ft subjected to an 18 mph wind. Express your answer in units of horsepower. Qualitatively compare the size of this structure with that of a gasoline engine capable of producing the same power.

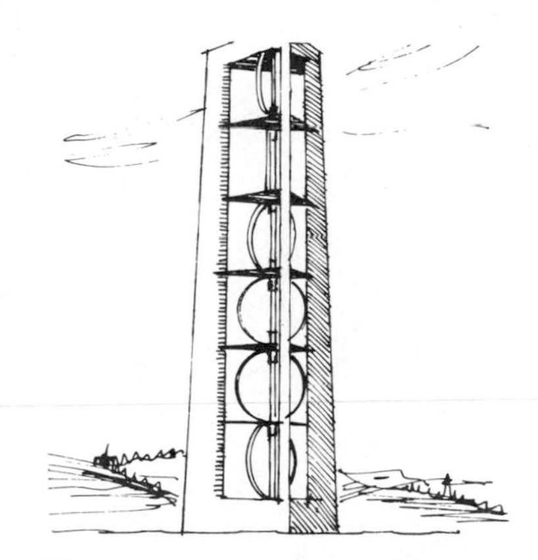

FIG. 3.77. Windmills may assume new and different forms.

3.88. Suppose you wish to power your home with a windmill. Investigation proves that breezes of 15 km/h are not uncommon at that particular location. Estimate the diameter of windmill required to simultaneously operate a TV set (240 W), an electric oven (3000 W), and eight 100-W light bulbs in a 15-km/h wind. Assume an overall efficiency of 40 percent in converting from kinetic energy of the air to electrical energy.

3.89. Estimate the power produced by a 10-m diameter wind machine operating at sea level in a wind at 20 km/h at a temperature of $-25°C$. Assume an efficiency of 40 percent. Repeat your calculation under identical conditions except the machine is operating at an elevation of 2000 m and the air temperature is 30°C. To what do you attribute the difference in power output for these two situations?

3.90. Windmills have long been used to pump water. Estimate the rate at which a 10-ft diameter windmill operating in a 15-mph wind could lift water to the surface of the earth from a depth of 100 ft. Assume a windmill efficiency of 40 percent and that water weighs 8.34 lb/gal. Express your answer in units of gallons per minute (gpm).

3.91. In the early 1940s a windmill having a swept-area diameter of 53 m was constructed on a hill known as Grandpa's Knob in Vermont. Calculate the efficiency of this device reported to produce 1250 kW of electrical power at a wind velocity of 15.3 m/s.

3.92. Estimate the swept area required for a 40-percent efficient wind turbine to produce 100 kW if operating at sea level in a 20-km/h wind having a temperature of 23°C.

3.93. Compare the energy density of a 20°C, sea-level wind blowing at 20 km/h with that in a jet-stream wind at 500 km/h. Assume that the air density in the jet stream is 0.460 kg/m^3.

3.94. Assume that a wind turbine is placed on a glider in a jet-stream wind where the air density is 0.460 kg/m^3 and the velocity is 500 km/h. What swept area would be required to produce 100 kW? Assume an efficiency of 40 percent.

3.95. Conventional horizontal axis windmills of steel construction generally have a mass to swept area ratio of about 75 kg/m^2. For vertical axis windmills of the type shown in Figure 3.73, this ratio is about 9.8. It is claimed that 3400 kg machines placed in high altitude jet stream could produce 1.7 MW. Calculate the mass to area ratio for these machines assuming the same jet stream characteristics as in the previous problem.

3.96. What are your personal feelings regarding the future of wind power? List three serious obstacles to economically competitive large-scale wind power systems.

CHAPTER FOUR

FORCES AND STRUCTURES

A structure is defined as an assemblage of material that provides a system of support. The material might be steel, aluminum, wood, earth, plastic, fiberglass, bone, or a combination of several materials. The planning, design, and eventual construction of structures account for several challenges in the general field of engineering. The engineer usually is responsible for such things as the selection of materials, the specification of size and shape, the interface with other structures, the identification of internal and external forces, the prediction of displacements and other concepts that relate to structural safety, economy, performance, and general appearance. Typical structures include bridges, buildings, vehicle frames, spacecraft, aircraft, power transmission towers, prosthetic devices, dams, pipelines, ships, and machines (Figure 4.1).

The study of structures penetrates several different areas of engineering, such as civil, mechanical, electrical, agricultural, manufacturing, and others. In some cases, engineers specialize in designing a particular type of structure such as a bridge. For example, an engineer in the aircraft industry might concentrate her or his entire career on the design and development of lightweight aircraft structures. Others encounter a variety of multidisciplinary-type problems that require a structural analysis as part of the total solution. A structural engineer might design temporary landing surfaces for aircraft one year, instrument hardware for the space shuttle the next year, and possibly an overhead structure for a football stadium the following year. Some engineers with a

FIG. 4.1*a*. An old suspension bridge built in 1908 spans the Colorado River.

FIG. 4.1*b*. A three-hinged arch bridge built in 1966 also spans the Colorado River. (Photos courtesy of Utah Department of Transportation)

structural specialty serve as part of a team effort and must contribute simultaneously to multiple projects. Many engineers select management positions and use their technical expertise to review and approve structural designs intended for several different applications.

Like most engineering problems, a structural design frequently has unique challenges that require independent engineering analysis and judgement. A framework might be stationary and require no dynamic analysis, yet another framework exposed to lateral excitation associated with an earthquake or wind would require a time dependent analysis and the dynamic response would be important. The decision to use either a simple static analysis or a sophisticated dynamic analysis to design a structure such as a spaceprobe for upper atmospheric measurements might depend on such parameters as the probe stiffness, spacecraft accelerations, and excitation frequencies during lift-off and landing.

In some designs the so-called dead weight of structures is small when compared with the loads supported and the designer conveniently neglects the weight of the structure. Some model balsa wood bridges that weigh less than 0.1 lb have carried over 1500 lb. The dead weight of the bridge material is negligible when compared with the applied load. In contrast, structures such as earth dams and fills for highways weigh significantly more than the external loads that they must withstand. The designer of a dam, a retaining wall, or perhaps a foundation would experience obvious difficulty if he or she neglected the dead weight of the material. These few cited situations are only descriptive of the many particulars that require evaluation in structural design.

The development of a structure from the initial conception to the final completion is often interdisciplinary and requires the interaction of multiple groups of people with different interests and responsibilities as pictorially represented in Figure 4.2. During the design sequence, the engineer must communicate and empathize with these different groups. In the initial planning stages, the engineer is often required to participate in the evaluation of the total impact of the proposed structure. The impact on the physical terrain, the financial return on the investment, social implications, recreational interests, aesthetics, the disruption of wildlife habitat, and safety are some of the common areas of concern. For example, the construction of a large dam would concern the owners of the land to be flooded. Agriculturists would have interest in potential irrigation systems and possible changes in ground water levels. Some fly fishermen would complain about the destruction of virgin streams, whereas other fishermen would welcome the opportunity for trolling on the lake. Recreational enthusiasts would anticipate the use of the lake, but search and rescue units might be apprehensive. Power generating compa-

FIG. 4.2. The planning, design, and construction of structures often requires the interaction of several groups of people.

nies would have interest in the water as a source of energy. Some would be concerned about failure of the dam and a disastrous flood, whereas others would consider the dam for flood control. Some naturalists would be opposed to the disruption of the natural environment, yet others would consider the lake a refuge for wildlife. Some businesses nearby would welcome the increased flow of people. These diverse political, social, environmental and economic inputs during the planning stage often need to be included by an engineer in the structural design.

Engineers represent science and technology, yet they must understand people and have empathy for their concerns. In preparation for this need, engineering curricula include courses in life sciences and social sciences and the humanities, in addition to the basic core of courses in the physical sciences. This exposure is intended to contribute to the social polish of the engineer and prepare him or her to cope with people and their ideas.

An engineer involved in the design and development of some structures must communicate with the general public and exchange technical information with specialists in other disciplines. The engineer's primary responsibility, however, is to apply the fundamentals of science and technology in order to design some assemblage of materials to perform the necessary task. The engineer synthesizes the many conceptual expectations and actually creates some solution in the form of a structure. After several have talked about a project, the engineer must specify how to build it. The engineer must then select materials and decide on the proper configuration to maintain structural integrity. The effect of temperature, external forces, magnitudes and directions of internal forces, displacements, and buckling are examples of things that the engineer must often consider. The ease of fabrication and construction, maintenance and structural life are other technical areas of concern.

In the remaining section of the chapter, you will be introduced to forces and structures by solving some related problems. Some basic fundamentals relating to forces, moments, and static equilibrium equations are presented in order to equip the student with the necessary rudimentary skills. These skills are then applied to some typical structural problems encountered in engineering.

4.1/Models of Structural Loads

An engineer designing a structure must evaluate several different alternatives and make judgment decisions that will greatly affect the design process and the eventual structure. One of these judgment decisions is the selection of the force system that appropriately represents the actual loads to which the structure will be subjected. The forces that occur in nature are frequently complex. Some generalizations and simplifying assumptions usually have to be made to simulate the naturally occuring force with a mathematically equivalent force. Some of the loads acting on a structure are relatively small compared to others and have little influence on the total design. These negligible forces need to be identified and eliminated in order to simplify the problem. Other forces might be applied in rather a complicated manner, yet the overall effect can be modeled rather simply. For example, in Figure 4.3 a 200-lb man is standing at the midpoint of a small bridge. His weight is distributed over an area, depending on the size of his feet. Perhaps he also stands with more pressure on his left foot as compared to his right foot. Nevertheless, the total contact area is small when compared with the bridge surface and the total 200 lb force can be represented as a static concentrated load of 200 lb acting at the midpoint of the bridge. A static concentrated force is represented typically by an arrow attached at the point of application. An arrow would be added for each additional load that could be simulated as a concentrated load with some magnitude, direction, and point of application.

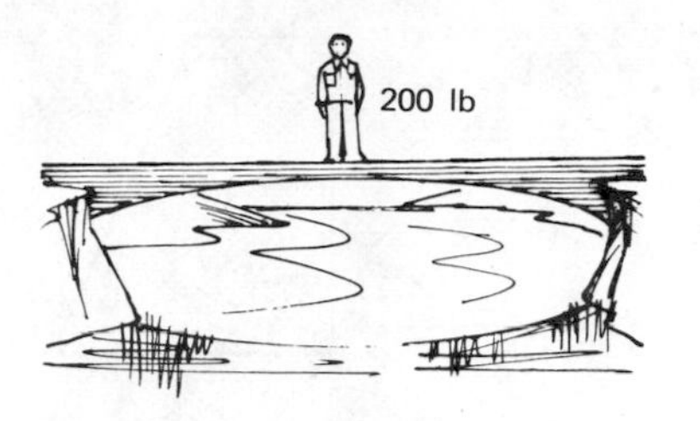

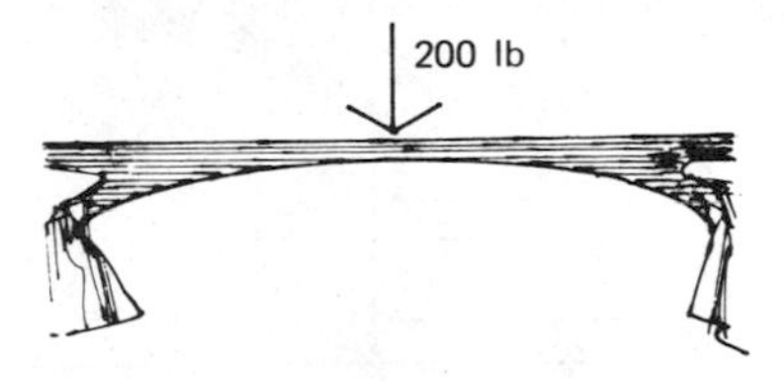

FIG. 4.3. The weight of a human may be modeled as a concentrated load.

Another commonly used model is the static uniformly distributed load as shown in Figure 4.4. For example, consider a large pipeline that crosses a creek and is supported by a truss. The weight of the fluid flowing in the pipe added to the dead weight of the pipe itself would represent a uniformly distributed load carried by the truss. A snow load on the roof, the weight of a crawler tractor on the ground surface, the lift on some aircraft wings, and the vertical bearing forces on foundations are other examples that could be ideally modeled as a static uniformly distributed load. The units are usually given as force per unit length or force per unit area.

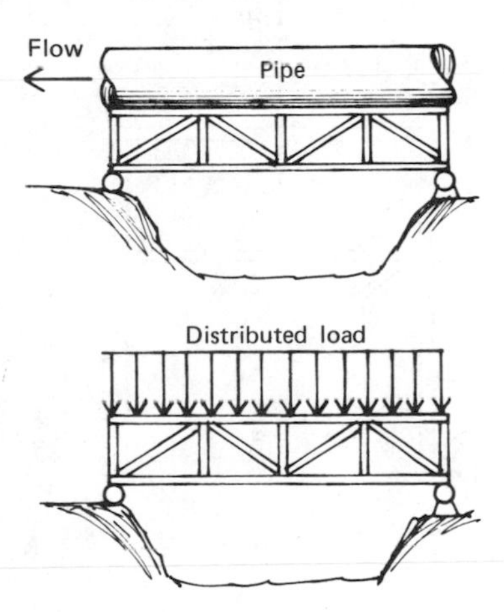

FIG. 4.4. A uniformily distributed load.

There are several nonuniformly distributed loads that occur in nature. An engineer must select some mathematical function that describes the distribution. The uniform load mentioned previously is a special case where the magnitude is constant and the resulting load distribution is a rectangle. Perhaps the next simplest case is the linearly distributed load that appears as a triangle. For example, the horizontal force of water against a dam increases linearly with depth, as shown in Figure 4.5. The soil pressure against a vertical retaining wall is also often simulated as a linearly distributed force. There are other more complex nonuniform force distributions that would require higher-order polynomials to properly model mathematically.

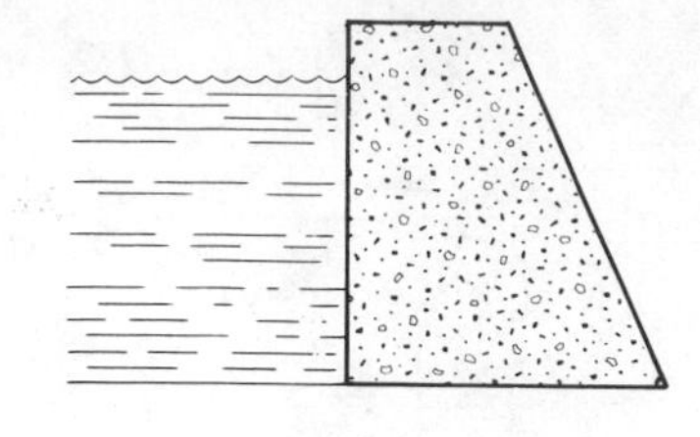

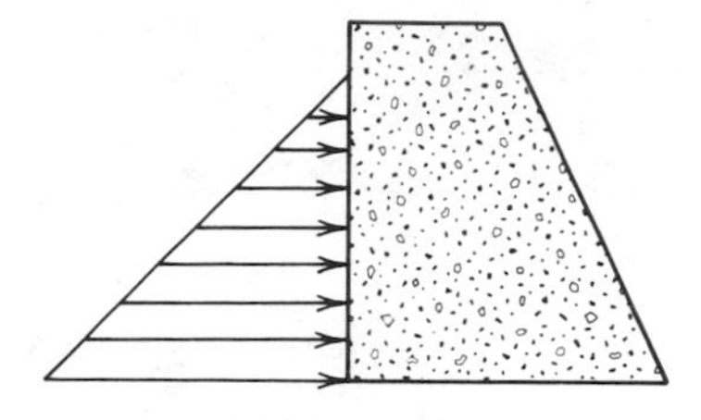

FIG. 4.5. A nonuniformily distributed load.

If the loads applied to a structure do not vary with time and if the structure is stationary, then the problem is defined as static. If the applied loads vary with time and/or the structure experiences appreciable acceleration, then the problem is dynamic and is considerably more complex. If inertial forces are negligibly small compared to other applied forces, engineers will often simplify the problem by solving an equivalent static problem. A typical periodic dynamic force is shown in Figure 4.6, where a motor with an unbalanced rotating shaft is mounted on a frame. As the shaft rotates, the magnitude and direction of the force vary as shown by the time dependent curve. A six-cylinder reciprocating automobile engine would also transmit a periodic force through the vibration isolation mounts to the automobile frame at a frequency three times that of the crankshaft's rotational speed.

Another type of dynamic load that an analyst must commonly model is the impact load. As shown in Figure 4.7, an air blast or shock wave on a building would be represented as a sudden force with a large amplitude and relatively short duration time. Automobile collisions, hammer blows, bats striking balls, and rocket thrusts are other examples of impact loads.

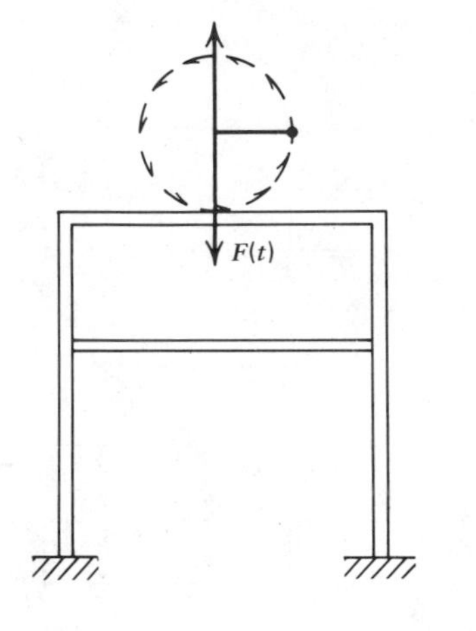

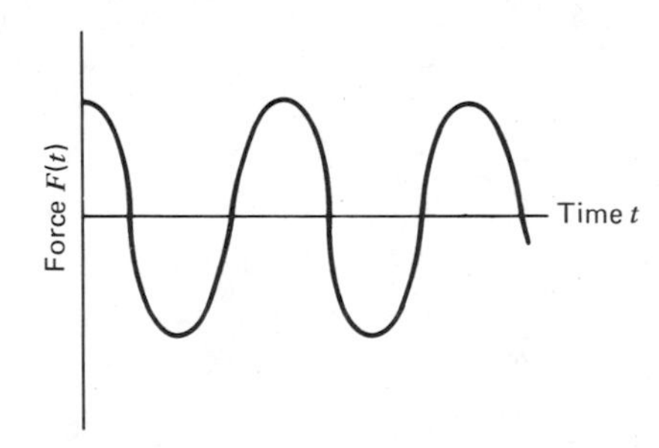

FIG. 4.6. A dynamic force may change magnitude and direction with respect to time.

There are some time dependent forces that have no definite pattern to frequency or amplitude. These so-called random forces do not repeat themselves and are not describable by a single continuous mathematical function. The analyst must then apply a statistical approach to properly simulate the force. For example, spacecraft would be subjected to random forces as the lift-off rockets are fired as shown in Figure 4.8. Lateral excitation forces during an earthquake are also examples of random forces.

Structures are frequently subjected to combinations of concentrated, distributed, static, and dynamic external loads. All these forces and possibly others could act simultaneously or independently on the structure. Since the exact magnitude and direction of all the forces are often unpredictable, some allowance for safety must be made. A margin of safety is provided by multiplying the predicted loads by a load factor. A load factor ensures the safety of the structure by compensating for such occurences as overloads, material flaws, fabrication mistakes, and other design oversights.

The determination of the magnitude of this load factor is a judgment decision made by the engineer, who considers such things as the exactness of the assumed loading system, the impact of a failure, economics, the validity of the structural analysis, and possibly other criteria. A load factor between 1.5 and 2 is common. In the past, the engineering profession has been reliable in the design of structures. However, there have been some buildings collapsing, bridges buckling, and dams washing away. Note that many of the relatively few failures that have occurred resulted from unusual natural hazards that exceeded the economic limitations that govern a design. Society could not afford to design all buildings to withstand all earthquakes, bomb blasts, hurricanes, tornadoes, and floods.

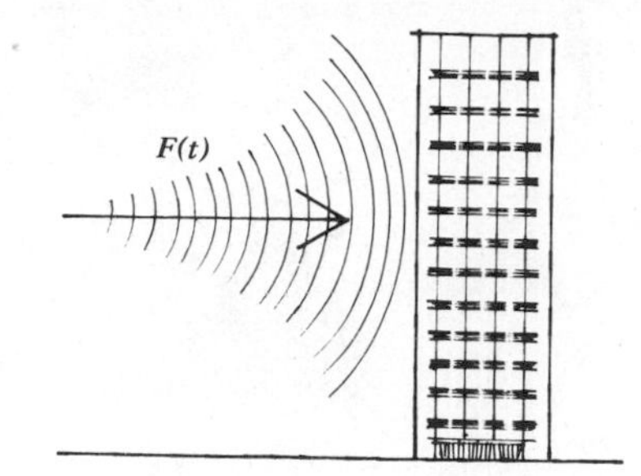

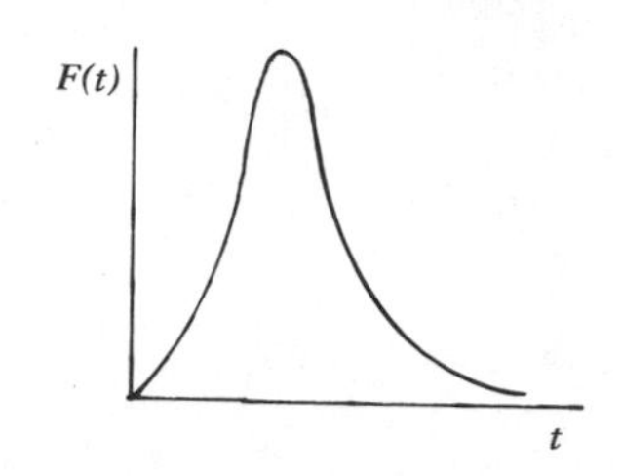

FIG. 4.7. An air blast may be modeled as an impact or shock load.

To demonstrate a typical judgment decision, assume that steel towers for the suspension of high-voltage electrical power lines must be designed. The engineer must eventually determine all the forces that act on the towers. One of these forces would result from transverse winds blowing on the cables. The magnitude of this force would depend on such factors as the frost accumulation on the cable, the air temperature, and the velocity of the wind. Assuming that a climatologist's report for the region estimates that a maximum 150 mph wind will occur once every 100 years, the engineer proceeds with the responsibility of making the best judgment on the magnitude of the force to be considered. If the towers were designed to withstand crosswinds on the cable of 150 mph in heavy frost conditions, then several towers may be significantly heavier than needed or placed too close together and our resources would be wasted. Perhaps the engineer would select a lighter, more realistic wind load for the area and design lighter towers. In the event of an unusually high wind and a possible tower failure, the failures would be repaired. The cost to repair a few towers might be significantly less than the additional cost of several heavy overdesigned towers. The other forces acting on the structure and such factors as public safety, construction requirements, and expenses resulting from power outages would also contribute to the engineer's judgment decision.

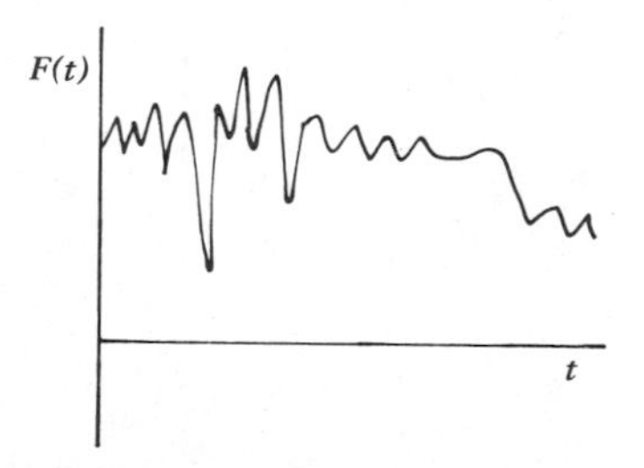

FIG. 4.8. Some forces vary randomly with respect to time.

4.2/Forces and Resolution

The term force describes either a push or a pull and represents the action of one body on another. A force can be described pictorially by an arrow. The arrow length represents the amount or *magnitude* of the force. The line along which the force acts is defined as the *line of action*. The *direction* of the force is then given by specifying the angle between the line of action and some reference line such as the x axis as shown in Figure 4.9a. The

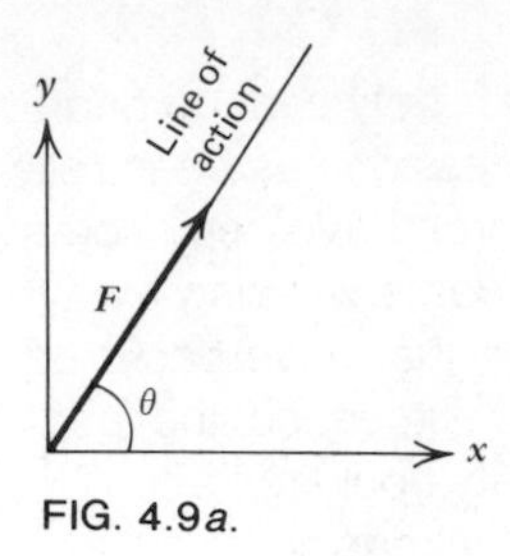

FIG. 4.9*a*.

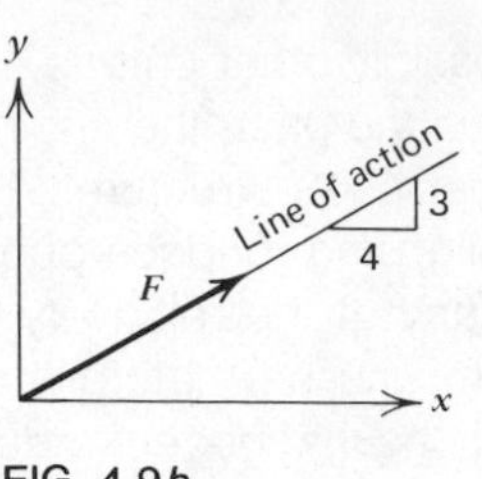

FIG. 4.9*b*.

arrowhead indicates the *sense* of the force along the line of action.

An alternative way to describe the direction of a force is to use the triangle method where the slope of the line of action is given by specifying the legs of a right triangle, shown in Figure 4.9*b*. In mathematical terms, a force is a vector quantity and is characterized by its magnitude, direction, and point of application. The magnitude of force is usually measured in such units as ounces, pounds, tons, newtons, or dynes. The orientation angle that indicates direction is measured in radians or degrees. The point of application is usually shown on a sketch by properly locating the arrowhead.

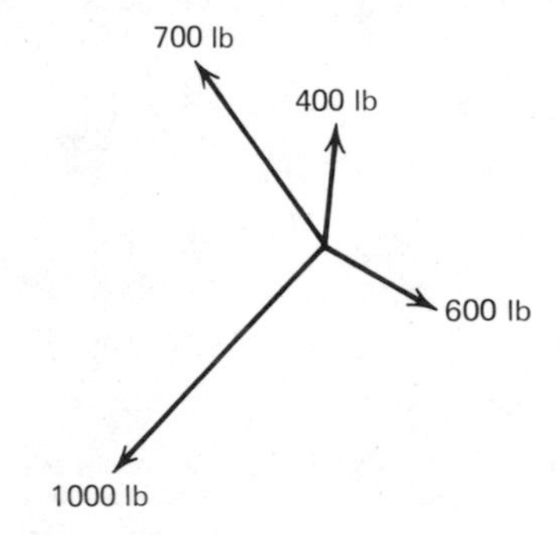

FIG. 4.10.

There are some terms relating to forces that need to be defined. Forces that act along the same line of action are *collinear.* If the lines of action of forces pass through a common point, then the force system is *concurrent.* A system of forces that act in a common plane are said to be *coplanar.* A system of coplanar concurrent forces is shown in Figure 4.10. The *principle of transmissibility* states that the external effect of a force on a body is the same for any point of application along the line of action as shown in Figure 4.11. As an example of the transmissibility principle, consider a long weightless rope attached to a tree. A pull of 50 lb anywhere along the rope would apply the same force to the tree.

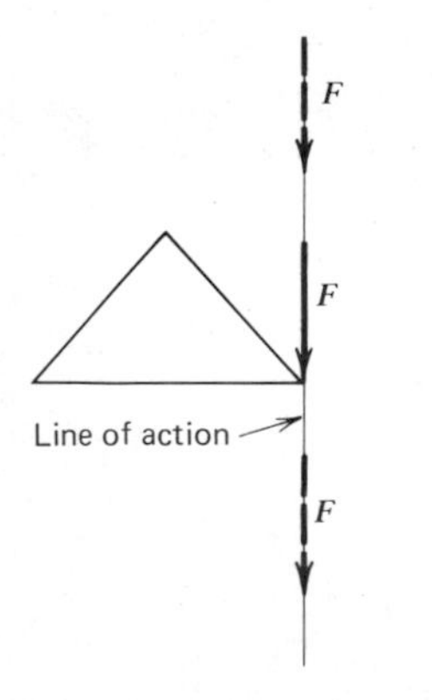

FIG. 4.11.

Forces can be added and subtracted; however, one must include both magnitude and direction. Forces can be added algebraically only if they are in the same direction. A 10-lb force in the vertical direction plus a 5-lb force in the horizontal direction is not equal to a force of 15 lb in either direction. In order to combine several forces, it is convenient to resolve them into rectangular components that have common directions and then add and subtract the appropriate magnitudes. The breakdown of a single force into components is called *resolution.* To find the rectangular components of a force, one considers the force to be the hypo-

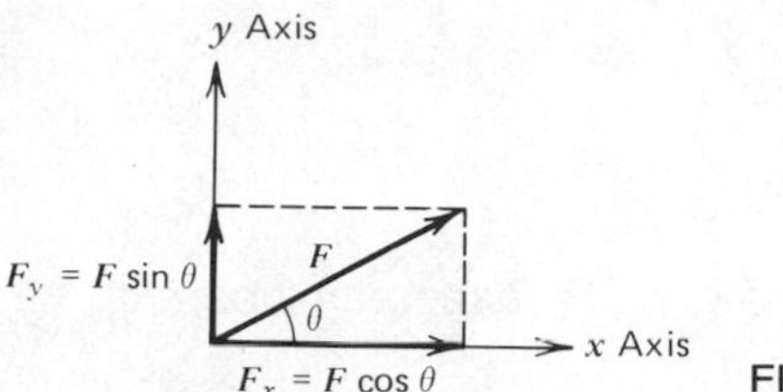

FIG. 4.12.

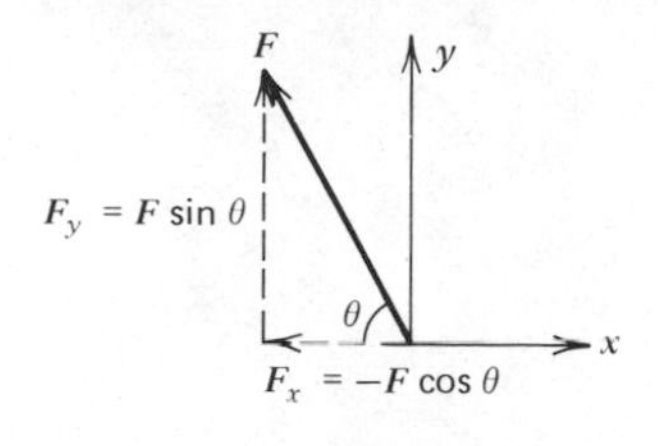

tenuse of a right triangle. The rectangular components are then determined using the trigonometric functions (see Appendix B) where the sides of the triangle represent the direction and magnitude of the components involved as shown in Figure 4.12.

The orientation and positive direction of the reference system of axes can be chosen arbitrarily. The reference system is frequently selected such that the vertical axis is the y axis and is positive upward. However, it is sometimes convenient to work with a rectangular coordinate system that has axes rotated at some angle from the horizontal. Such an inclined reference system sometimes simplifies the addition of forces. A positive sign is attached to the component if the component is in the direction assumed to be positive for the reference axis. A negative sign indicates that the component is in the opposite direction of the reference axis. Examples are shown in Figure 4.13.

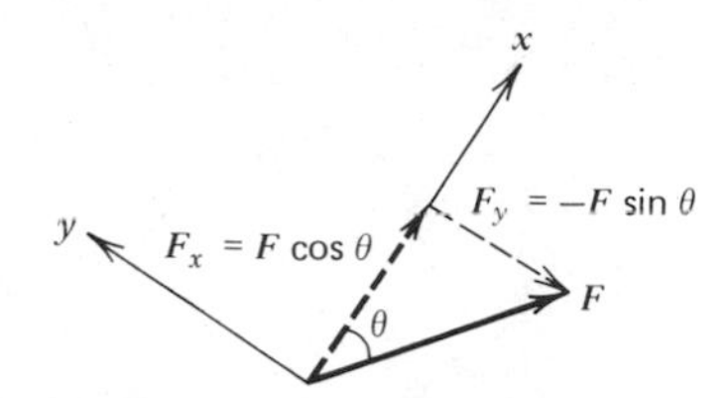

FIG. 4.13.

Several concurrent forces can be vectorially added together to find a single equivalent force called the *resultant.* The sum of the x and y components are considered as the legs of the triangle of which the hypotenuse is the resultant force. In a concurrent force system, the single resultant force could replace the system of individual external forces as shown in Figure 4.14. Both the magnitude and direction of the resultant force must be determined.

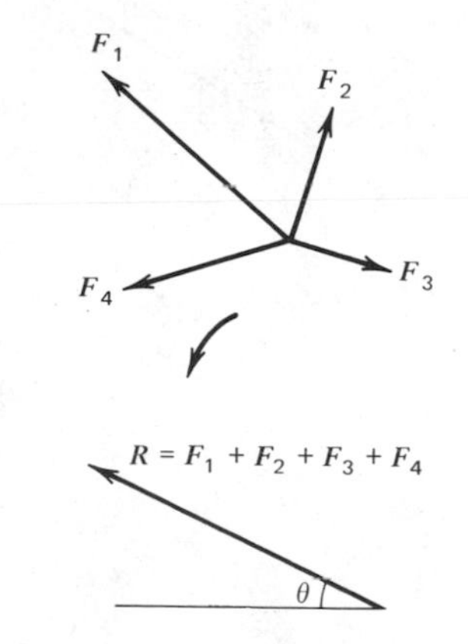

FIG. 4.14.

EXAMPLE PROBLEM 4.1

Resolve the force into components in the horizontal and vertical directions.

Solution

The force of 1000 lb is considered to be the hypotenuse of a right triangle. The horizontal or x component is found by writing

$$\cos 30 = \frac{F_x}{1000}$$

$$F_x = 1000 \cos 30° = 866 \text{ lb}$$

The vertical component F_y is also found using trigonometry.

$$\sin 30 = \frac{F_y}{1000}$$

$$F_y = 1000 \sin 30 = 500 \text{ lb}$$

The results can be checked using the Pythagorean theorem:

$$1000 = \sqrt{(866)^2 + (500)^2}$$

Note that both F_x and F_y are in the positive directions as shown. In resolving forces into components, the arrowheads are selected such that one traveling along the components eventually arrives at the tip of the force being resolved.

EXAMPLE PROBLEM 4.2

Resolve the force into components in the x and y directions.

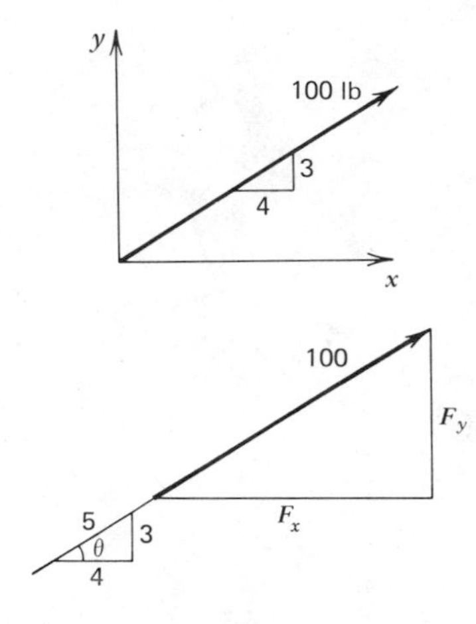

Solution

The components can be found by treating the 100-lb force as the hypotenuse of a right triangle. The ratio of the sides can now be used instead of trigonometric functions and angles. The horizontal component is

$$F_x = \frac{4}{5}(100) = 80 \text{ lb}$$

The vertical component is

$$F_y = \frac{3}{5}(100) = 60 \text{ lb}$$

Note that

$$\cos \theta = \frac{4}{5} \qquad \sin \theta = \frac{3}{5}$$

EXAMPLE PROBLEM 4.3

Add the two forces and determine the resultant.

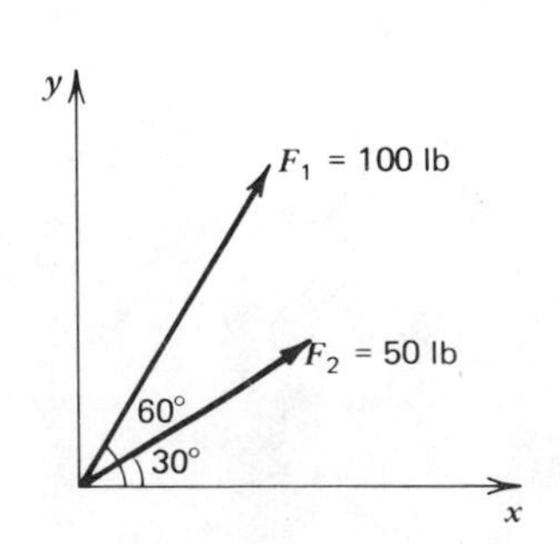

Solution

The forces are first resolved into components that can then be added algebraically. The x and y components of the 100-lb force are identified as F_{1x} and F_{1y}, respectively, and are determined as follows:

$$F_{1x} = 100 \cos 60° = 50 \text{ lb}$$

$$F_{1y} = 100 \sin 60° = 86.6 \text{ lb}$$

The x and y components of the 50-lb force are determined as follows:

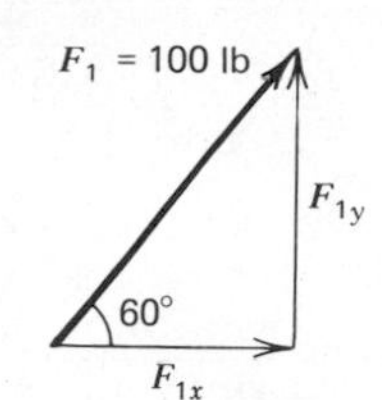

$$F_{2x} = 50 \cos 30 = 43.3 \text{ lb}$$

$$F_{2y} = 50 \sin 30 = 25 \text{ lb}$$

The components of the resultant force are determined by adding the components of the individual forces in the x and y directions.

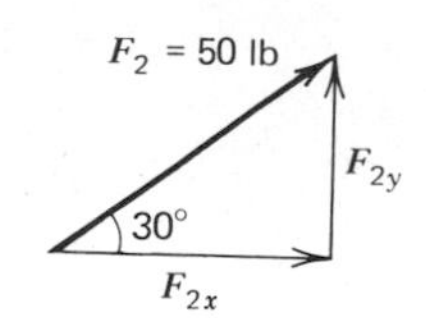

$$R_x = F_{1x} + F_{2x} = 50 + 43.3 = 93.3 \text{ lb}$$

$$R_y = F_{1y} + F_{2y} = 86.6 + 25 = 111.6 \text{ lb}$$

The magnitude of the resultant is then found by using the Pythagorean theorem.

$$R = \sqrt{93.3^2 + 111.6^2}$$

$$R = 148.9 \text{ lb}$$

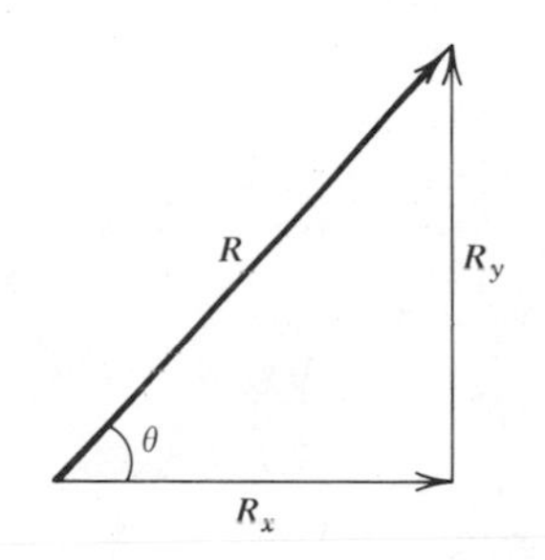

The line of action of the resultant can be referenced to either the x or y axis. The angle θ referenced to the x axis is determined as follows:

$$\theta = \tan^{-1} \frac{116}{93.3}$$

$$\theta = 51.2°$$

EXAMPLE PROBLEM 4.4

Determine the resultant force for the concurrent force system. Magnitudes and directions of forces are shown.

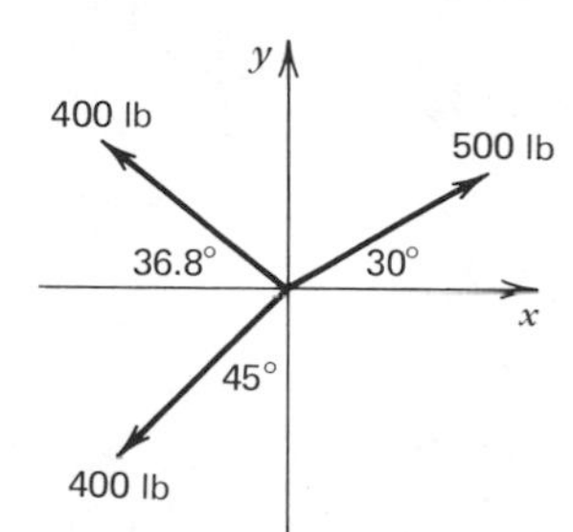

Solution

One begins by resolving the individual forces into components in the x and y directions. The magnitude of each force must be multiplied by the appropriate trigonometric function. Force components that point in the negative sense along the x and y axes are identified as negative. The results are summarized in the table. The x component of the resultant force is the sum of all the x components and and is −171 lb. Summing in the vertical direction, the y component of the resultant force is 206 lb. The magnitude of the resultant force can now be determined as:

Force	F_x	F_y
500	433	250
400	−320	240
400	−284	−284
R	−171	206

$$R = \sqrt{206^2 + 171^2} = 268 \text{ lb}$$

The direction of the line of action can be found as follows:

$$\theta = \tan^{-1}\left(\frac{206}{171}\right) = 50.3°$$

$$\theta = 50.3°$$

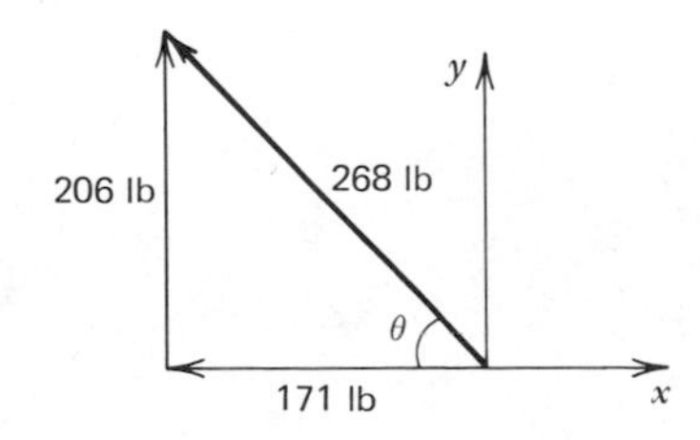

Note that the horizontal component of R is negative and the vertical component is positive; therefore R is to the left and up as shown.

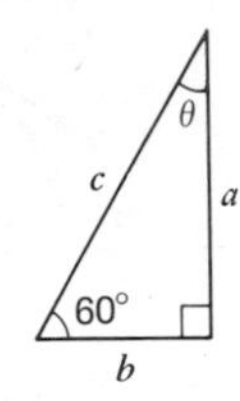

PROBLEMS

4.1. Determine sides b and c and the unknown angle θ for the right triangle. Side $a = 150$.

4.2. Determine the unknown angles θ and λ and the side b for the right triangle.

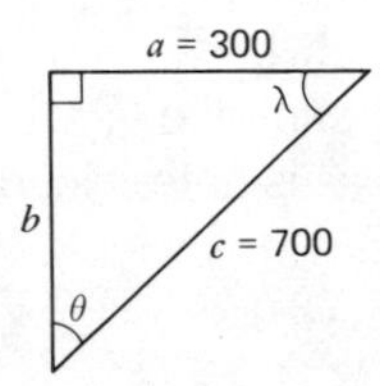

300 lb

$\theta = 38.6°$

4.3. Determine the horizontal and vertical components of the 300-lb force.

4.4. Determine the horizontal and vertical components of the 1500-lb force applied to the truss.

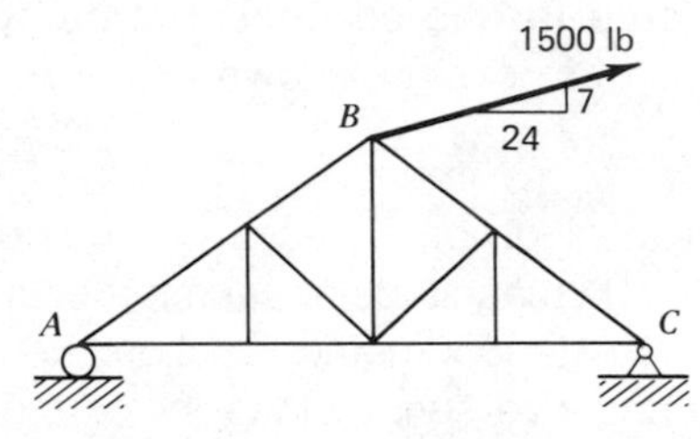

4.5. The tension in a guy wire is 1000 lb. Determine the vertical component that attempts to pull the anchor out of the ground at point C.

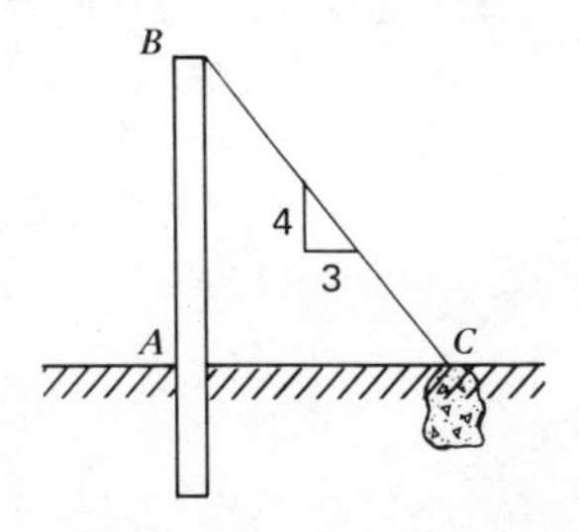

4.6. Determine the resultant force of the concurrent force system.

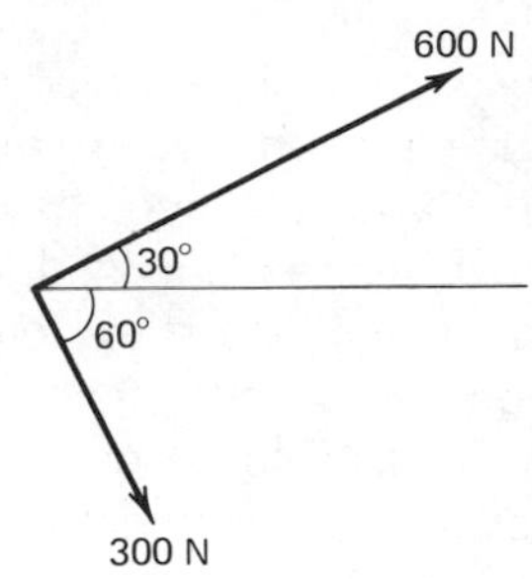

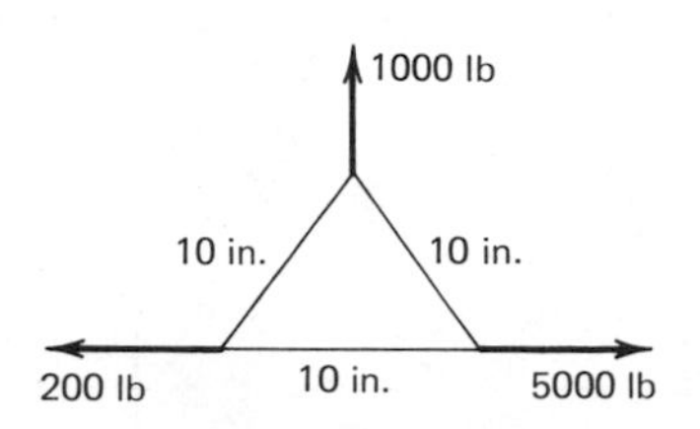

4.7. Determine the resultant force of the concurrent force system.

4.8. Determine the resultant force of the concurrent force system.

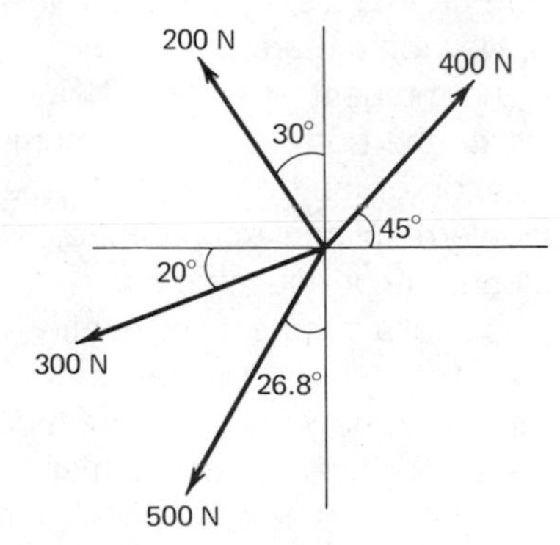

4.9. A rope tow is being used to pull skiers up a snow-covered hill, which is 500 ft long and has a slope of 20 degrees measured from the horizontal. Roughly estimate the minimum tension in the cable required to pull 20 skiers up the hill that weigh 180 lb each. Neglect the friction between the skis and the snow.

4.3/Moment of a Force

If a force is applied to a rigid body initially at rest, the body tends to either translate, rotate, or move with combined translation and rotation. For example, as shown in Figure 4.15, a force applied at the midpoint of a free uniform bar slides the bar such that every point moves on equal distance and the bar is said to translate. If the same force is applied at some other point, then the bar both

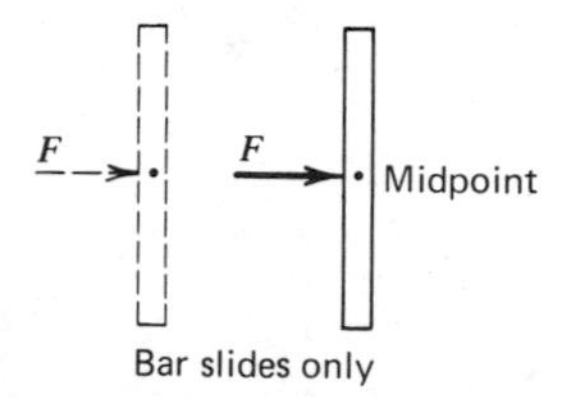

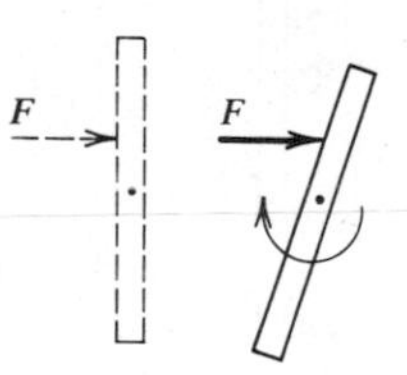

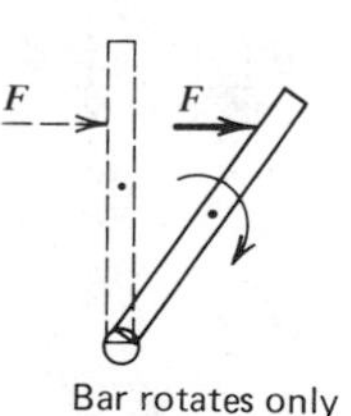

FIG. 4.15.

translates and rotates. The amount of rotation depends on the point of application of the force. If a point on the bar is fixed, then the applied force causes the bar to rotate only.

A force was defined earlier as a push or a pull; however, it is now apparent that a force can also cause rotation. Such rotation occurs when you push on a door or when you pull on the handle of a wrench to turn a nut. Most everyone has experienced that the turning ability depends not only on the magnitude of the force, but also on the distance away from the door hinge or the nut. In engineering parlance, the term moment is used to describe this twisting or rotation that a force can produce. The moment of the force F about the point A is defined as

$$M = Fd$$

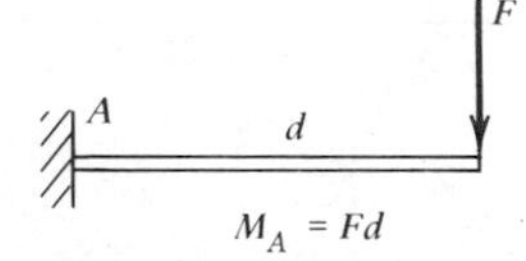

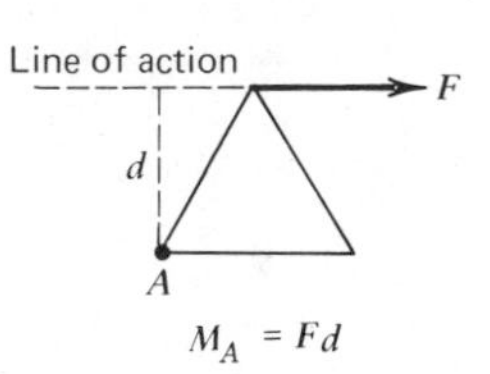

FIG. 4.16.

where d is the perpendicular distance between the point A and the line of action of the force F as shown in Figure 4.16. The units of a moment are the product of force and distance and are therefore expressed as foot pounds, Newton meters, and so on.

It is important to note that the moment of a force does not depend on the angle through which the body rotates. A moment can be developed even if no rotation of the body occurs. Again, a force pulling on a wrench handle results in a moment or twist applied to the nut even if the stubborn nut doesn't move.

An algebraic sign is attached to a moment to identify the twisting action as either clockwise (cw) or counterclockwise (ccw). The sign convention is arbitrary; however, it is often convenient to refer to ccw moments as positive and cw moments as negative. In two-dimensional problems, the total moment about some arbitrary point is the algebraic sum of all the individual moments, as shown in Figure 4.17.

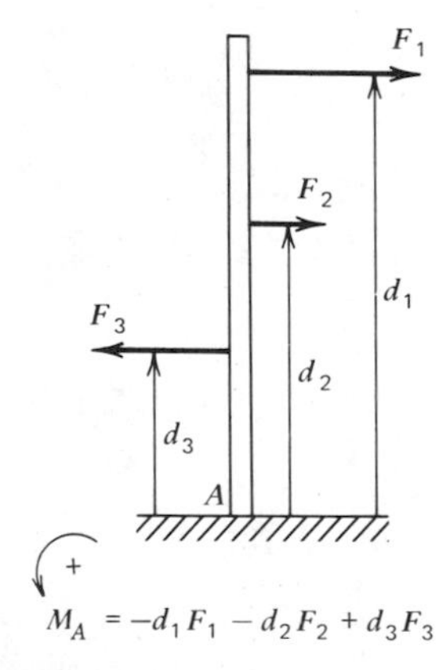

FIG. 4.17.

It is often convenient to replace all the forces that act on a body with one equivalent force called the resultant. This single resultant force must provide the same push or pull and the same moment or twist as all the individual forces combined. If the force system is concurrent, then the total moment is zero and the resultant must equal only the vector sum of all the individual forces and its line of action must pass through the common point of intersection. In the case of nonconcurrent force systems, the resultant must equal the vector sum of all the individual forces. Furthermore, it must also provide a moment about any arbitrary point equal to the sum of the moments of the individual forces about the same point. Consequently, the resultant of a nonconcurrent force system requires that both the magnitude and point of application be found. The point of application is determined by equat-

F_1 d_2 F_2 d_1 A F_3 — Nonconcurrent forces

$R = F_1 + F_2 + F_3$, A, $M_A = d_1F_1 - d_2F_2 + d_2F_3$ — Force and moment

R, A, d, $d = \frac{M_A}{R}$ — Resultant force

FIG. 4.18.

ing the total moment of the individual forces about some point *A* to the moment of the resultant with respect to the same point *A* as shown in Figure 4.18.

EXAMPLE PROBLEM 4.5

Determine the total moment of all the forces about points *A*, *B*, and *C* in the truss shown in the diagram.

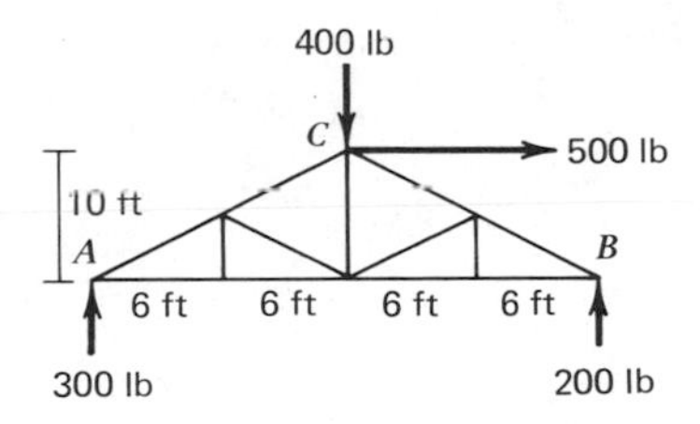

Solution

The moment about point *A* can be determined by multiplying the magnitude of each force by its lever arm with respect to point *A* and then combining the products algebraically. The positive direction is arbitrarily selected as ccw. The lever arm is the perpendicular distance from the point *A* to the line of action of the force. For example, the lever arm from *A* to the 400-lb force applied at point *C* is not *AC*, rather the perpendicular distance along the lower chord of the truss, which is 12 ft.

$$M_A = -(400)(12) - (500)(10) + (200)(24)$$

$$M_A = -5000 \text{ ft}\cdot\text{lb}$$

The moment about point *B* is determined in a similar manner. Note that a minus sign indicates that the moment is cw. Also, the moment of the 200-lb force about point *B* is zero since the perpendicular distance from *B* to the line of action of the force is zero.

$$M_B = -300(24) + 400(12) - 500(10)$$

$$M_B = -7400 \text{ ft}\cdot\text{lb}$$

The moment about point *C* is

$$M_C = -300\,(12) + 200(12)$$

$$M_C = -1200 \text{ ft}\cdot\text{lb}$$

Remember that the lever arm extends perpendicular from the point about which moments are being taken to the line of action of the force, which is not always the point of application.

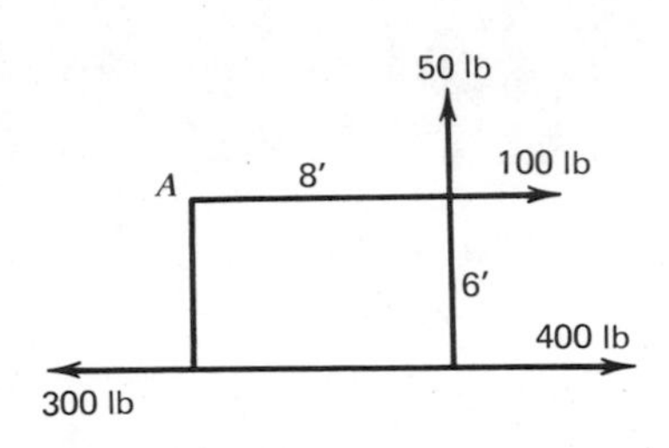

EXAMPLE PROBLEM 4.6

For the rigid body loaded as shown, determine (a) the total moment about point *A* and (b) the resultant force.

Solution

The total moment about *A* is the sum of the moments of all forces about point *A*. The ccw direction is arbitrarily assumed to be positive.

$$M_A = (50)(8) - (100)(0) + (400)(6) - (300)(6)$$

$$M_A = 1000 \text{ ft}\cdot\text{lb}$$

The magnitude, direction and point of application of a single resultant force can be found for the nonconcurrent system. The magnitudes of the components of the resultant force can be found by algebraically combining the components of the individual forces as shown in the following table.

Force	F_x	F_y
F_1	0	50
F_2	100	0
F_3	400	0
F_4	−300	0
R	200	50

$$R = \sqrt{(50)^2 + (200)^2} = 206$$

$$\theta = \tan^{-1}\frac{50}{200} = 14^\circ$$

The direction of the line of action of the resultant is determined by finding the inverse tangent of R_x and R_y.

The point of application of the resultant force is found by summing moments of all the individual forces about some point, say *A* and equating to the moment of the resultant force times some unknown distance *d*.

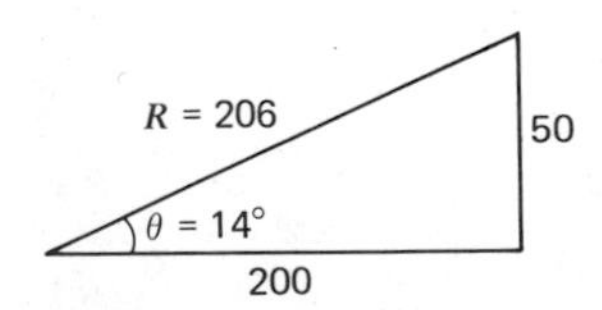

$$M_A = 1000 \text{ ft}\cdot\text{lb} = Rd = 206d$$

$$d = 4.85 \text{ ft}$$

Therefore the line of action of R is located 4.85 ft in a direction perpendicular and such that the moment is plus (ccw).

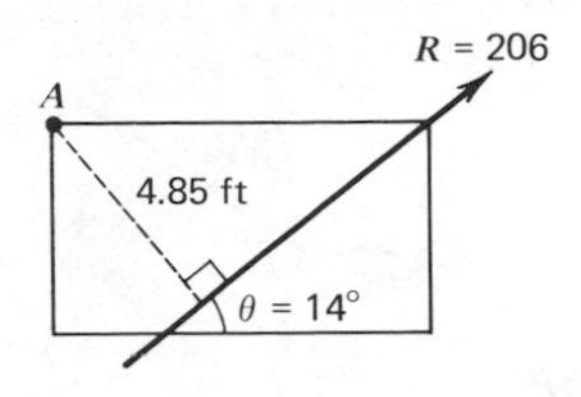

PROBLEMS

4.10. A mechanic is using an end wrench to remove a nut. A force of 120 lb is applied as shown. Determine the moment applied by the mechanic with respect to point A.

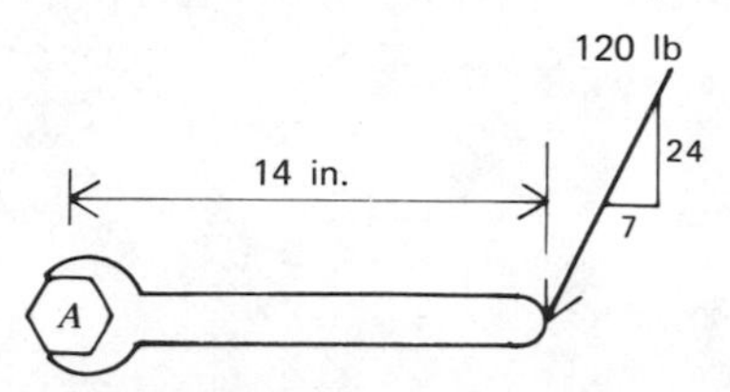

4.11. The total lift force on an airplane wing is given as L = 25 000 lb. The wing structure weighs 5000 lb, and a fuel tank attached at the wing tip weighs 1200 lb. Determine the total moment about the wing root at point A.

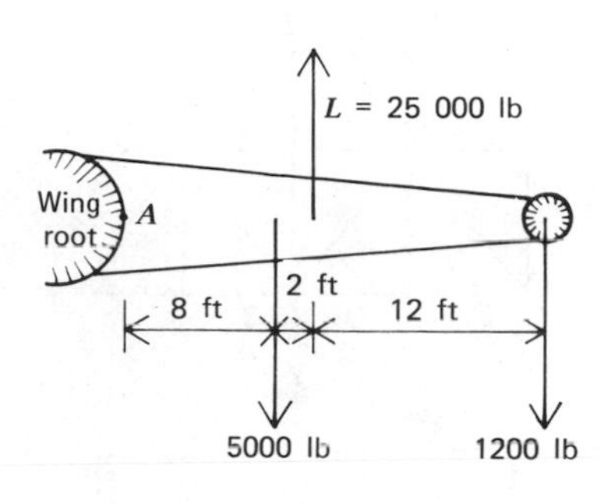

4.12. The horizontal wind load on a transmission tower is lumped into a concentrated force of 300 lb as shown. The cables transmit a force of 800 lb. Determine the moment about point B at the tower base.

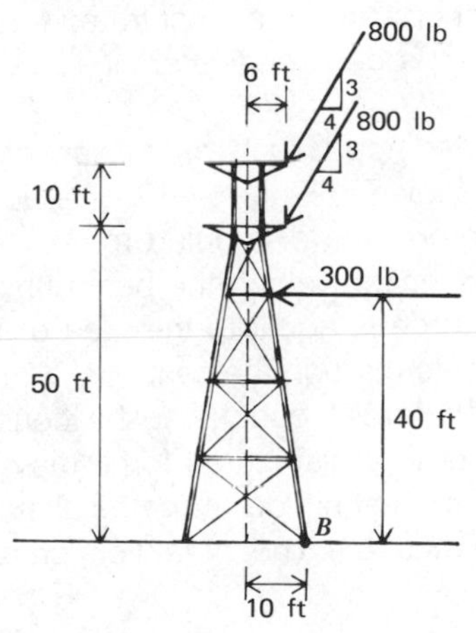

4.13. A large crane lifts 5 yd^3 of concrete in its bucket. The concrete weighs 3800 lb/yd^3. Determine the moment at the base of the 75 ft boom at point A due to the weight of the concrete when the boom is at 70° from the horizontal.

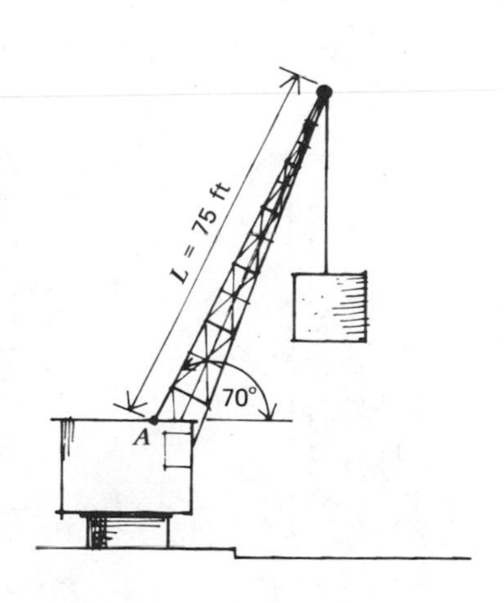

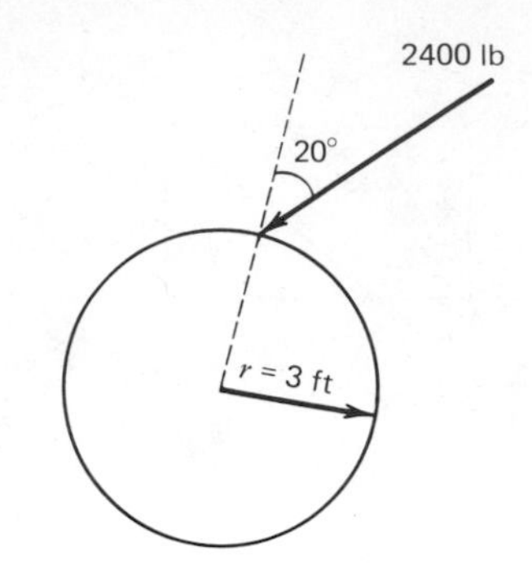

4.14. A fragment of solid material strikes a spherical satellite on the circumference. The impact is equivalent to a static force of 2400 lb as shown. Determine the magnitude of the moment of the force about the center of the satellite.

4.15. Determine the total moment about points *A* and *B*. (Hint: Resolve the 500-lb force into components before taking moments.)

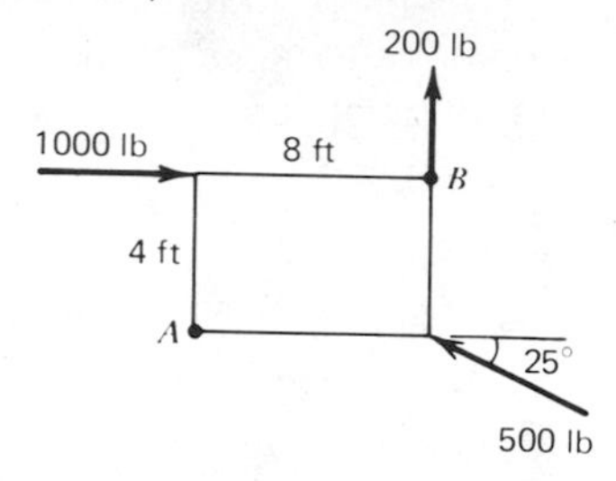

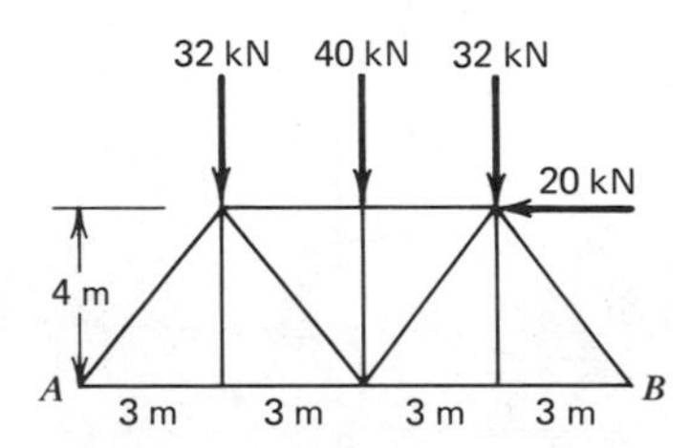

4.16. Determine the total moment about points *A* and *B*.

4.4/Distributed Loads

A concentrated load acts on a relatively small area that can be described as a point. Distributed loads extend across an area or along a line. The load distribution could be uniform, triangular, or vary according to some higher-order polynomial. Since a resultant force for both concurrent and nonconcurrent concentrated forces has been discussed, one can now proceed and determine a resultant force for distributed loads.

A typical uniformly distributed load such as a blanket of snow, is applied to a beam as shown in Figure 4.19. The beam is assumed to be uniform; therefore, the distributed force, w, is given in terms of pounds per linear foot along the beam (lb/ft). The magnitude of the resultant force is equal to the area of the distributed force rectangle and acts through the centroid.

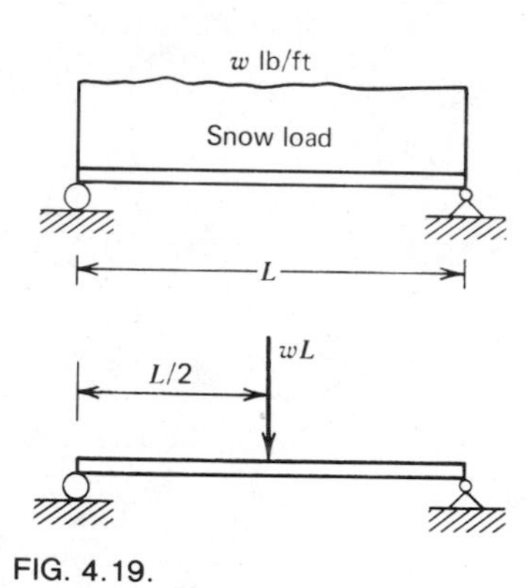

FIG. 4.19.

More complicated distributed forces could also be considered in a similar manner. In general, a distributed force may be replaced by a single resultant. The magnitude of the resultant is the area under the load curve and the line of action passes through the centroid.

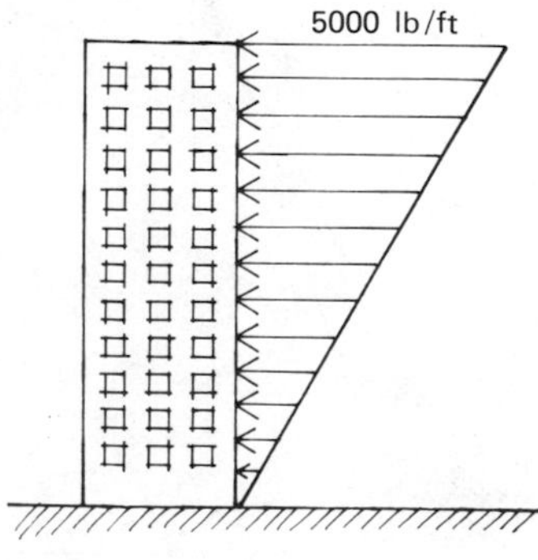

EXAMPLE PROBLEM 4.7

A high-velocity wind is blowing against the side of a building 80 ft high. The wind force against the building on a vertical strip 1-ft wide is assumed to vary linearly from zero at the bottom to

5000 lb/ft at the top. Determine the resultant wind force and the moment produced by the wind about an axis along the ground surface.

Solution

The magnitude of the resultant wind force is equivalent to the area of the force triangle.

$$F = \left(\frac{1}{2}\right)(80)(5000) = 20{,}000 \text{ lb}$$

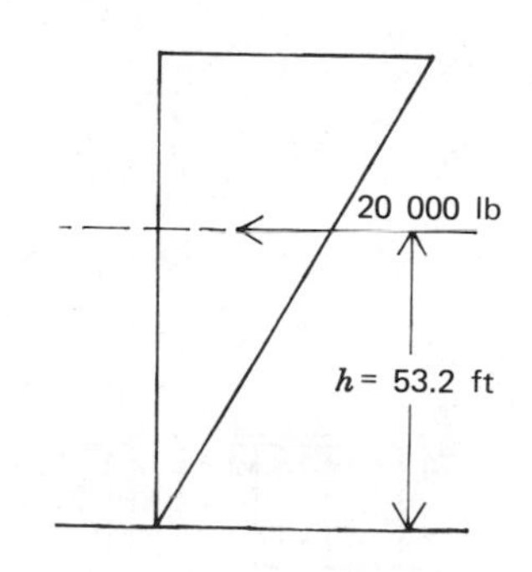

The resultant is located at the centroid of the triangle.

$$h = \left(\frac{2}{3}\right)(80) = 53.2 \text{ ft}$$

The moment of the wind about the ground axis is:

$$M = (53.2)(20{,}000) = 1.06(10)^6 \text{ ft}\cdot\text{lb}$$

EXAMPLE PROBLEM 4.8

A beam is subjected to a distributed load as shown. Determine the magnitude and point of application of a single concentrated force equivalent to the distributed load.

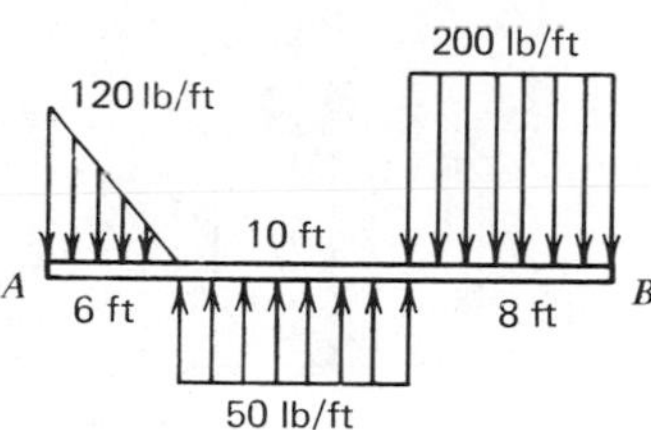

Solution

The magnitude of a concentrated load for each section can be determined as the area of the load diagram. The point of application of the concentrated load is at the centroid of each diagram.

$$F_1 = \left(\frac{1}{2}\right)(120)(6) = 360 \text{ lb downward}$$

$$F_2 = (50)(10) = 500 \text{ lb upward}$$

$$F_3 = (200)(8) = 1600 \text{ lb downward}$$

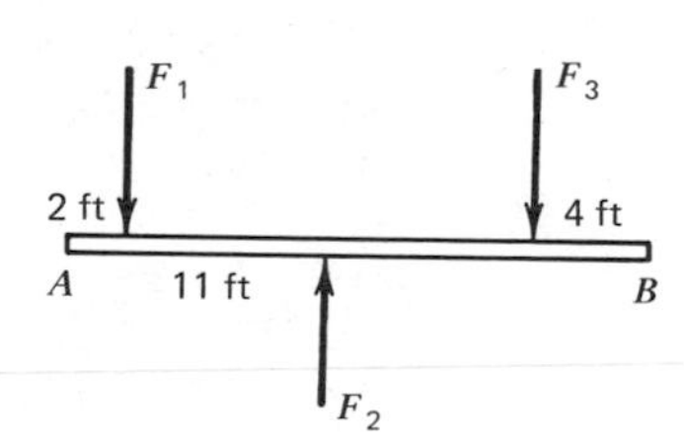

The magnitude of the system resultant is:

$$R = F + F_1 - F_2 + F_3 = 360 - 500 + 1600$$

$$R = 1460 \text{ lb}$$

The point of application of the resultant force is found by equating moments about point A. Assume that ccw is positive:

$$-dR = -2F_1 + 11F_2 - 20F_3$$

$$-d(1460) = -2(360) + 11(500) - 20(1600)$$

$$d = 18.64 \text{ ft}$$

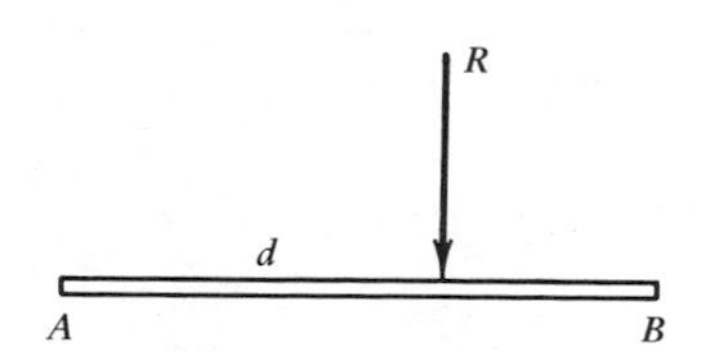

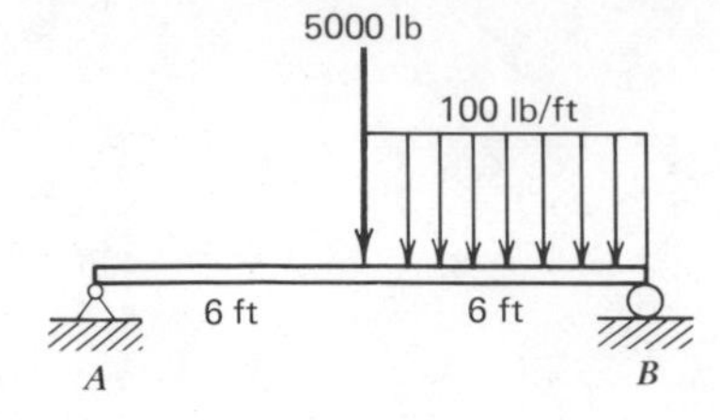

PROBLEMS

4.17. A beam is subjected to a concentrated load of 5000 lb and a distributed load of 100 lb/ft as shown. Determine the magnitude and location of the resultant force.

4.18. The force distribution on a 1-ft strip from a retaining wall is shown. Determine the moment of the resultant force about *A*.

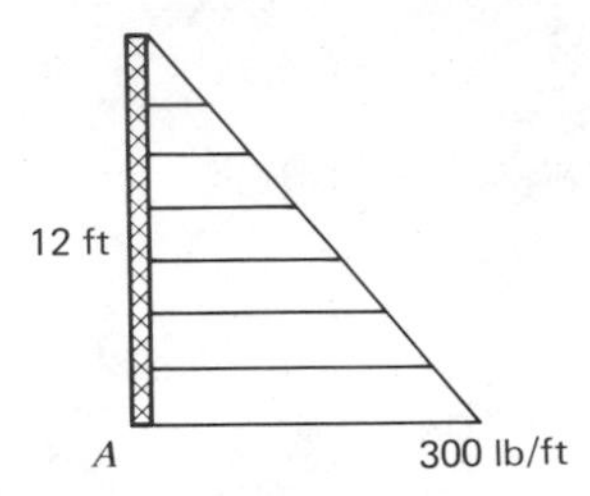

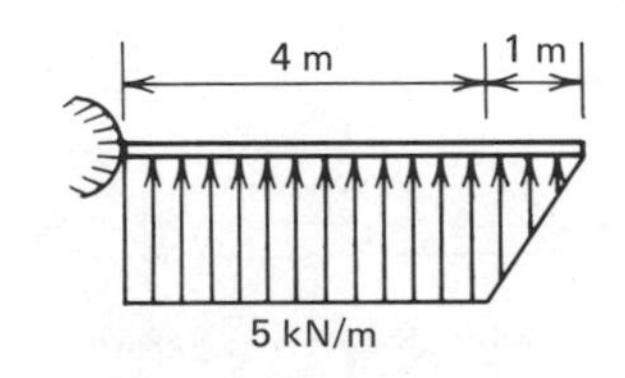

4.19. The lift force on an airplane wing is approximated as shown in the diagram. Determine the magnitude and location of a single force equivalent of the distributed force.

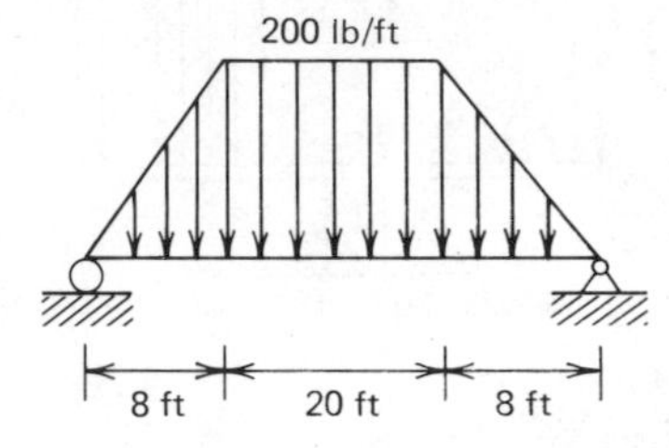

4.20. The dead weight of a bridge is distributed along the span as shown. Determine the magnitude and point of application of a single concentrated load equivalent.

4.5/Equilibrium

Three fundamental laws that relate the motion of a particle to the applied forces were first introduced by Sir Isaac Newton in the seventeenth century. These laws can be stated as follows:

> *I. A particle continues in its state of rest or of uniform motion in a straight line if there is no unbalanced force acting on it.*
>
> *II. The acceleration of a particle is proportional to the resultant force acting on it.*
>
> *III. To every action, there is an equal and opposite reaction.*

As stated, Newton's laws apply to a particle. A particle is a special object in mechanics that is defined to have mass but no size. Sometimes a particle is referred to as a point mass even though it might represent something as large as the moon or as small as the head of a pin. In particular, a particle cannot rotate and cannot be subjected to any moments since there are no lever arms. Forces acting on a particle represent a concurrent force system.

The first law of Newton defines static *equilibrium.* If the resultant force acting on a particle is zero, the particle is in equilibrium and remains at rest. If the particle is initially moving at some velocity, then under equilibrium conditions the particle continues to move with no change in velocity. Static equilibrium is also evident from the second law by equating the force to zero and recognizing that the acceleration must also then be zero. In concurrent force systems, equilibrium requires only that the resultant of all external forces acting through the point be zero as expressed by the equation,

$$\Sigma F = 0$$

Forces that act on a rigid body may not all pass through a common point. Such a nonconcurrent force system would produce moments that would tend to rotate the body. Consequently, a rigid body is in equilibrium only if the resultant force and moment are both equal to zero. The necessary and sufficient conditions for equilibrium of a rigid body can be expressed mathematically in the form of the equations:

$$\Sigma F = 0$$

$$\Sigma M = 0$$

The moment equation can be applied with respect to any point either on or off the body itself. In coplanar force systems, the summation of forces in any two orthogonal directions and the summation of moments results in three independent equations. Some examples of rigid bodies in equilibrium are shown in Figure 4.20.

As one can readily observe, there are several structures that appear to be stationary and consequently must be in static equilibrium. The resultant of all the forces and moments of forces must necessarily be zero, since no apparent motion occurs. Although most structures do experience some measurable, internal deformation when subjected to external loads, the deformations are slight and occur very slowly. The analysis of forces is usually completed by assuming the structure to be a rigid body and in static equilibrium.

A structural analysis often includes the interaction of adjacent bodies. For example, a building sets on a foundation, a bridge rests on abutments, an aircraft is suspended by air currents, a vehicle frame rests on a suspension system, and a fence post is

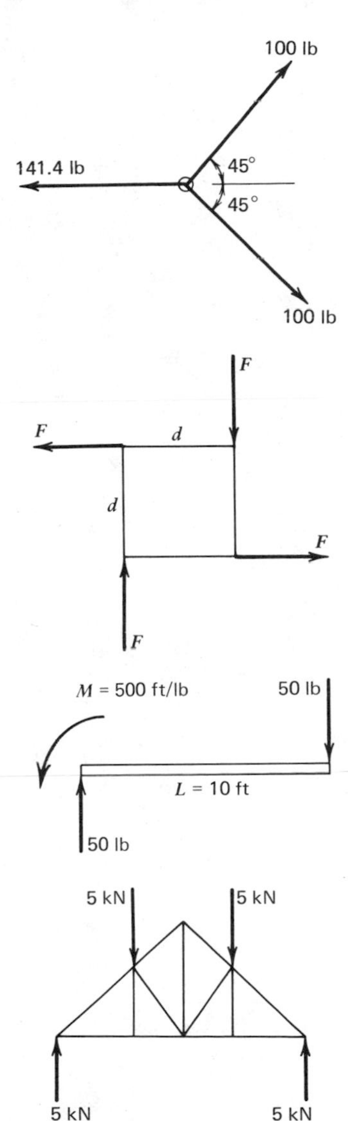

FIG. 4.20.

anchored in the ground. In order to apply the equations of equilibrium, the particular body or structure of interest must be removed from its surroundings. The interaction of the surrounding bodies must be represented as reaction forces and moments. Eventually, a free body diagram (FBD) must be drawn that isolates the body and includes all the reaction forces, external forces, and moments that act on the body itself.

The interaction of one structure with another structure or the interaction of internal structural members usually takes place at a joint or a connection. The force system that simulates the interaction exactly would be difficult to determine and depends on several physical factors such as bearing surface, friction, and joint mobility. The analysis of each joint would require unwarranted special attention and the analysis would frequently be cumbersome. In most engineering work, the actual structural connections are catagorized into general types of connections that theoretically simulate the joint movement and are representable by standard reaction force systems. Some of the basic connections and the corresponding reaction forces are given in Figure 4.21.

The roller support provides a constraint in only one direction. The reaction is equivalent to a single force of unknown magnitude but in a direction perpendicular to the sliding surface. For example, the rollers between a bridge deck and the horizontal plane of the abutment provide support in the vertical direction, but allow horizontal motion to accommodate expansion and contraction due to temperature changes. The bridge roller reaction can be replaced by a single force acting either up or down in the vertical direction.

The hinge or pin connection prevents translation in any direc-

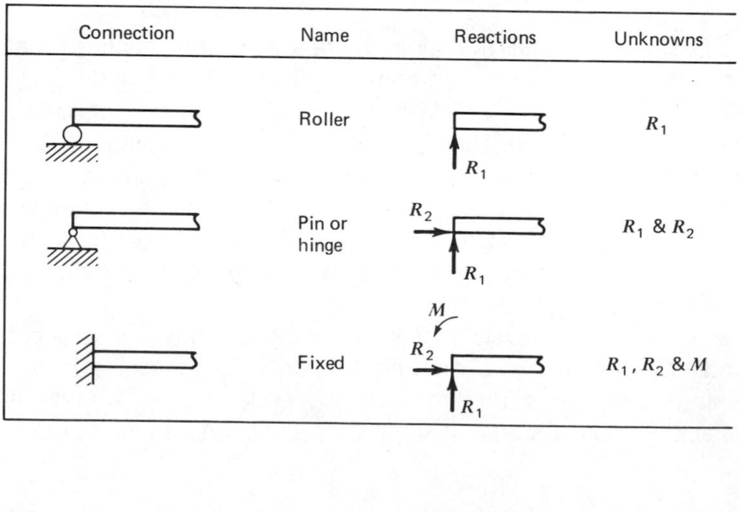

Connection	Name	Reactions	Unknowns
	Roller	R_1	R_1
	Pin or hinge	R_2, R_1	R_1 & R_2
	Fixed	M, R_2, R_1	R_1, R_2 & M

FIG. 4.21.

tion but does allow rotation. The total reaction at a pinned joint can be represented in component form by specifying the two perpendicular components of the resultant reaction. The ends of a wooden roof truss toenailed to bearing plates of a wall, the free edges of a box, or the bolted connections of steel members in a radio tower could be treated as hinged joints.

The fixed joint prevents both translation and rotation. The reactions include the two unknown force components similar to the hinged joint, but the moment is also unknown. A beam end embedded in concrete or a post firmly anchored in the ground could be treated as a fixed connection. The degree of fixity of a joint is usually a judgment decision that an engineer must make. Sometimes a joint has some moment rigidity but is not perfectly fixed and the engineer must decide whether the hinged or the fixed joint would be appropriate.

When rollers, pins, and fixed joints are replaced with the appropriate reaction forces and moments, their direction is not always obvious. For example, the structure shown in Figure 4.22 is supported by both a roller and a pinned connection. Any of the reaction systems shown could be assumed at the onset of the solution. In other words, the directions of the reactions can be selected arbitrarily in the beginning. As one applies the equilibrium equations to determine these unknown reactions, the sign will be positive if the proper direction was originally selected. A negative value will result if the opposite direction was arbitrarily selected. Therefore, don't fret about assuming the proper direction immediately. Simply draw on the reactions in some arbitrary direction and solve the problem using the equilibrium equations. If one of the unknowns turns out negative, say for example -50 lb, then the reaction is 50 lb in a direction opposite to that chosen on the FBD initially.

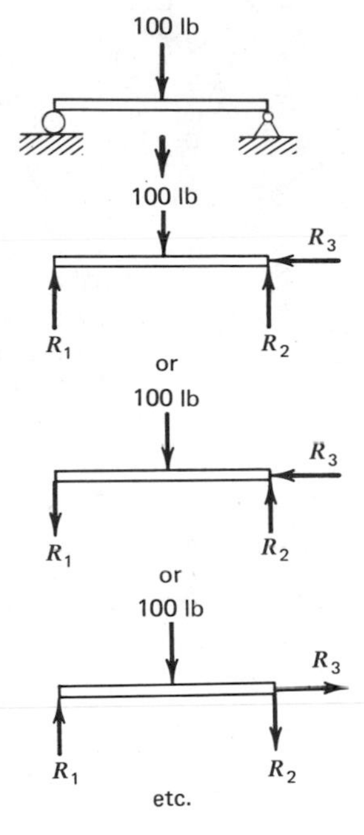

FIG. 4.22.

In general, the unknown forces in coplanar force systems can be determined in the following manner:

1. Draw a free body diagram. Include pertinent dimensions and all the external forces that act on the body. The connections must be replaced with the equivalent force system.
2. Set the summation of all the horizontal components of forces equal to zero.
3. Set the summation of all the vertical components of forces equal to zero.
4. Set the summation of the moments of all the external forces equal to zero with respect to any point in the plane of the body.

The order of application of the equilibrium equations is arbitrary. These conditions allow the determination of up to three unknowns in coplanar equilibrium problems.

EXAMPLE PROBLEM 4.9

A 1000-kg mass is suspended by two cables as shown in the diagram. Determine the force in each cable. The acceleration due to gravity is $g = 9.824\ \mathrm{m/s^2}$.

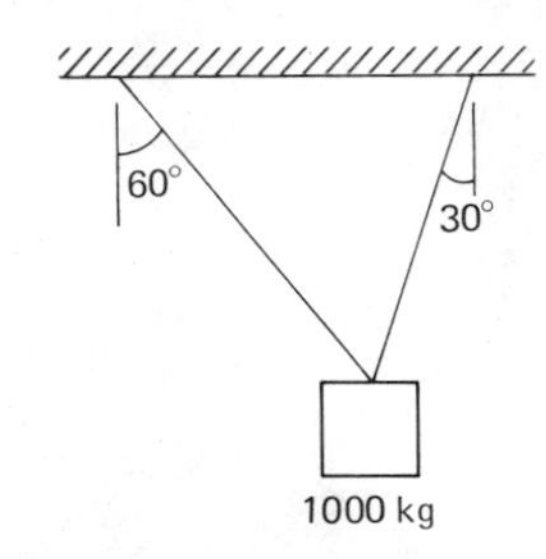

Solution

Draw a FBD. Represent the unknown tension forces in the cables as T_1 and T_2. The downward force on a 1000-kg mass is computed as

$$F = mg = (1000\ \mathrm{kg})\ (9.824\ \mathrm{m/s^2})$$

$$F = 9824\ \mathrm{kg \cdot m/s^2} = 9824\ \mathrm{N}$$

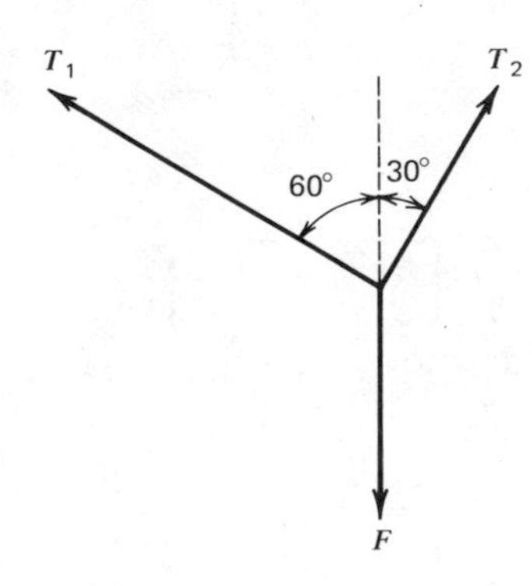

The resultant force of the concurrent force system must be zero since the system is in equilibrium. Therefore, the summation of the horizontal and vertical components of all the forces must equal zero.

Unless otherwise noted, the horizontal direction is represented by x and the vertical direction by y. The x direction is positive to the right and the y direction is positive upward.

$$\Sigma F_x = 0 \qquad -T_1 \sin 60° + T_2 \sin 30° = 0$$

$$\Sigma F_y = 0 \qquad -T_1 \cos 60° - T_2 \cos 30° + 9824 = 0$$

The equations can be rewritten as

$$-0.866\,T_1 + 0.5\,T_2 = 0$$

$$0.5\,T_1 + 0.866\,T_2 = 9824$$

Multiplying the first equation by 0.5, the second equation by 0.866 and then adding

$$T_2 = 8507.2\ \mathrm{N}$$

Upon substitution of T_2, it follows that

$$T_1 = 4911.7\ \mathrm{N}$$

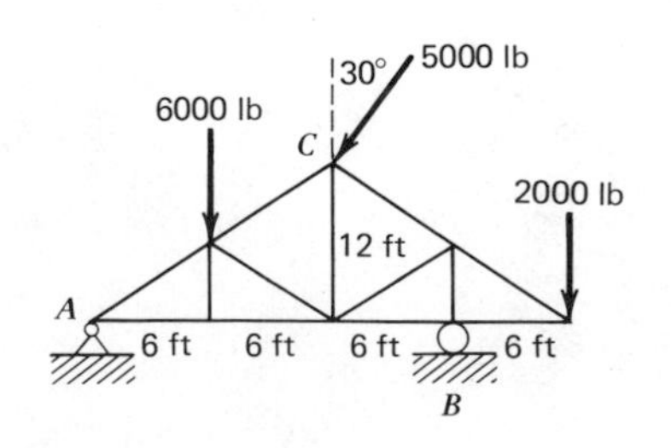

EXAMPLE PROBLEM 4.10

Determine the reactions at the supports on the truss at points A and B for the loads given.

Solution

Draw a free body diagram (FBD), which shows the reaction

forces of the pin at A and the roller at B. The directions of R_1, R_2, and R_3 are selected arbitrarily as shown. The other external forces are also shown on the FBD.

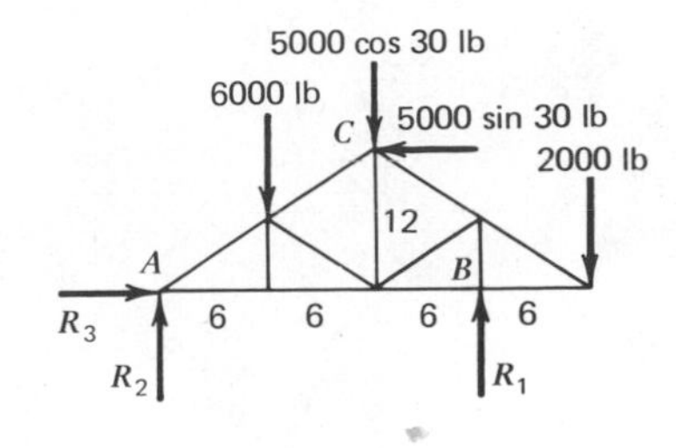

The equations of equilibrium are then applied to the total structure. One could begin by summing forces in the horizontal direction.

$$\Sigma F_x = 0$$

$$R_3 - 5000 \sin 30 = 0 \qquad R_3 = 2500 \text{ lb}$$

Since the truss is in equilibrium, the summation of moments about any point must equal zero. To eliminate some unknown forces in the equation, select point A and equate moments to zero.

$$\circlearrowleft^{+} \Sigma M_A = 0$$

$$-6(6000) - 12(5000)(.866) - 24(2000)$$
$$+ 12(5000(.5) + 18R_1 = 0 \qquad R_1 = 5886.7$$

Finally $\Sigma F_y = 0$

$$R_2 + R_1 - 6000 - 5000(0.866)$$
$$- 2000 = 0 \qquad R_2 = 6443.3$$

The results can be verified by summing moments about some other point, say point C.

$$\Sigma M_C = 0$$

$$12(R_3) - 12(R_2) + 6(6000) + 6R_1 - 12(2000) = 0$$

EXAMPLE PROBLEM 4.11

The wing of an aircraft that is in steady flight is shown. The structural weight of the wing is 5000 lb and is assumed to be concentrated at the center of gravity G. The total lift force is also assumed to be concentrated as shown. Determine the reactions on the wing at the point where the wing is attached to the body of the aircraft.

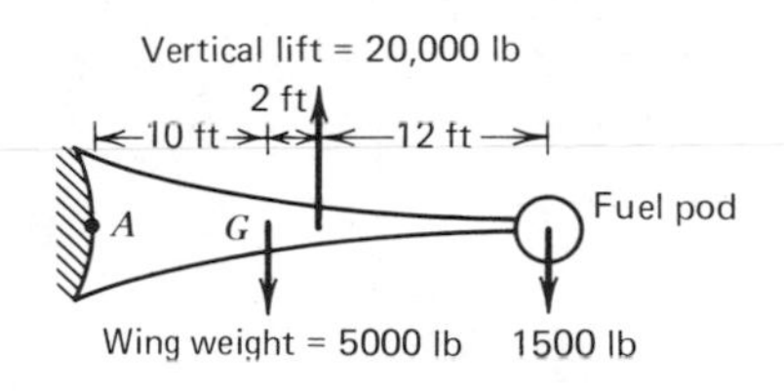

Solution

The equilibrium equations apply to bodies that have zero acceleration (constant velocity) as well as bodies at rest. A FBD showing the reactions at the fixed joint A and other forces can be drawn as shown.

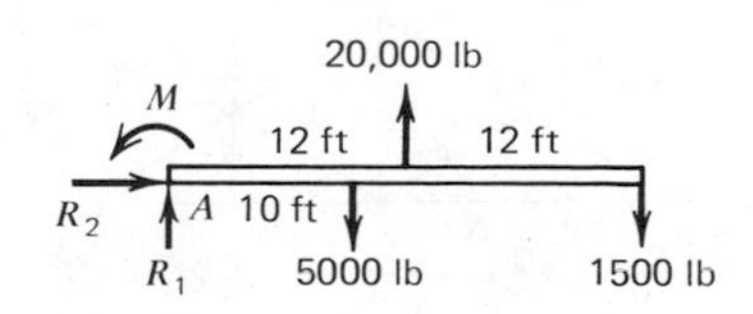

The equilibrium equations are then applied. First the moments about any point, say A, must be zero.

$$\circlearrowleft^{+} \Sigma M_A = 0$$

$$M - 10(5000) + 12(20\,000) - 24(1500) = 0$$

$$M = -154\,000 \text{ ft}\cdot\text{lb} \qquad M = 154\,000 \text{ ft}\cdot\text{lb}$$

The horizontal forces must sum to zero.

$$\Sigma F_x = 0$$

$$R_2 = 0 \qquad R_2 = 0$$

The vertical forces must sum to zero.

$$\Sigma F_y = 0$$

$$R_1 + 20\,000 - 5000 - 1500 = 0$$

$$R_1 = -13\,500 \text{ lb} \qquad R_1 = 13\,500 \text{ lb} \downarrow$$

PROBLEMS

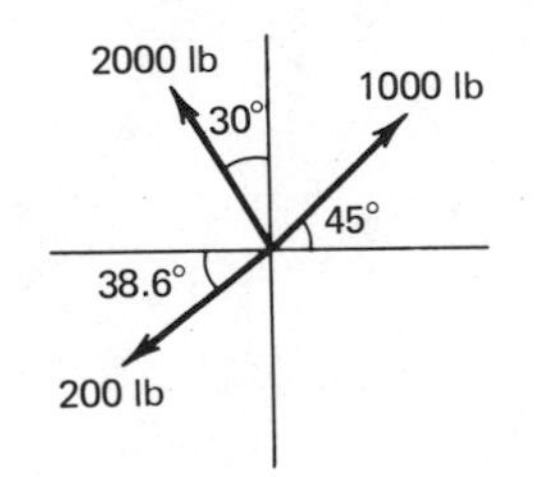

4.21. Determine the external force required for the concurrent force system to be in equilibrium.

4.22. A swimmer drops a short distance onto the end of a diving board. Assume the total force on impact to be 500 lb downward. Determine the reactions at the fixed end of the board.

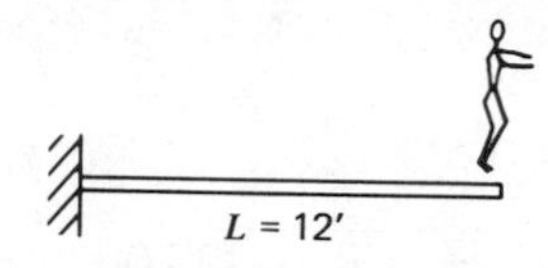

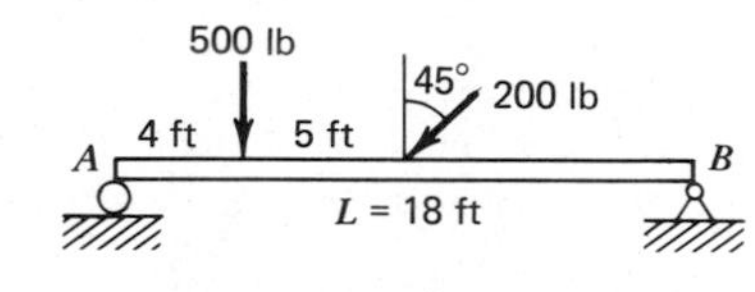

4.23. Determine the beam reactions for the beam subjected to concentrated loads as shown in the diagram.

4.24. A floor joist must carry a uniformily distributed load of 50 lb/ft. Estimate the beam reactions at the ends where the floor joist is "toenailed" onto the wall.

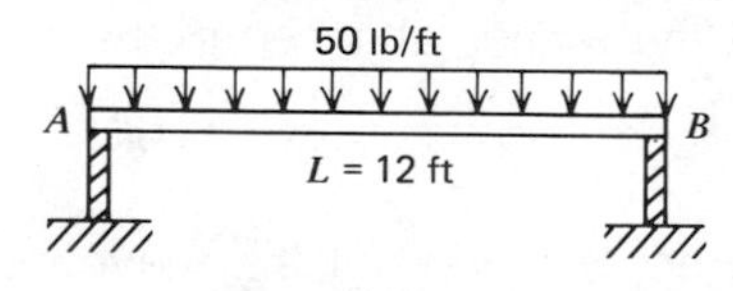

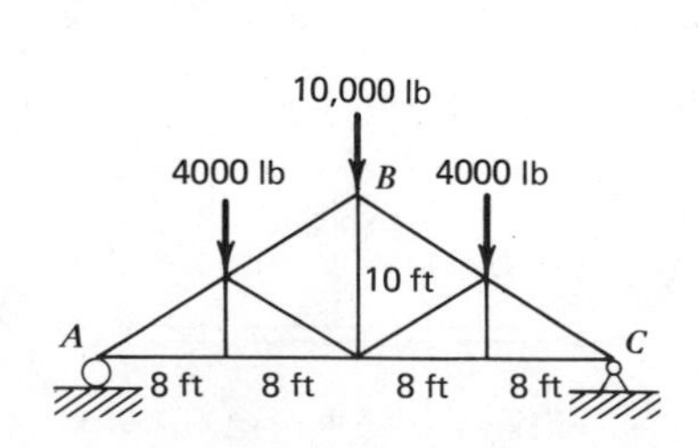

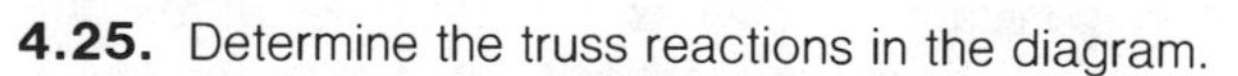

4.25. Determine the truss reactions in the diagram.

4.26. The block W weighs 1000 lb and is supported by two cables as shown in the diagram. Determine the tensions T_1 and T_2.

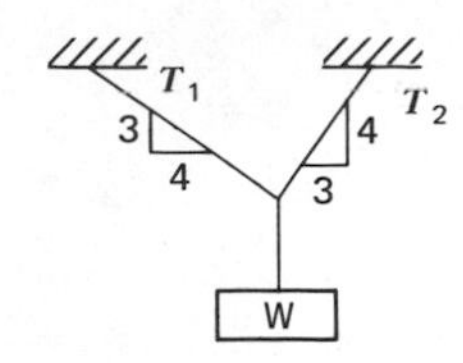

4.27. A reentry vehicle is flying at a constant velocity for the position shown in the diagram. Determine the lift L, and the drag D.

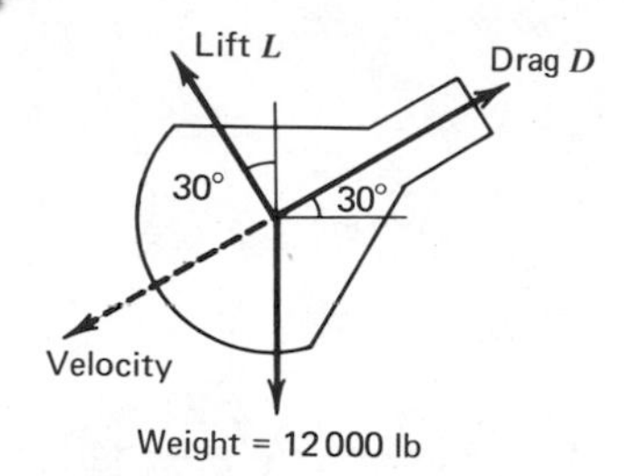

4.28. Determine the reactions for the truss loaded and supported as shown in the diagram.

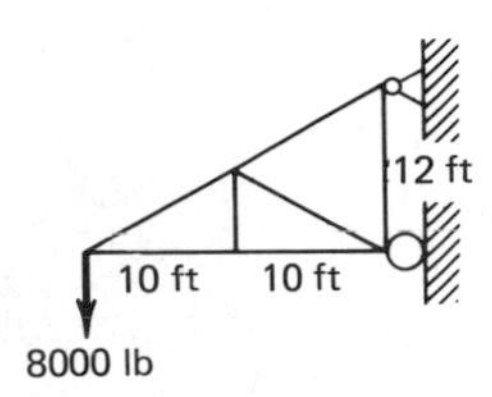

4.29. A corner fence post is firmly anchored in the ground and supports four strands of barbed wire at 1-, 2-, 3- and 4-ft elevations from the ground. Each wire is tensioned at 80 lb. Determine the reactions of the ground on the post.

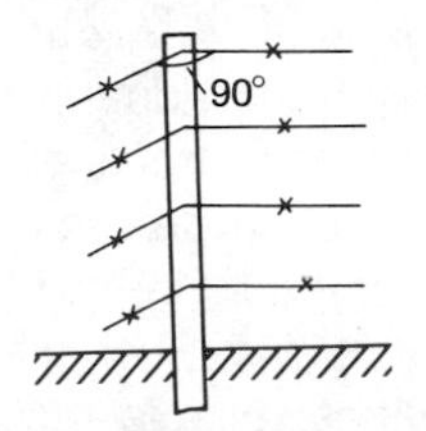

4.30. An automobile weighs 2000 N, has a wheelbase of 3.5 m, and rests on a horizontal road. The normal reaction on each front wheel is 650 N and 350 N on each rear wheel. Determine the horizontal distance to the center of mass measured from the rear wheels.

4.31. A 10-m support tower for a ski lift is to be positioned such that the moment at the anchored base is zero. The tower assembly mass is 3500 kg and can be considered as concentrated at the midpoint. The acceleration due to gravity is 9.81 m/s^2. The tension in the cable is constant at 66 kN at the angles shown. Determine the angle θ as measured from the vertical for zero moment.

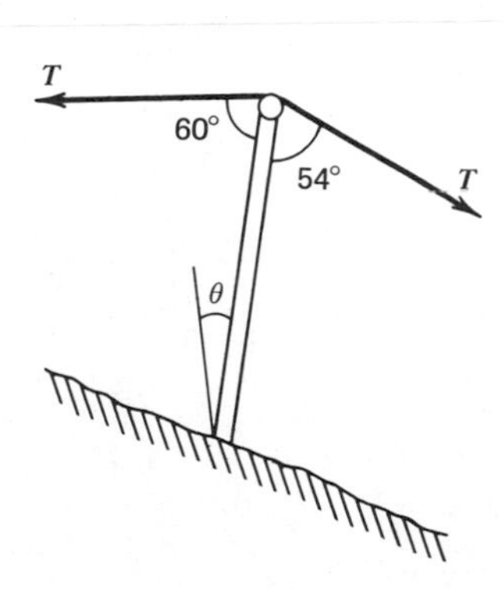

4.32. Determine the truss reactions in the diagram.

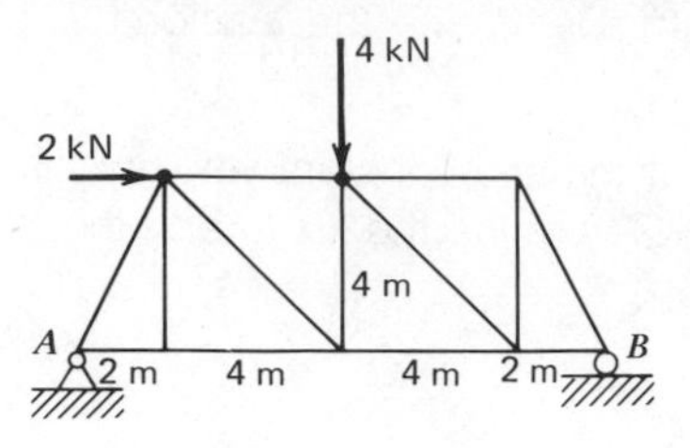

4.6/Trusses

There are several different types of structures that are identified by such names as beams, plates, frames, trusses, and shells. The static analysis of both internal and external forces associated with the structures is accomplished through the application of the equilibrium equations. A knowledge of the magnitude and distribution of the internal loads and moments of the structure is required for proper design. The analysis of the forces associated with a truss is typical of many structural problems and will be presented in some detail.

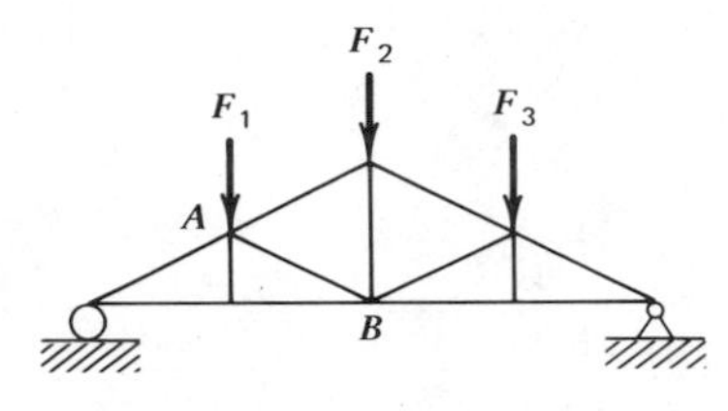

FIG. 4.23.

A truss is a structure composed of straight individual members that are connected together only at their extremities as shown in Figure 4.23. Although the joints in actual trusses are often welded, nailed, or riveted, an engineer models the joint as a pin connection, which implies zero moment rigidity. Such a simplification is universally accepted and provides reliable results. The external loads are also assumed to be applied only at the joints. Distributed and other concentrated loads that might appear to be applied along the members of a truss are transferred to the joints through stringers, crossbeams, or similar devices. The weight of the individual members of a truss is usually small compared to the external loads and is therefore neglected.

It has been shown that the equations of equilibrium can be applied to static structures such as a truss to determine reaction forces and moments. If the complete structure is at rest, then the resultant external forces and moments must be zero. Furthermore, the members and joints of the truss experience no appreciable relative motion and must also be in equilibrium. If the internal forces on the members and joints were not balanced, then the members would accelerate, much to our dismay. The same conditions of equilibrium that apply to the whole structure must also apply to all joints, members, or groups of members and joints.

In order to analyze internal forces that exist in members, one simply makes an imaginary cut, removes the member or joint, considers the internal forces as external forces, and then applies the equations of equilibrium. There is something unique about a truss member, however, that greatly simplifies the problem. The only possible direction of the force is collinear with the member itself. This is easily verified and is a consequence of the hinged joint which is used to define a truss. In Figure 4.24, the member AB is removed from the truss in Figure 4.23. Since the ends of member AB are pinned, the reactions are given as usual in the form of two perpendicular components at each end. By definition of a pinned joint, the moments are zero. Equating the moments of the forces about point B on member AB equal to zero shows that $R_1 = 0$. Equating moments about point A on member AB proves that $R_3 = 0$. In general, there is no non-zero component of force perpendicular to the member, hence, the internal force can only be collinear.

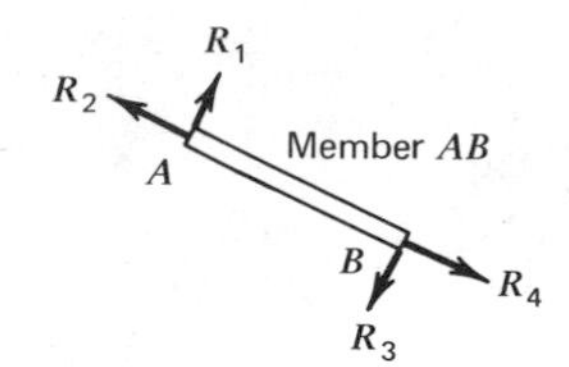

FIG. 4.24.

Although internal forces are collinear with the member itself, they may either push or pull. Forces that tend to pull on the members are called tension forces. Forces that push on a member are called compression forces as shown in Figure 4.25. Note that the internal load in a member is constant in magnitude and sign along the total length between joints.

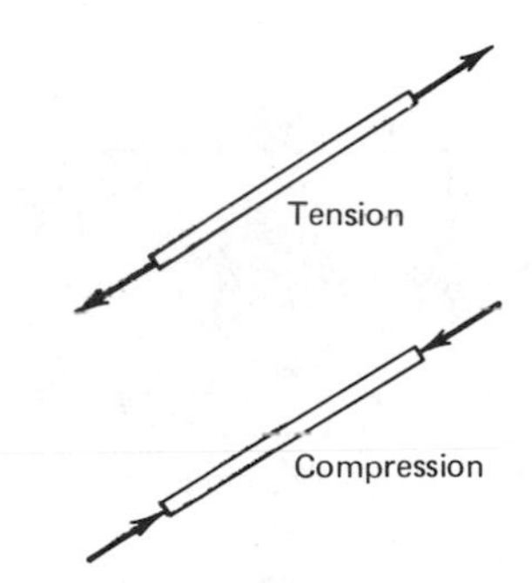

FIG. 4.25.

In addition to the members of a truss being in equilibrium, the joints must also be in equilibrium. If one were to make some imaginary cuts and isolate a joint as shown in Figure 4.26, the resulting concurrent force system must be in equilibrium. In drawing the free body diagram of the isolated joint, the lines of action of the internal forces are known to be along the members. A member in tension pulls on the joint and a compression member pushes on the joint (Figure 4.26). Initially, one can arbitrarily assume the force to be either tension or compression. If one assumes tension initially and applies the equations of equilibrium to find the magnitude of the force, a positive result indicates that the initial assumption of tension is correct. A negative answer simply says that the force is opposite to the direction chosen.

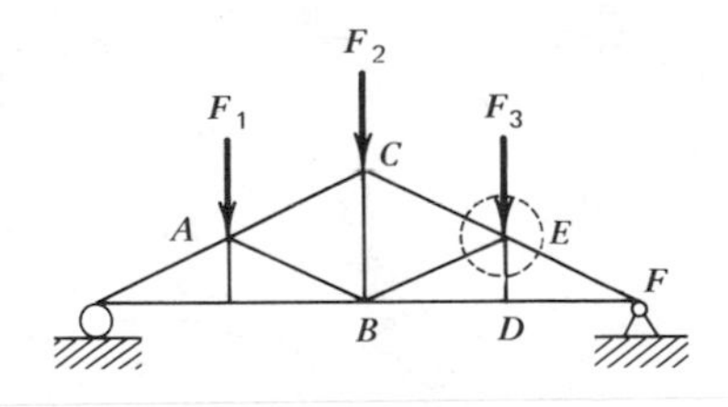

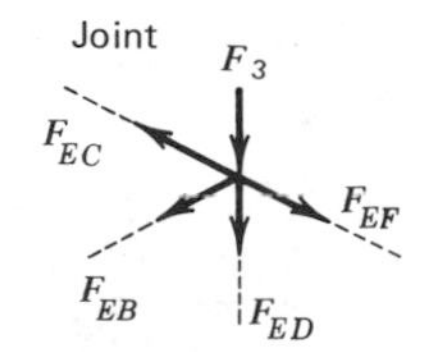

FIG. 4.26.

In the analysis of the force system, the objective is to find the reactions and the force in the internal members of the truss. As mentioned, the reactions are found by first treating the complete structure as a rigid body and applying the equations of equilibrium. The internal forces can then be found by drawing FBD's of each member, joint or section as necessary and applying the equations of equilibrium. Some example problems follow.

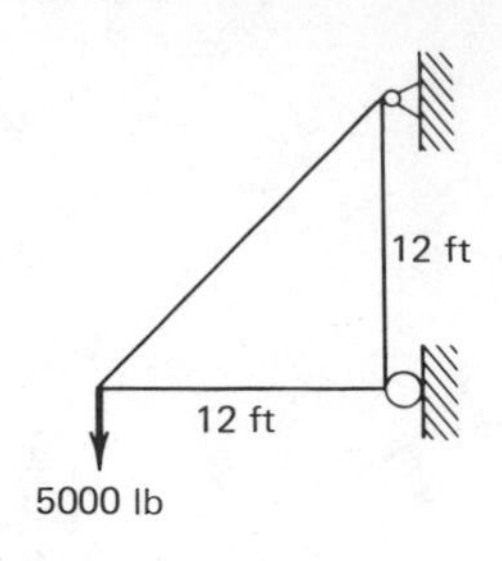

EXAMPLE PROBLEM 4.12

Determine the internal force in each member of the truss as shown in the diagram.

Solution

One can begin by drawing a FBD as follows that shows the external forces and replaces the pin and roller connections with the appropriate reaction forces R_1, R_2, and R_3. The directions are assumed and the magnitudes can be found by applying the equilibrium equations for the total truss. It is convenient to begin by summing moments about point B.

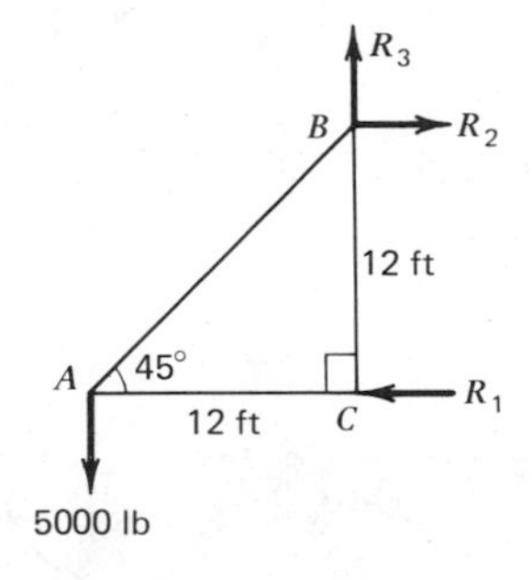

$$\circlearrowleft + \Sigma M_B = 0$$

$$(12)5000 - 12R_1 + 0$$

$$R_1 = 5000 \qquad R_1 = 5000 \text{ lb} \leftarrow$$

The summation of horizontal forces must equal zero.

$$\Sigma F_x = 0$$

$$-R_1 + R_2 = 0$$

$$R_1 = R_2 \qquad R_2 = 5000 \text{ lb} \rightarrow$$

The summation of vertical forces equals zero.

$$\Sigma F_y = 0$$

$$-5000 + R_3 = 0$$

$$R_3 = 5000 \qquad R_3 = 5000 \text{ lb} \uparrow$$

The external reactions of the truss R_1, R_2, and R_3 have now been determined. One can continue the analysis and find the internal forces carried by each member. To begin, arbitrarily select a joint, say joint A, and draw a FBD of the forces that act on the joint. The force in member AB must act along the line AB; however, the sense is unknown. Assume that the force in AB is tension; therefore it pulls on the joint. Assume F_{AC} to be compression and it, therefore, pushes on the joint. The determined magnitudes of F_{AB} and F_{AC} will be negative if the wrong sense is assumed. The equilibrium equations are now applied to the joint.

$$\Sigma F_y = 0$$

$$-5000 + F_{AB} \sin 45° = 0$$

$$F_{AB} = 7072 \qquad F_{AB} = 7072 \text{ lb } T$$

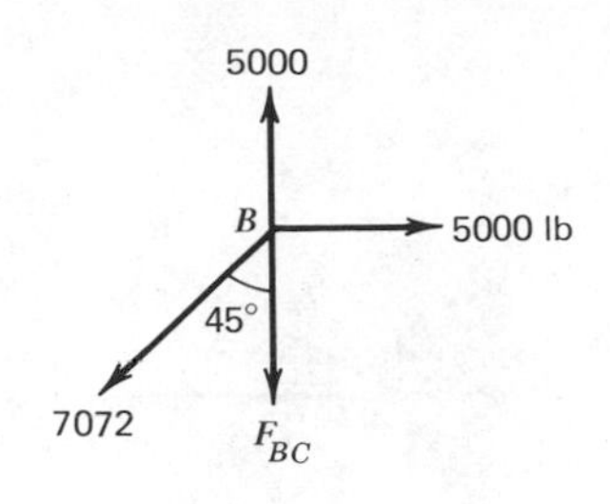

$$\Sigma F_x = 0$$

$$-F_{AC} + 7072\,(\cos 45°) = 0$$

$$F_{AC} = 5000 \qquad F_{AC} = 5000 \text{ lb } C$$

The forces acting at joint B are shown. The directions are collinear with the members. The force F_{AB} was determined as 7072 lb tension at joint A; therefore F_{AB} is also shown as tension at joint B. Both the magnitude and sense of F_{BC} must be determined. F_{BC} is assumed to be tension. The equilibrium equations are now applied.

$$\Sigma F_y = 0$$

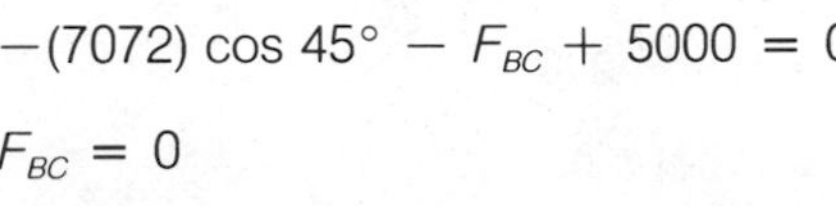

$$-(7072)\cos 45° - F_{BC} + 5000 = 0$$

$$F_{BC} = 0$$

EXAMPLE PROBLEM 4.13

Determine the internal forces in the members of the truss as shown in the diagram.

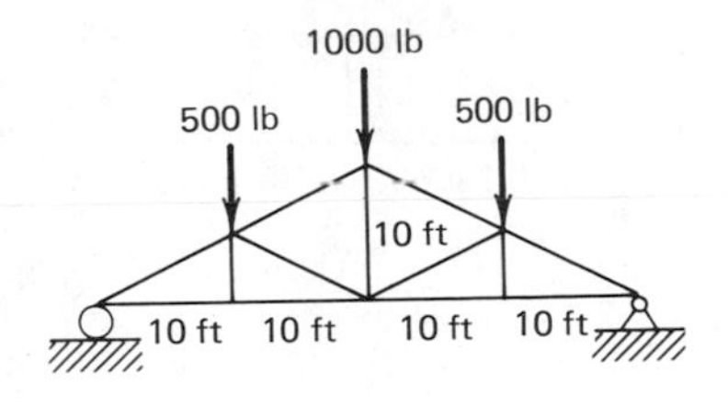

Solution

A FBD of the complete structure is drawn as follows that shows the external and reaction forces. The reactions are then found using the equations of equilibrium.

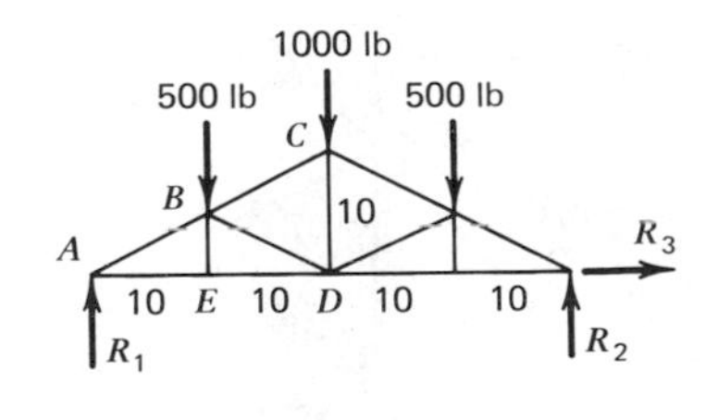

$$\Sigma F_x = 0$$

$$R_3 = 0 \qquad R_3 = 0$$

$$\circlearrowleft^{+}\Sigma M_A = 0$$

$$-500(10) - 1000(20) - 500(30 + R_2(40) = 0$$

$$R_2 = 1000 \text{ lb}\uparrow$$

$$F_y = 0$$

$$-500 - 1000 - 500 + 1000 + R_1 = 0$$

$$R_1 = 1000 \text{ lb}\uparrow$$

The internal forces are found by applying the equations of equilibrium to the joints. Beginning with joint A, the forces are found as follows:

$$\Sigma F_y = 0$$

$$1000 + F_{AB} \sin 26.5° - 0$$

$$F_{AB} = -2241 \text{ lb}$$

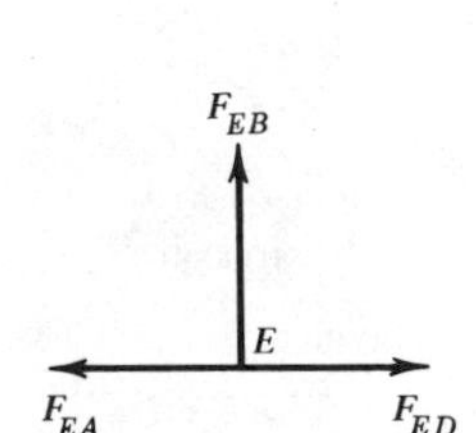

Since F_{AB} is negative, the direction of the force is opposite to that shown initially. Since tension was assumed, the member is in compression.

$$F_{AB} = 2241 \text{ lb C}$$

$$\Sigma F_x = 0$$

$$F_{AB} \cos 26.5° + F_{AE} = 0$$

$$(-2241)\cos 26.5° + F_{AE} = 0$$

$$F_{AE} = 2006 \text{ lb} \qquad F_{AE} = 2006 \text{ lb } T$$

Draw a FBD of joint E and determine F_{BE}.

$$\Sigma F_y = 0$$

$$F_{BE} = 0$$

$$\Sigma F_x = 0$$

$$-F_{AE} + F_{ED} = 0 \qquad F_{ED} = 2006 \text{ lb } T$$

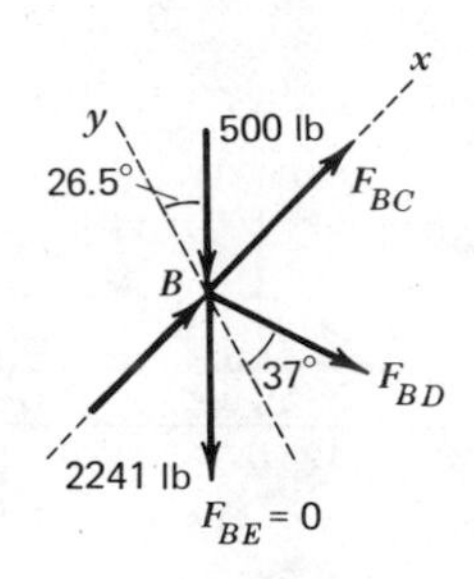

Draw a FBD of joint B and determine F_{BD} and F_{BC}. For convenience, orient the x and y axes as shown.

$$\Sigma F_y = 0$$

$$-500 \cos 26.5° - F_{BD} \cos 37° = 0$$

$$F_{BD} = -560.2 \qquad F_{BD} = 560 \text{ lb } C$$

$$\Sigma F_x = 0$$

$$2241 + F_{BC} - (500)\sin 26.5 + F_{BD} \sin 37° = 0$$

$$F_{BC} = -1681 \qquad F_{BC} = 1681 \text{ lb } C$$

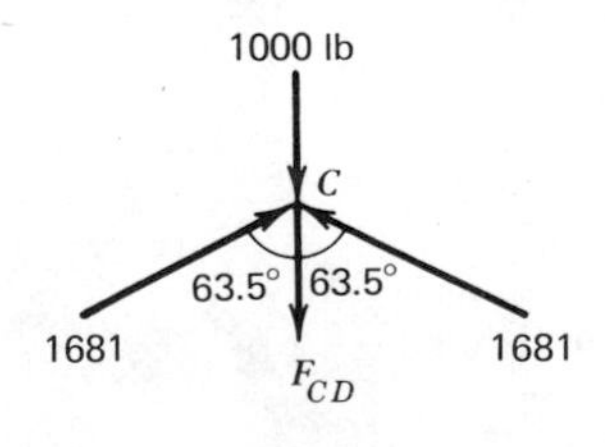

Draw a FBD of joint C and determine F_{CD}. Note that the structure is symmetric.

$$\Sigma F_y = 0$$

$$-1000 - (2)(1681) \cos 63.5 - F_{CD} = 0$$

$$F_g = 500 \qquad F_{CD} = 500 \text{ lb } T$$

Check the results by considering joint D.

$\Sigma F_y = 0$

$500 - (2)(560)\sin 26.5° = 0.25 \approx 0$

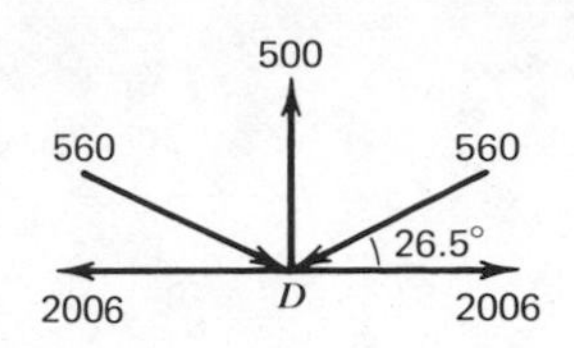

PROBLEMS

4.33. Determine the force in each member of the truss shown.

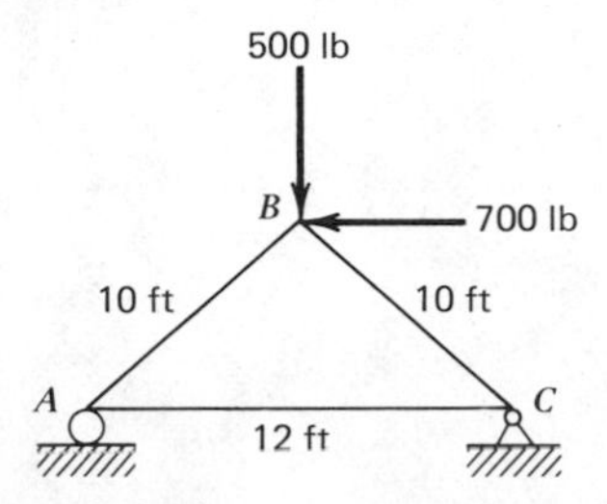

4.34. Determine the forces in members *AB*, *AC*, and *DC* of the truss shown.

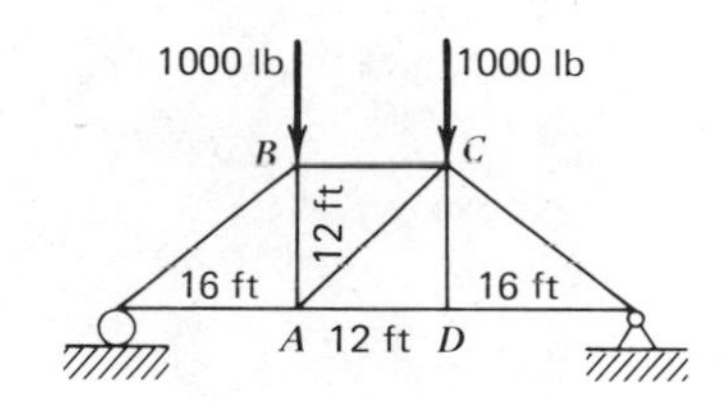

4.35. Determine the force in each member of the truss as shown.

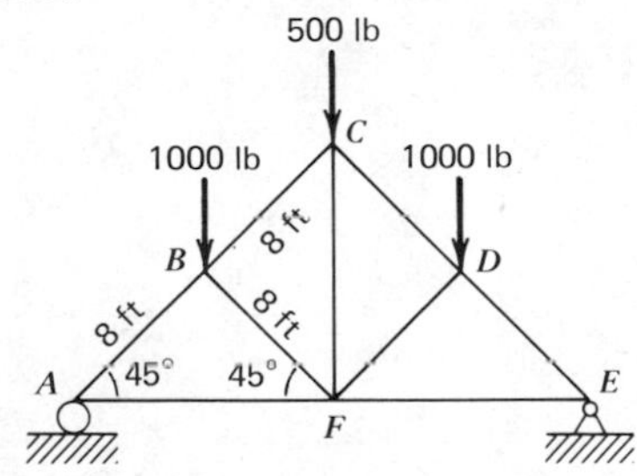

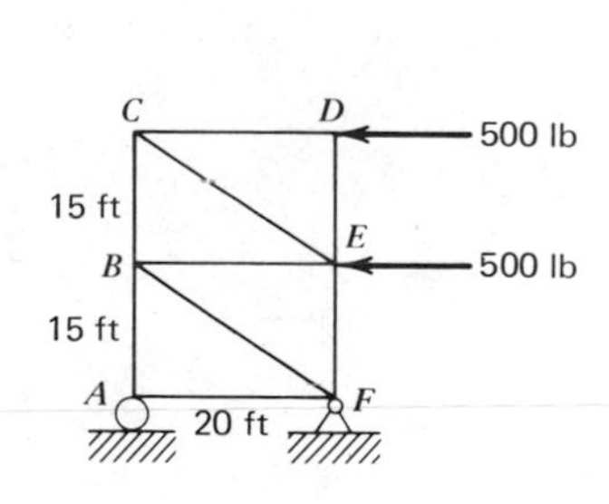

4.36. Determine the force in each member of the truss as shown.

4.37. Determine the force in members *AB* and *BC* in the diagram.

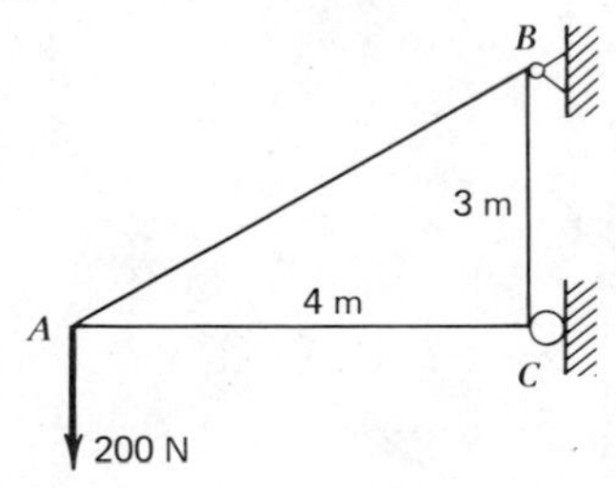

4.38. Determine the force in members *AB*, *BC*, and *CD* as shown in the sketch.

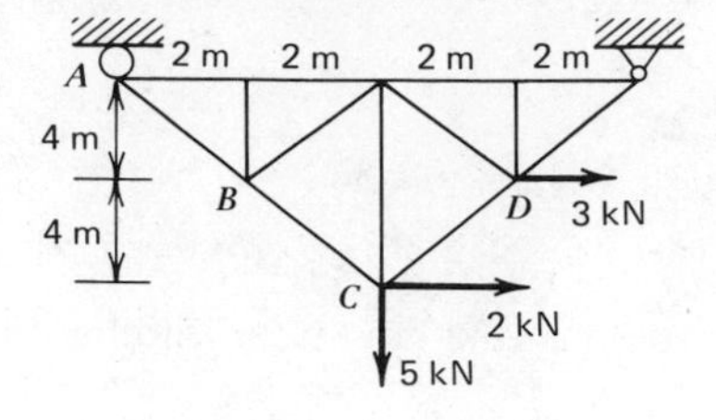

CHAPTER FIVE

MATERIALS AND ENGINEERING

PART A/STRUCTURE AND PROPERTIES

5.1/Introduction

In the stone age, humans started to improve their standard of living when they first began to gather rocks of various size and shape. Some rocks were stacked to provide shelter, some were thrown at animals to obtain food, some were used to grind grains, and some were probably gathered for their artistic attraction. The early years of material technology experienced little shaping or fabrication of raw materials. A caveman simply looked until earth provided the right size rock (Figure 5.1). Civilizations followed that shaped the raw materials as they made bricks of a standard form with straw or bamboo for reinforcement. Other generations found rare metals such as gold and silver to make weapons of war and tokens of love. Copper workings have been discovered that date back to about 6000 B.C. A gold helmet recovered from the tomb of Mes-Kalam-Dug was supposedly made before 3000 B.C.

FIG. 5.1. The use of materials began many centuries ago.

Prior to the nineteenth century, engineers were limited in their material selection primarily to naturally occurring materials such as rock, clay, earth, iron, rare metals, leather, and wood. During the nineteenth century, an accelerated interest in materials and material technology began. For example, in 1839 Goodyear treated sticky natural rubber with sulphur and provided an elastic material. Scarce amounts of aluminum became available at an exhorbitant price of about $20 a pound in 1857. The Bessemer converter and the open-hearth furnace initiated the steel age in the late 1850s. Synthetic textiles that resulted in rayon became available at the close of the nineteenth century.

Since 1900, scientists have broadened the choice of commercial type metals by a factor of 10^3, while polymers and composites have increased even by a greater amount. Phenolic resins such as Bakelite were introduced around the beginning of the twentieth century. Investigations of the now-accepted covalent structure of polymers gained acceptance in the 1930s and precipitated the gigantic list of synthetic materials such as nylon, polyethylene, polyvinylchloride (PVC), dacron, teflon, and others. Nickel, chromium, molybdenum, carbon, vanadium, tungsten, and other elements have been alloyed with steel to achieve improved material properties much as a cook adds spices in soup to improve the flavor. Textiles, ceramics, and metals have been added to plastics; steel has been added to concrete; and several other mixtures have initiated the vast selection of materials available today.

TABLE 5.1. Some Ingredients of the Earth's Crust

Primal Substances	*Percent*
Oxygen	46.50
Silicon	27.60
Aluminum	8.13
Iron	5.00
Magnesium	2.09
Titanium	0.44
Carbon	0.009
Nickel	0.008
Copper	0.007
Uranium	0.0004
Silver	0.00001

The basic ingredients of our broad selection of materials have been available for hundreds of years. The availability of some primal substances contained within the earth's crust is given in Table 5.1. Oxygen is obviously very plentiful. Silicon, a primary raw material for the production of cement, bricks, glass, and other ceramics, is very abundant. There is more natural aluminum than any other metal. There is not an excessive amount of carbon, which is well known as the basic building block of fuels, drugs, plastics, and synthetic textiles. Copper and silver are relatively scarce.

It so happens that the recovery of the basic materials from raw earth requires energy and technology. As shown in Table 5.2, many of our modern materials require significant energy inputs. Aluminum is the most abundant metal by far; however, one doesn't find commercial aluminum as a bright sparkling rock resting on the surface of the earth. Roughly 1 shovelful of every 12 shovels of clay is aluminum, yet it requires energy and know-how to extract aluminum from clay. The common ore bauxite (aluminum oxide), is a less expensive source of aluminum, but the ore is also very stable and requires energy to be smelted.

TABLE 5.2. Energy Input Values

Material	*Energy Input (Mj/kg)*
Aluminum	276.3
Copper	142.3
Polyethylene	125.6
PVC	108.8
Stainless steel	67.0
Structural steel	58.6
Cast steel	33.5
Cement	12.6

The general field of material science and technology includes such assignments as the extraction of basic elements from the earth's crust, the separation and purification of elements from other raw materials, the chemistry of materials, the processing of

FIG. 5.2. A scientist studies the microstructure of materials to determine why materials behave as they do.

materials into useful shapes, the identification of material properties, the selection of materials for particular applications, and the discovery of new concepts. A community of scientists representing fields such as physics, chemistry, and metallurgy work together with engineers. However, the scientist and engineer have different roles.

The materials scientist analyzes and predicts material behavior by studying such things as the molecular structure that is observable through a microscope (Figure 5.2). The scientist attempts to understand how materials conduct electricity, why materials corrode, and other "hows" and "whys" of material behavior. The scientist is concerned with atoms, atomic bonding, crystal structure, grain boundaries, and other *microscopic* phenomena. The scientist attempts to describe the conduction of heat and electricity, tensile strength, corrosion, melting, elongation, hardness, surface conditions, and other material properties in terms of microscopic interactions. New and better materials can be synthesized by manipulating the atomic structure of the basic elements to achieve desired results.

An engineer takes the materials as developed by scientists and builds products (Figure 5.3). The observable characteristics of a material that represent the resultant effect of many atoms are commonly referred to as *macroscopic* characteristics. The engineer is primarily concerned with macroscopic behavior, particularly with measurable material properties, such as melting points, strength limits, conductivity, elasticity, surface conditions, and weight. The engineer selects materials according to their properties and the performance required. For our generation, the Stone Age and the rare metal age are both history. The steel age is fading. Although the United States now consumes steel at the rate of approximately 150,000,000 tons per year, it appears as though the tonnage of plastic will exceed that of steel before the year 2000. The volume of plastic already exceeds the volume of steel used per year. As illustrated in Figure 5.4, engineers must be familiar with the complete spectrum of metals, polymers, rubbers, ceramics, textiles, and other materials in order to take advantage of weight, strength, cost availability, maintenance, and other material characteristics that are required in competitive design.

FIG. 5.3. An engineer applies the knowledge obtained by scientists to build products.

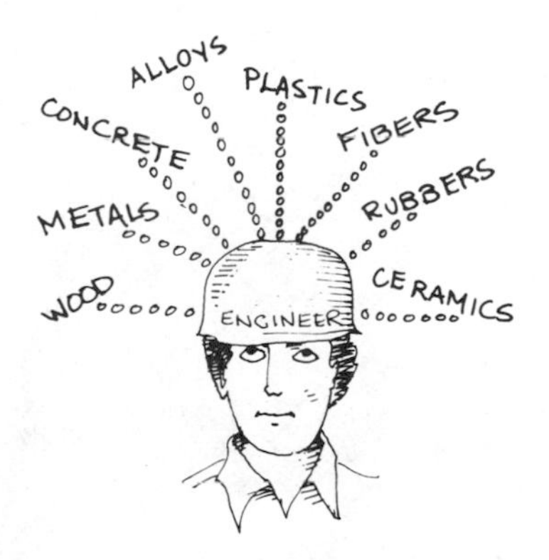

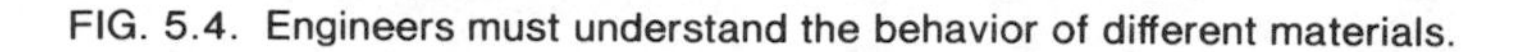

FIG. 5.4. Engineers must understand the behavior of different materials.

5.2/The Internal Structure of Materials

The basic constituents of materials have puzzled minds for years. Ancient Greek philosophers reasoned that water was the primal substance from which all material was made. Later, they decided that all matter was composed of four basic things—fire, water, earth, and air. Ancient Chinese philosophers proposed that all materials were made from five primal elements—fire, water, earth, air, and wood. More recent scholars claim that matter consists of communities of atoms that have discrete particles in motion. Atoms are so small that approximately 10^{22} are contained in each cubic centimeter of engineering material.

Rutherford Atom

Electrons

Nucleus

+

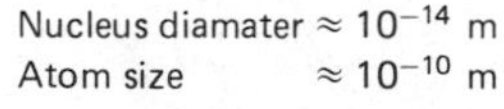

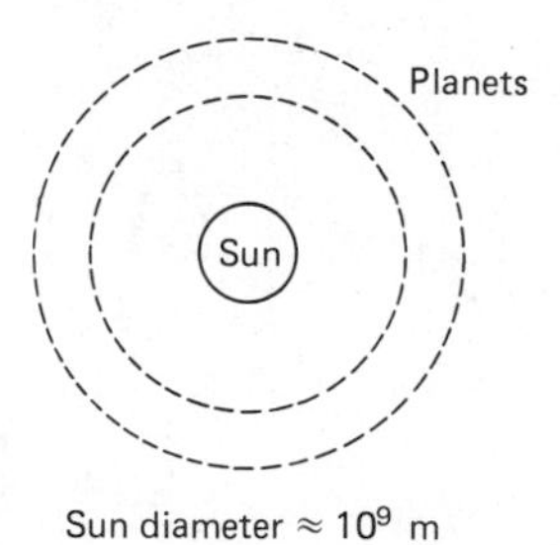

Sun diameter ≈ 10^{9} m
Solar size ≈ 10^{12} m

FIG. 5.5. Rutherford modeled the atom as a tiny solar system.

5.2.1/Atomic Structure

In 1911, Ernest Rutherford bombarded gold foil with alpha particles from radium and used the scattering pattern to develop a model of the unseen structure of atoms. He modeled the atomic structure as a tiny solar system, shown in Figure 5.5. Rutherford postulated that tiny electrons orbit around a centralized nucleus that consists of protons and neutrons much as planets orbit around the sun. Although Niels Bohr, Louis de Broglie, and other scientists have refined the model, the Rutherford model is still descriptive of modern theories of atomic structure.

Not all of these atoms or tiny solar systems are identical. Scientists have identified over 100 primal substances called elements that consist of homogeneous groups of atoms. Atoms are different because they consist of different numbers of neutrons, protons, and electrons. In electrically neutral atoms, the number of protons is equal to the number of surrounding electrons. In the lighter elements, except hydrogen, the number of protons is also equal to the number of neutrons in the nucleus. The number of neutrons exceeds the number of protons in the heavier elements. The weight of an atom is almost proportional to the number of protons and neutrons concentrated in the nucleus, since either of the two weighs roughly 2000 times that of an electron.

The electron cloud configurations that surround a nucleus are rather complex. Every atom, regardless of the number of electrons, contains several possible orbital paths or shells that correspond to different electron energy levels as shown for the boron atom in Figure 5.6. In the lighter elements, only the first few energy levels are actually occupied by electrons. If the atom is excited through the addition of energy, then the electrons tend to

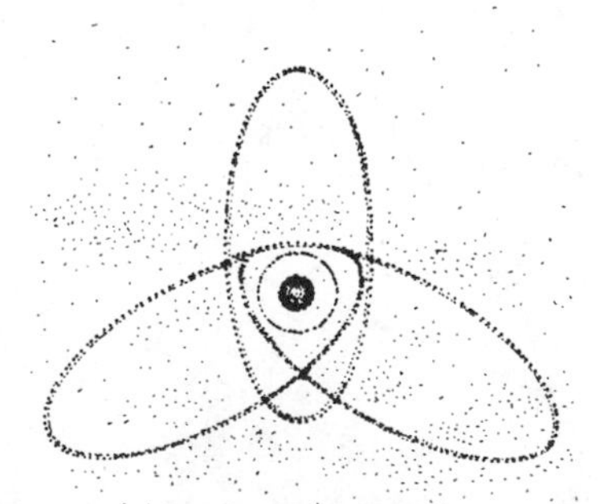

FIG. 5.6. Electron orbits in the Boron atom.

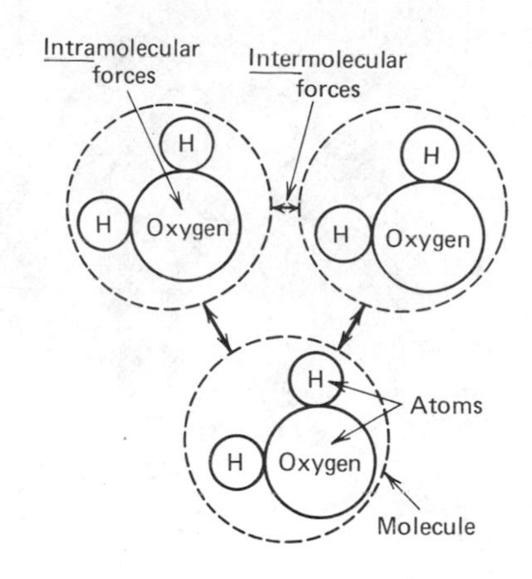

FIG. 5.7. Intramolecular forces bind atoms and intermolecular forces bind molecules.

move outward into the higher energy orbits. These outer electron shells become more fully occupied as the atomic number increases. The volume of the atom doesn't necessarily increase as the number of orbiting electrons increases. The nucleus also has additional protons that pull the electron clouds closer together.

5.2.2/Molecular Structure

Atoms combine together and form a unit called a molecule. Molecules combine to form materials. Every molecule of a particular material contains the same kind and number of atoms bonded together in the same arrangement. For example, the water we drink is made from millions of molecules that are held together by *inter*molecular forces and each molecule consists of two hydrogen and one oxygen atom bonded together by *intra*molecular forces as shown roughly in Figure 5.7.

The intramolecular forces include three primary types of bonding between atoms: ionic, covalent, and metallic bonds. These primary bonds describe the different ways in which the electrons of atoms may be combined such that atoms are attached or bonded together.

The ionic bond results from a transfer of one or more electrons from one atom to another as shown in Figure 5.8. The atom becomes a positive ion and is usually smaller than the neutral atom since the attraction of the nucleus is distributed over fewer electrons. In ionic bonding, the element that loses electrons is commonly called a metal, whereas the element that gains electrons is the nonmetal. For example, in the formation of sodium chloride (NaCl), sodium is the metal and loses one electron to chlorine.

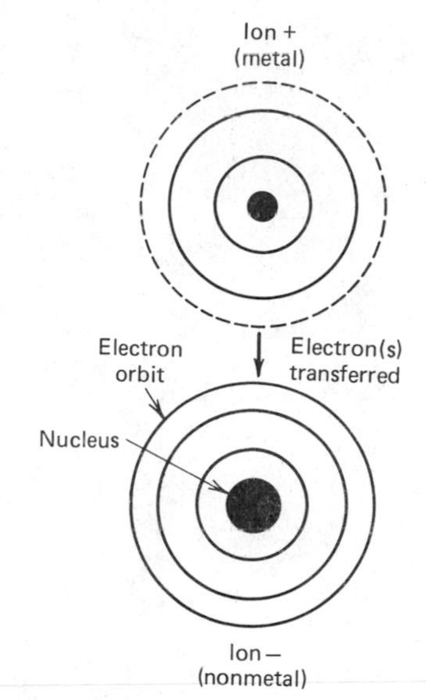

FIG. 5.8. Ionic bond.

The covalent bond is that bond by which electrons remain with their nucleus but are shared between atoms, as shown in Figure 5.9. In some cases, each atom contributes electrons equally. Sometimes one atom contributes all the electrons shared. The combination of carbon with other atoms is often a covalent type bond as found in many fuels, plastics, drugs, living substances, and other organic materials. The covalent bond is very strong as evidenced in diamonds.

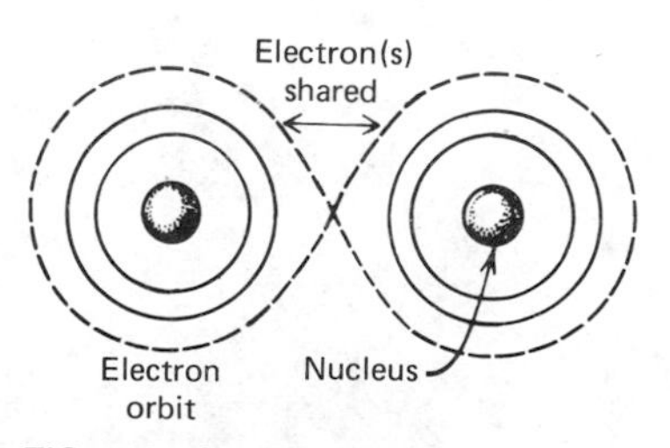

FIG. 5.9. Covalent bond.

The third type of primary bond is the metallic bond. In the metallic bond, the "borrowed" electrons are not bound to partic-

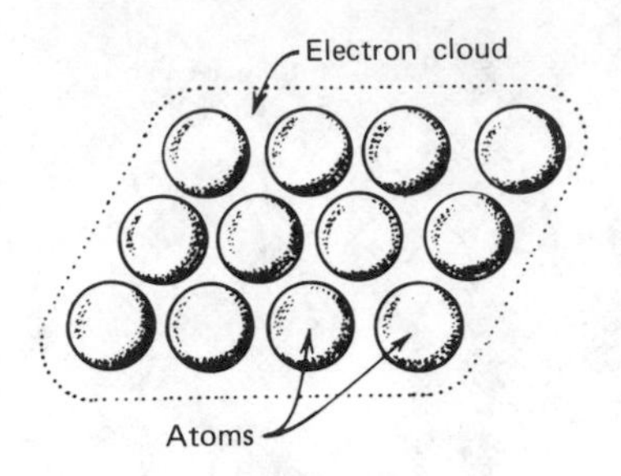

FIG. 5.10. Metallic bond.

ular atoms as they are in covalent bonds. Some of the atoms provide electrons that are used by the complete community of atoms. An array of positively charged metal ions is surrounded by an electron cloud as shown in Figure 5.10. These electrons are highly mobile and can be visualized as a free electron gas. This mobility of electrons accounts for the high thermal and electrical conductivity of metals.

The molecules themselves are attracted to each other by intermolecular forces that are commonly called secondary bonds or van der Waal's forces. The magnitude of these forces varies significantly and have an influence on the mechanical behavior of the material.

Molecules can be created, rearranged, disassembled, or otherwise changed during a chemical reaction and different bulk materials formed. The atoms that make up molecules cannot be created nor destroyed, nor can they be altered. The burning of paper, corrosion of metals, degradation of wood and the melting of ice result in molecular changes, but the individual atoms still persist.

5.2.3/Metals

The vernacular term *metal* often refers to a material such as steel that has high-strength characteristics, conducts heat and electricity, and is machinable, hard, and lustrous. More precisely, the word metal refers to a class of approximately 80 elements that appear on the periodic table. The metals tend to lose electrons in chemical reactions, whereas the nonmetals gain electrons. Some of these elemental metals such as aluminum, magnesium, and titanium are used for structural materials, whereas others are used for fertilizers, liquid coolants, and table salt. In this section, the general classification of materials called metals refers to solids that consist of either pure elemental metals or metal alloys. An alloy is a mixture of two or more metals and may also include other additives or impurities.

Metallic materials appear to be continuous lumps of rigid masses. Armed with a high-powered microscope, however, one could observe that materials really consist of billions of small particles that are spinning, vibrating, and moving about in a somewhat orderly fashion. A metal structure that supports multiple floors loaded with people, equipment, and other paraphernalia must depend on the integrity of many molecules to attract one another with intermolecular forces around 10^{-13} lb. These numerous intermolecular forces are extremely small; however, they can impart instantaneous accelerations greater than 10^{14} times the

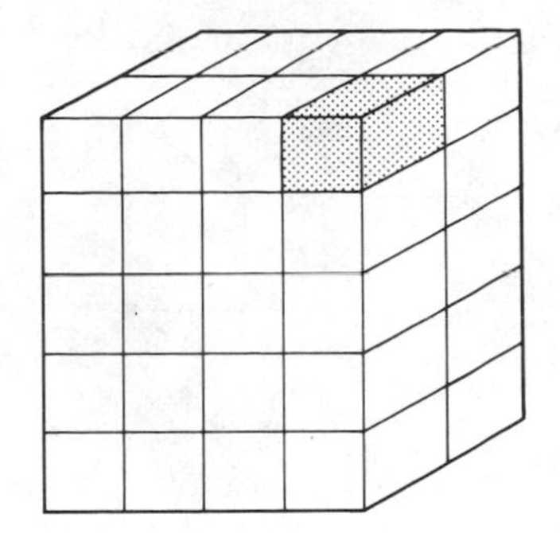
FIG. 5.11. The microstructure of metals consists of an arrangement of unit cells.

acceleration due to gravity on atomic particles. It is exciting to know that our material world depends on tiny atomic universes and their adherence to physical law and order.

The microstructure of metals consists of atoms that are arranged in a regular three dimensional arrangement of unit cells that are repeatedly connected together as shown in Figure 5.11. This ordered arrangement of atoms is referred to as crystallinity. It is one of the most general characteristics that distinguishes metals from other materials. These special atomic lattices are called Bravais crystal structures and there are 14 different types. The body-centered cubic (bcc) and the face-centered cubic (fcc) are two of the most common crystal structures.

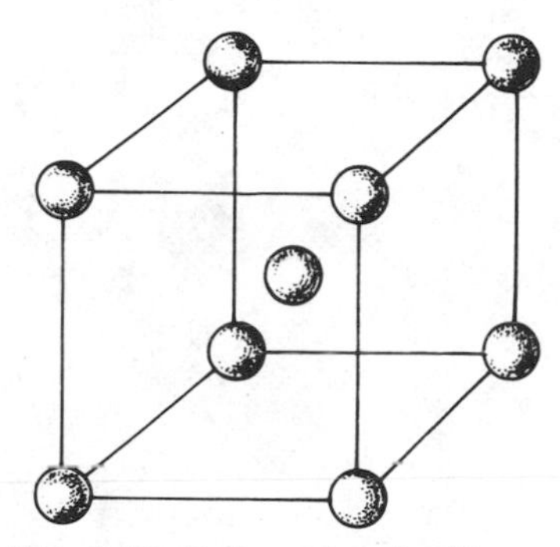
FIG. 5.12. Body-centered cubic (bcc).

The body-centered cubic arrangement of atoms is shown in Figure 5.12. The cube is the unit cell that is repeated in a three-dimensional array. There is one atom at each of the corners of the cube, plus one atom at the center. The atoms at the corners are common to adjacent cubes in the lattice. The radii of the atoms have been reduced and the atoms have been pictured somewhat as marbles. It so happens that all corner atoms are in contact with the center atom yet not with each other, as shown in Figure 5.13. The bcc group of metals includes iron, some steels, chromium, tungsten, and vanadium. These metals all have high melting points.

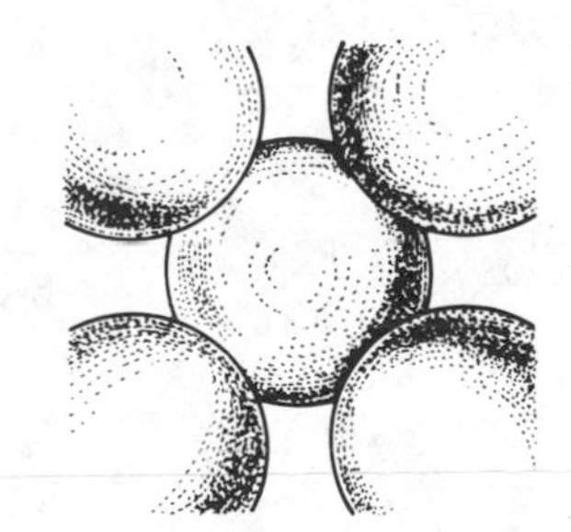
FIG. 5.13. Corner atoms are in contact with the center atom in a bcc unit cell.

The face-centered cubic (fcc) is shown in Figure 5.14. One atom occupies each corner of the unit cube, and one occupies each face at the center. Some of the fcc materials are aluminum, copper, gold, lead, nickel, and silver.

If a substance may exist in more than one stable crystalline form, it is said to be polymorphic or allotropic. Many metals are allotropic and may change form in response to temperature increases or decreases. Iron changes from a bcc structure to an fcc structure if heated to a temperature of 910°C. If the temperature continues to increase, the iron changes back to a bcc at its melting point. Also, titanium exists as a close-packed hexagonal crystal below 882°C and at temperature above, it is a bcc. The allotropic metals change crystalline structure as a function of temperature; therefore, heat treating is a common industrial technique used to strengthen a material that exhibits allotropy. By heating an allotropic metal, however, one might also weaken some crystalline structures.

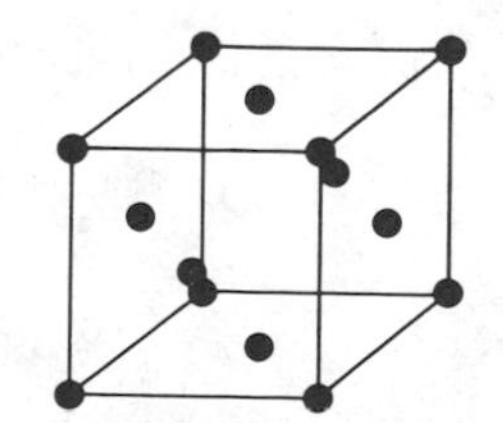
FIG. 5.14. Face-centered cubic (fcc).

What would happen if one of the corner atoms in the bcc iron were replaced with chromium or some impurity? Such a sub-

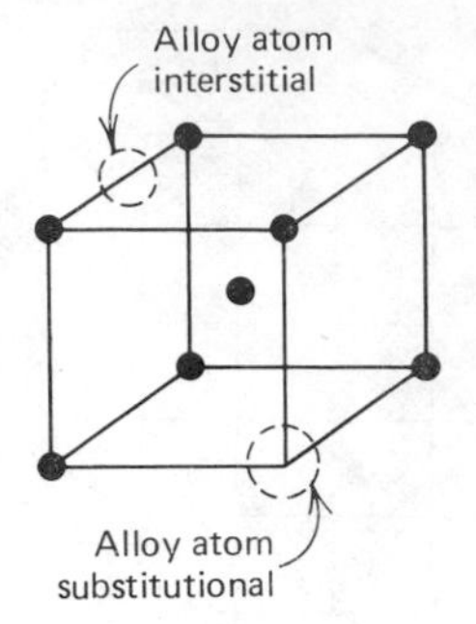

FIG. 5.15. Interstitial and substitutional methods of alloying metals.

stance that has metallic properties and is composed of more than one element is called an alloy. The predominant metal is called the base metal or the solvent. The other elements are called alloying elements and may be intentionally added or perhaps unintentionally present as impurities. If the atoms of the alloying solute replace and fill positions in the crystal lattice normally occupied by the base metal, then the alloy system is substitutional, shown in Figure 5.15. Interstitial solids result when the solute atoms from the alloy occupy positions between the atoms of the base metal.

A more rigorous treatment of the microstructure of metals is usually covered in more advanced material engineering courses. Atomic movements in crystalline solids in response to mechanical loads along with other concepts are usually encountered later by engineering students. The intent of this discussion was to introduce some atomic concepts and perhaps motivate students in their desire to understand the "whys" of the behavior of metals.

5.2.4/Ceramics

The general classification of materials commonly referred to as *ceramics* includes several different types. Rock, mineral ores, clay, cement, glass, brick, refractories, insulation, abrasives, tile, porcelain, and welding fluxes are some common ceramics. The use of these ceramics began in the Stone Age and has continued into the twentieth century. Rocks, sand, and cement are still used abundantly in the construction of highways, buildings, pipelines, cutting tools, and dams. The nonmetallic linings used in high-temperature furnaces are called refractories and must withstand extreme thermal, chemical, and pressure environments. Some of the glasses now used in building construction provide structural support, inhibit heat flow, transmit visible light, and absorb ultraviolet radiation. Ceramics are used in paints, toothpastes, lubricants, packaging, electronic transducers, computers, and human organ transplants.

The molecular structure of ceramic materials consists of mixtures of metallic and nonmetallic elements. The primary bonding between their atoms can be both ionic and covalent. The metal atoms have relatively little attraction for their outermost electrons and lose them in ionic bonding to the stronger attraction of the nonmetal nuclei. The nonmetals bond to each other through the covalent bond. These bondings provide a strong attraction for electrons and account for the high melting points, low corrosive action, and overall chemical stability typical of most ceramics.

Many of the ceramics are noncrystalline and are referred to

as glass by materials scientists. Some of the ceramic materials do have their atoms arranged in a repeating lattice formation and are therefore crystalline in structure. The ceramic lattices consist of atoms and ions that are significantly different in size and are therefore more complex than the uniform lattices typical of most metals.

The clay minerals are examples of ceramics that have ordered microstructures. There are two common crystalline units that serve as the basic building blocks for clay. One unit consists of silicon and oxygen arranged in a tetrahedron. The other unit is made from aluminum and oxygen in the form of an octahedron. The microstructure of clays consist of various combinations of these crystalline units bonded and stacked together in the form of a flaky pancake. These flakes are surrounded by an aqueous solution as shown symbolically in Figure 5.16. The bonding of molecules within the flakes of clay particles is strong. The attraction between adjacent flakes depends on the thickness of the water film. In wet clays, the water isolates the particles and the attraction is small. Consequently, the flakes slide easily and the wet clay is slippery.

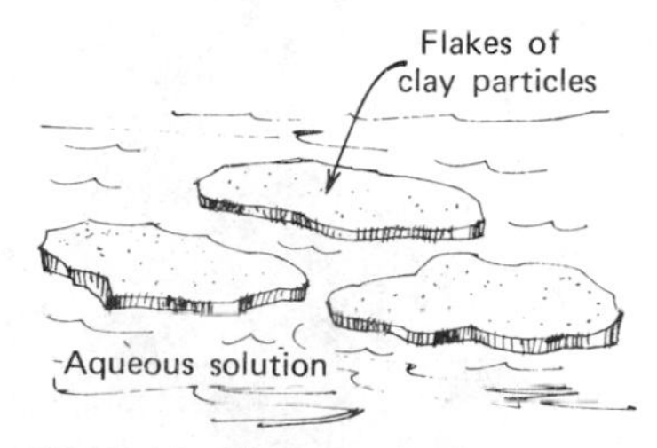

FIG. 5.16. Microscopic clay particles are surrounded by water.

What happens to the microstructure in the drying of clays? Initially, the wet clay contains water in two different regions. The water suspended within the unit cells in the particle layers is called pore water. The water films that separate the particle layers and promote the slippage or plasticity of clay are called shrinkage water. When the clay is dried, the shrinkage water is lost first. The clay shrinks, the flaky particles become closer together and the stronger attraction forces increase the strength of the clay. Continued increases in temperature cause chemical reactions in the microstructure.

Many different materials with contrasting properties can be made by structuring the molecules of clay in different ways. The silica in the mud that sticks on one's shoes can be mixed with some water and cement and then plastered on walls. If heated to around 2200 F, silica compounds would result in bricks that could be used for building construction or the lining of high temperature (3000 F) furnace walls. Common window glass is 75 percent silica. The same silica that is part of an opaque brick is also used for precision optical transparent lenses. Synthetic stones used on counter tops, vitreous enamels for sinks, pottery for cooking such as the trade name "Pyrex" and filters for water purification all include silica bonded together in some microstructure of atoms. The manipulation of the tiny universes from mineral clays can result in many useful products.

5.2.5/Polymers and Organics

Substances that are based around the element carbon are called *organic* materials. They may exist in the form of solids, liquids, or gases. Wood, petroleum, natural gas, natural rubber, coal, collagen in connective tissue, proteins, and hair are examples of organic materials that occur in nature. Drugs, adhesives, detergents, plastics, nylon, insecticides, rubber, and some foods are typical synthetic organics that have been developed through chemistry.

The use of synthetic organic materials has been increasing at a phenomenal rate since about the 1940s. Wooden floors, moldings, gun stocks, window casings, furniture, skis, fence posts, telephones, and handles are being replaced with plastic. Synthetic fibers are used in carpets, tires, dental floss, fishing lines, baling twine, rope, trampolines, clothing, blankets, and draperies. The homemaker is being surrounded with plastic dishes, bowls, utensils, eggbeaters, flower pots, picture frames, and toys. Metal gears, pipes, and structural frames are now being made from plastic. Ceramic insulators and optical lenses must also compete with their plastic counterparts. Twentieth-century engineers need to be aware of the advantages and disadvantages of these relatively new synthetics in order to select the best material in product design.

H
H — C — H
H
Methane

FIG. 5.17. Methane molecule.

Carbon is often considered to be the most basic element in the chemistry of life. As mentioned, it plays the dominant role in the microstructure of plastics and other organic materials. The reason carbon is so fundamental relates to the four electrons in the outer orbits that are available for primary bonding. Carbon seldom participates in ionic bonding with other atoms since the gain or loss of four electrons would be unlikely. Carbon prefers covalent bonding. Some of the four electrons may be shared mutually by other carbon atoms to produce a strong, repetitive chain of carbon-carbon molecules. Other electrons may covalently bond atoms of hydrogen, chlorine, oxygen, nitrogen, and others to the carbon chain.

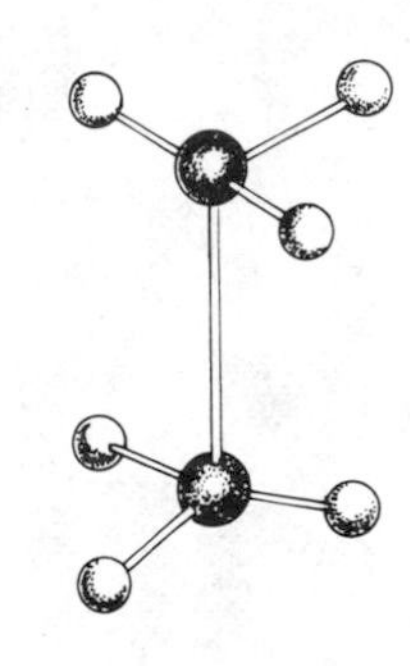

FIG. 5.18. Ethane molecule.

A molecule of methane gas consists of a carbon atom sharing its four electrons with hydrogen as shown in Figure 5.17. In methane, all four valency electrons are satisfied and the molecule is said to be saturated. A molecular model of ethane is shown in Figure 5.18, where three hydrogens are attached to each carbon. In the case of ethylene, only two of the four valency electrons are covalently bonded to hydrogen atoms as shown in Figure 5.19. The other two carbon electrons are shared with another carbon atom. The vinyl compounds substitute an atom such as chlorine

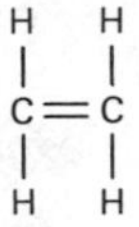

```
H   H
|   |
C = C
|   |
H   H
```

Ethylene

FIG. 5.19. Ethylene molecule.

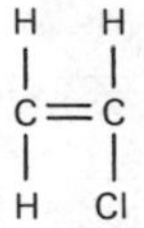

```
H   H
|   |
C = C
|   |
H   Cl
```

Vinyl chloride

FIG. 5.20. Vinyl chloride molecule.

for one of the hydrogen, shown in Figure 5.20. Such molecules that contain multiple bonds between carbon atoms are said to be unsaturated. The double bond is represented pictorially by two lines between the carbons, which indicates that two electrons are shared.

Synthetic organic materials are made by chemically connecting long chains of carbon molecules together. The fundamental link or single molecule in the chain that is repeated is called the monomer or just plain "mer." The process of attaching these monomers together is called polymerization and the completed chain is referred to as a polymer. Consequently, the term *polymer* refers to a broad class of carbon based monomers that are connected together. For example, the unsaturated ethylene molecule available from crude oil as a by-product has a double bond between the carbons. The well-known polymer polyethylene, which is commonly used in packaging, is formed by persuading one of the electrons in the carbon double bond to transfer over and attach to an adjacent monomer of ethylene. The net result is a new chain of carbons with no double bonds, as shown in Figure 5.21. The polymerization of the monomer vinyl chloride results in polyvinylchloride as used extensively in pipe, shown in Figure 5.22. Nylon, butadiene, and over a million other different polymers can be made by chaining different monomers together. Although it sounds easy, it took several years to develop the chemical expertise to persuade electrons and atoms to participate in polymerization.

The covalent bonding in organics tends to improve the stability of the material. The chemical compounds in molecules of polymers seem to be relatively inert compared to many metals. The polymers that have double carbon bonds, such as the synthetic rubbers, seem to be more prone to deteriorate than do the saturated carbon compounds such as polyethylene. Some plastic containers seem to exist forever, yet products such as chest waders deteriorate within a short time when exposed to sunlight.

The microstructure of these polymer chains may be visualized as a snarled pile of string or perhaps as a bowl of cooked

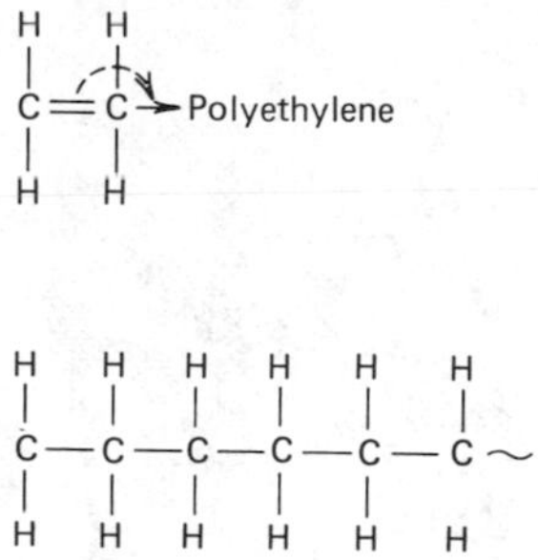

```
H   H
|   |
C = C → Polyethylene
|   |
H   H

H   H   H   H   H   H
|   |   |   |   |   |
C — C — C — C — C — C ~
|   |   |   |   |   |
H   H   H   H   H   H
```

Polyethylene polymerization

FIG. 5.21. Polyethylene polymerization.

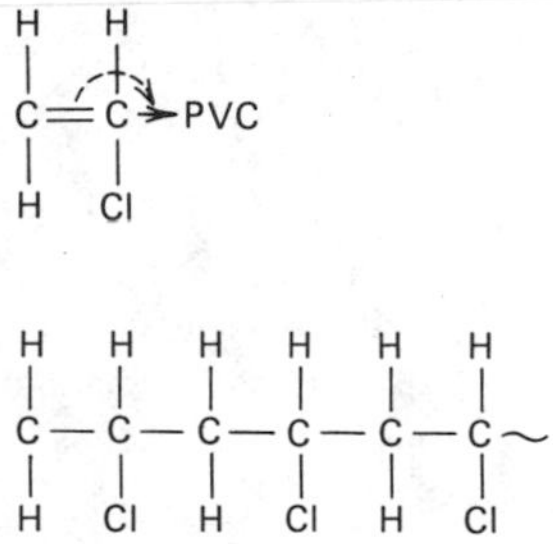

```
H   H
|   |
C = C → PVC
|   |
H   Cl

H   H   H   H   H   H
|   |   |   |   |   |
C — C — C — C — C — C ~
|   |   |   |   |   |
H   Cl  H   Cl  H   Cl
```

Polyvinylchloride polymerization

FIG. 5.22. Polyvinylchloride polymerization.

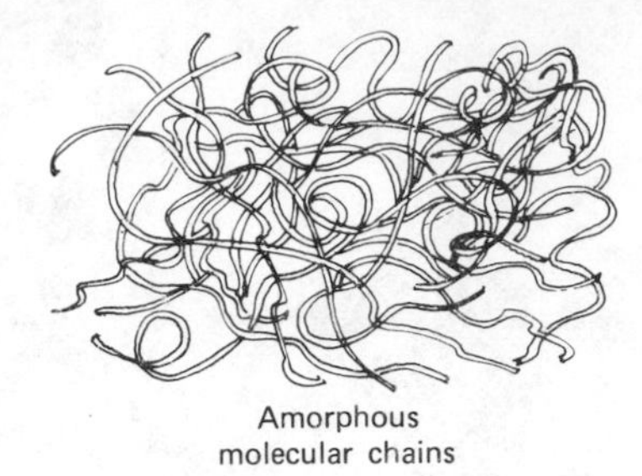

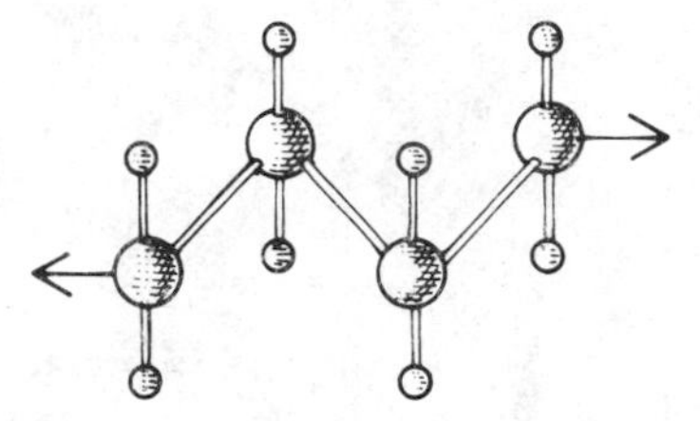

FIG. 5.23. Microstructure of polymers.

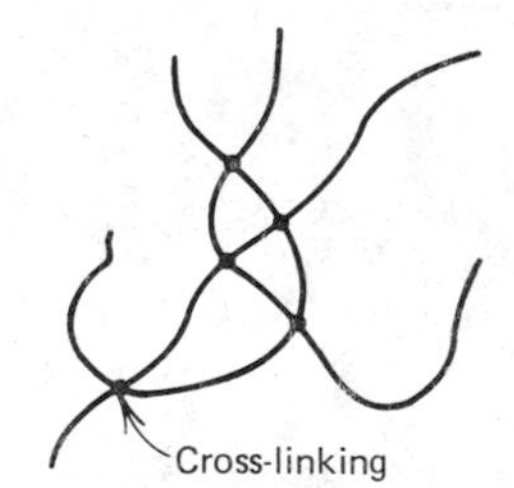

FIG. 5.24. Kinks in molecular chains provide some elasticity.

FIG. 5.25. The fluidity of plastics may be reduced through cross linking.

spaghetti. If the chains are random throughout, then the formation is said to be amorphous. A polymer that has molecular strands aligned in some repeating fashion is said to be crystalline, as shown in Figure 5.23. The degree of crystallinity affects the strength and elasticity of the material. Nylon is one of the most crystalline polymers, and has a tensile strength up to 200 000 psi and a modulus of elasticity of 1 000 000 psi. Some crystallinity in the microstructure can be achieved by stretching out the tangled chains.

The long strands of molecules are three dimensional and have a backbone of carbon with other atoms such as hydrogen arranged around the outside. Many of the molecular chains have kinks that provide elastic stretching similar to that of a coiled spring as shown in Figure 5.24. The elastic behavior of the bulk material is related to the spring action of the polymer chains. Highly ordered polymers have molecular chains that behave like parallel springs. Such polymers are more elastic than those such as polystyrene that have complicated arrangements of chains. Many of the polymers experience so-called stress relieving as the spaghetti-type chains reorient themselves under load to become more effective in handling the internal forces.

There are weak intermolecular forces between chains that tend to inhibit their relative motion. The fluidity of many plastics is due to the relative sliding of these polymer chains. In some cases, they slide past one another much as the layers in a fluid. Also, they do not always slide back when the load is released and some permanent deformation results. This sliding can be reduced by cross linking the chains at intermediate points as shown in Figure 5.25. Cross linking is essentially a local spot weld that is created by covalent bonding between chains. It reduces the sliding action and adds rigidity to the bulk material. Hard rubber, for example, is the result of cross linking through the addition of sulphur. The mechanical properties such as strength and resistance to creep can be greatly improved through cross linking. In polyethylene, cross linking can increase the tensile strength about 30 percent, and the elasticity is about doubled.

The hardness of a polymer depends on the attraction between molecular chains. This attraction is influenced by the type of atoms that are covalently bonded around the periphery of the carbon backbone. Polyvinylchloride has a single chlorine atom in the monomer, yet the atom provides an additional polar attractive force to the chain that makes PVC a rigid material. It is used extensively in plastic structures such as pipe. The microstructure is stable and provides a long life.

PROBLEMS

5.1. Explain the role of a materials scientist as compared with an engineer.

5.2. Describe the Rutherford model of the atom.

5.3. How does an element differ from an atom?

5.4. Briefly explain the three primary bonds.

5.5. Why are metals good conductors?

5.6. Why do ceramics usually have high melting points?

5.7. Explain what is meant by polymerization.

5.8. Sketch a typical microstructure of a polymer.

5.9. Describe the microstructure of a metal alloy as compared to a pure metal.

5.10. Describe why clay is slippery when wet yet stabilizes on drying.

5.11. Name at least five products made from sand.

5.12. What is the most abundant metal available from the earth?

5.3/Mechanical Properties of Materials

Mechanical properties describe the response of materials subjected to applied forces and displacements. The properties defined in this section apply to metals, ceramics, and polymers. The numerical values that quantify material characteristics may be different, but the definitions of the properties themselves are the same.

5.3.1/Stress

The strength of a material is usually specified in terms of *stress,* which is defined as the internal force per unit area. In Figure 5.26, an external axial force F is applied to each end of a nonuniform bar. The bar is in static equilibrium. At any section along the bar, the normal stress σ in the material is defined as:

$$\sigma = \frac{F}{A}$$

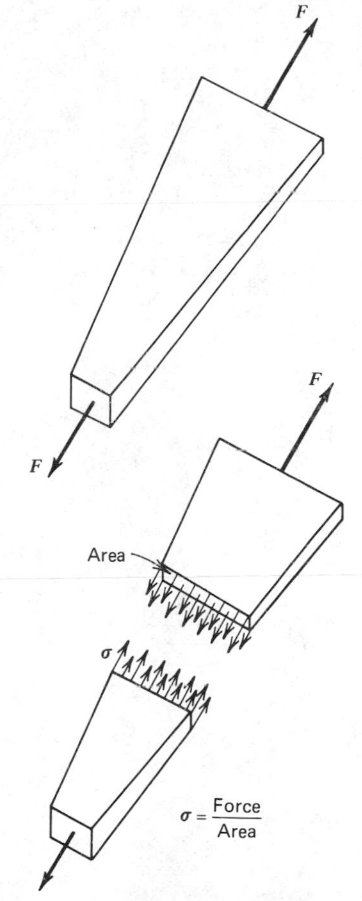

FIG. 5.26. Stress is defined as force per unit area.

where F is the force and A the cross-sectional area. In axial loadings, the direction of stress is perpendicular or normal to the cross-sectional area and is uniformly distributed as shown. The stress is *tensile* if the member is being "stretched" and *compressive* if the bar is "pushed" together. The units of stress may be pounds per square inch (psi), pounds per square foot (psf) or Newtons per square meter (N/m^2).

The total internal force at any cross section is the product of stress times area and must equal the applied force F at the end of the bar in order to satisfy equilibrium conditions. Since the force is constant, the stress must vary inversely as the cross-sectional area changes along the length of the bar. A smaller area results in a larger stress.

Would a bar experience stress if twisted? Do bending moments induce stress? The answer to both questions is yes. Stress can be induced into materials through forces that tend to twist, bend, shear, compress or elongate a material. The interaction of all these stresses and formulas that relate their magnitudes to applied loads are covered in later courses in engineering. The normal stress defined here is only a first step, yet material strength is most frequently identified by specifying the maximum normal stress allowable.

5.3.2/Strain

Materials deform under load. A term is needed to describe the elongation or compression of some unit length of material. Engineering *strain*, ϵ, is defined as

$$\epsilon = \frac{\Delta L}{L}$$

where ΔL is the change in length of the original length and L is the original length of the bar before the load was applied, as shown in Figure 5.27. If the change ΔL is measured in the same units as the length L, then the ratio is dimensionless. A strain might be specified as $500(10)^{-6}$ in./in., $500(10)^{-6}$m/m or perhaps 0.05 percent. A plus strain indicates elongation, and a negative value of strain indicates compression.

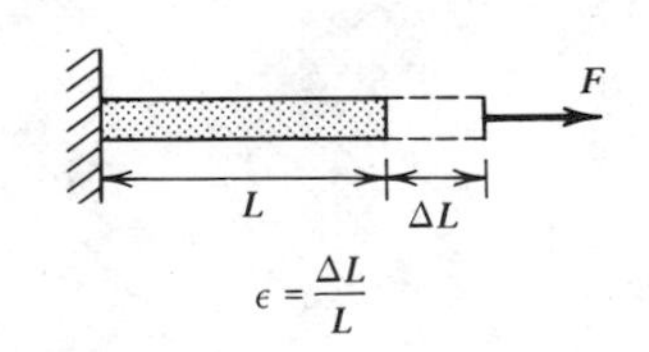

FIG. 5.27. Strain.

5.3.3/Modulus of Elasticity

All materials do not deform in the same manner when subjected to identical loading systems. Some stretch more than others under equal tension. Up to a certain point, defined as the *elastic limit,* most materials behave as an elastic spring as shown in Figure 5.28. The material develops an internal strain that is proportional to the internal stress in the material. This proportionality constant is called the *modulus of elasticity* and is given the symbol E. For elastic materials subjected to a one-dimensional stress, the relationship between stress and strain can be expressed as

$$\sigma = E\epsilon$$

which is commonly called Hooke's law.

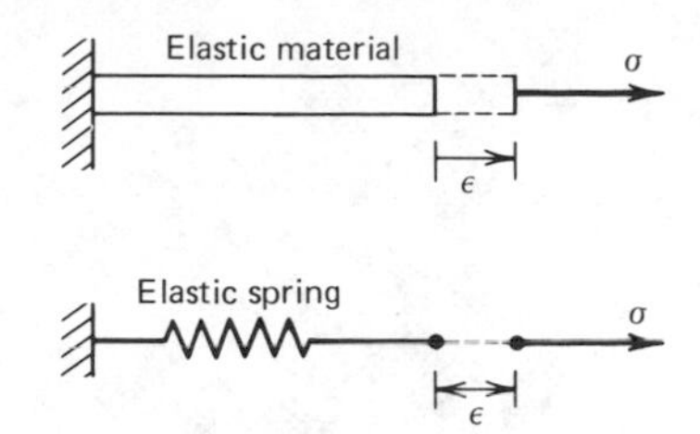

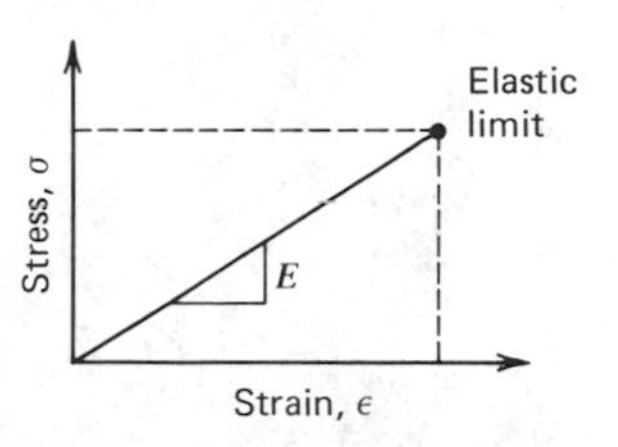

FIG. 5.28. Elastic materials behave like springs.

5.3.4/Mechanical Properties of Metal Alloys

Pure iron is known as ferrite and has a tensile strength of around 40 000 psi. The addition of only 0.50 percent carbon to the iron increases the tensile strength to slightly over 100 000 psi. Such a sprinkle of carbon on the crystalline lattice of iron adds to material strength much as a dash of salt adds to the flavor of a baked potato. The strength capacity is more than doubled. The iron-carbon alloy is known as commercial steel and there are over 20 000 different mixtures available. Some of these steels have tensile strengths in the neighborhood of 600 000 psi, which is about 15 times the strength of pure iron.

Plain-carbon steel contains small traces of manganese, phosphorus, sulfur, and silicon; however, carbon is the most dominant alloy. Low-carbon steels (less than 0.3 percent C) are used to make automobiles, pipe, bridges, buildings, appliances, and other structures. They are ductile, easily welded, relatively inexpensive, heat treatable and machineable. Medium-carbon steels (0.35 to 0.55 percent C) have higher tensile strength, are less ductile, are harder, and are used for farm implements, gears, shafts, and machines. High-carbon steels (0.6 to 1.5 percent C) have even higher tensile strengths and are used in axes, hammers, knife blades, piano wire, and saws. As one continues to add carbon, the steel becomes harder and more brittle. Steel with over 2 percent carbon is called cast iron. It is brittle, not recommended for welding, is difficult to machine, yet is used extensively in casting.

The stress-strain curve for most carbon steel alloys is given in Figure 5.29. The modulus of elasticity is the slope of the elastic portion of the curve and remains constant up to the yield point of the material. At the elastic limit or yield point, the steel specimen suddenly begins to deform more readily as shown by the drop in the stress-strain curve. As the load is continued beyond the yield point, the atoms in the crystalline microstructure of the metal are

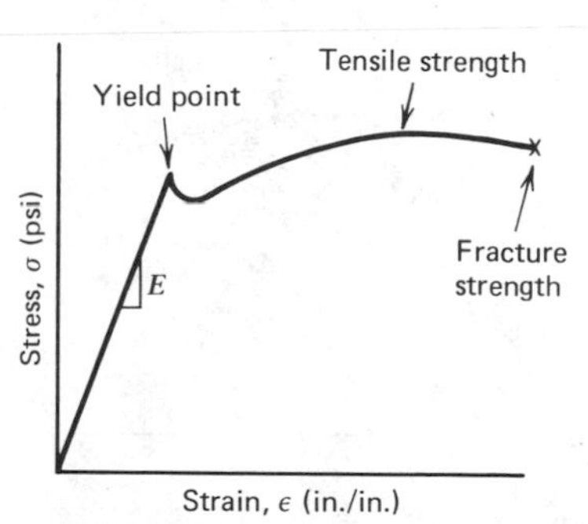

FIG. 5.29. Stress-strain curve for carbon-steel alloys.

distorted and begin to experience traffic jams as they dislocate. This effect is called work or strain hardening and actually strengthens the material. The material also becomes more brittle with strain hardening. Permanent or plastic deformation results from the distortion in the crystalline lattice. If the load is continued, the stress-strain curve reaches a maximum stress identified as the *tensile strength* (ultimate stress). Material separation eventually occurs at the *fracture point*. The modulus of elasticity for steel is $30(10)^6$ psi. The alloying of steel doesn't change the slope of the elastic portion of the stress-strain curve significantly.

Carbon alloying improves the strength of iron, yet the resistance to corrosion is reduced. Steel rusts more easily than iron. The addition of chromium and nickel to iron reduces the corrosive action, and the material is commercially recognized as stainless steel. Stainless steel has high strength and toughness, has good resistance against corrosive environments, and is lustrous in the eyes of consumers. On the negative side, however, stainless steel is twice as expensive as low carbon steel, it is heavier, and it won't keep a cutting edge as well as other steels unless specially treated.

The addition of such elements as tungsten, vanadium, molybdenum, and chromium into iron produce tool steels that have exceptional hardness, strength, abrasion resistance, and toughness. Tool steels can maintain desirable properties and operate at red-hot temperatures for a short length of time. A small amount of columbium will increase the tensile strength of steel an additional 15 000 psi. An almost negligible amount of hydrogen or oxygen suspended interstitially in iron will promote cracking and other harmful effects.

Aluminum is a lightweight metal that is used extensively in structures. Although pure aluminum is very ductile and a good conductor, its tensile strength is only around 10 000 psi. Through alloying, aluminum tensile strengths of over 75 000 psi can be achieved, which exceeds the strength of the heavier low carbon steels. The stress-strain curve for nonferrous metals, such as aluminum, usually have an initial elastic range as shown in Figure 5.30. The yield point is not clearly defined since there is not a sharp peak. The yield point is then defined as the stress that will produce a specified amount of permanent deformation, usually around a strain of about 0.002. This means that such a nonferrous material loaded to the yield point would have a residual strain of 0.002 in./in. if the load were removed. The maximum tensile strength is beyond the elastic limit. Aluminum is easy to machine, form, weld, and is frequently used in space vehicles, airplanes,

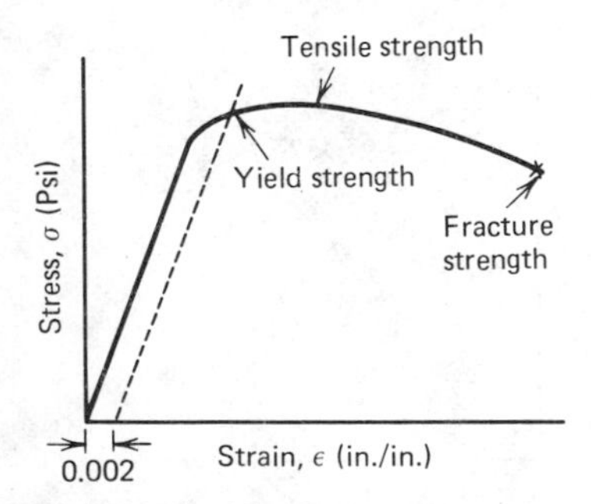

FIG. 5.30. Stress-strain curve for nonferrous metals.

cans, shipping containers, and other areas where weight is a limiting factor. In fact, an aluminum bulldozer has been made that can be transported by helicopter.

There are many other metals available for engineering applications that have good mechanical properties. Magnesium is a lightweight metal than can be used successfully at 600 F as a structural member. Titanium can be alloyed to have strengths greater than 200 000 psi and weighs only about 0.6 times the weight of steel. Titanium is resistant to corrosion at room temperatures. Beryllium is also available with tensile strengths around 270 000 psi; however, it is rather brittle. There are thousands of metal alloys with different characteristics. Table D.1 in the Appendix indicates some common metals and their properties.

5.3.5/Mechanical Properties of Ceramics

Ceramics are usually brittle and hard. The brittleness is partially explained due to the lack of mobility of the individual atoms. Ceramics consist of molecular compounds that have atoms of unequal sizes and local atomic slippage is impeded as compared to metals as shown in Figure 5.31. The tiny particles of the microstructure are trapped and cannot slide over the larger particles. Furthermore, ceramic materials often include ions from ionic bonding that repel any relative movement of the atoms. Instead of local yielding and ductile action, ceramics break suddenly and are brittle.

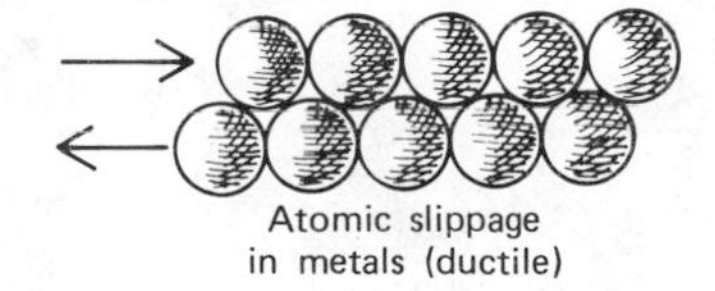

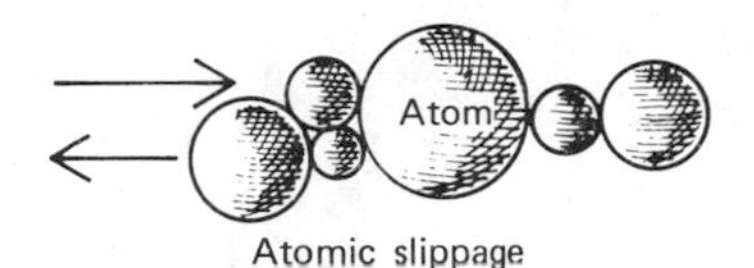

FIG. 5.31. Ceramics are usually brittle due to the lack of molecular mobility.

The strength limits of ceramics are often different for tension and compression. Most ceramics will carry considerably more load in compression than in tension. Any unfortunate person who shoots a plate glass window with a BB gun will probably observe a small hole on the front side of the glass and a much larger cone shaped hole in the back as shown in Figure 5.32. The person might ask why. The answer is simple—the impact of the bullet initiates a compression wave that passes through the glass. Upon reaching the back side, the compression wave rebounds as a tension wave. The glass then fails in tension and the trapped momentum of the particle motion carries the glass away.

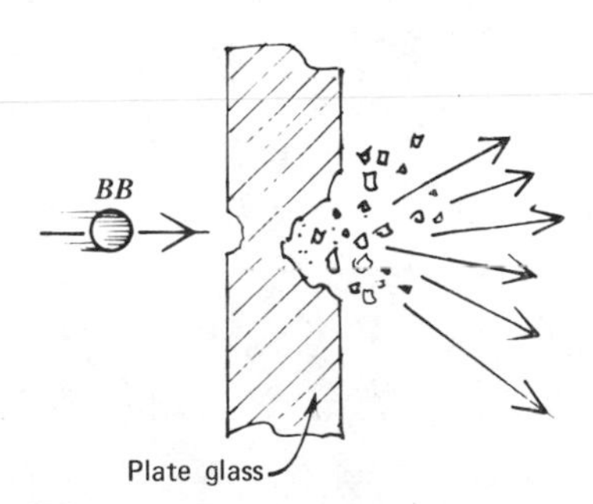

FIG. 5.32. The spalling of glass.

The ionic and covalent bonds that hold ceramic compounds together are strong. One would expect high-strength characteristics in both tension and compression. The difference in strength capacities is somewhat described from the microcracks that are induced during the solidification of the ceramic materials. The molten temperatures of ceramics are usually higher than metals. The melt temperature of silica is 3100 F, titanium carbide 6370 F,

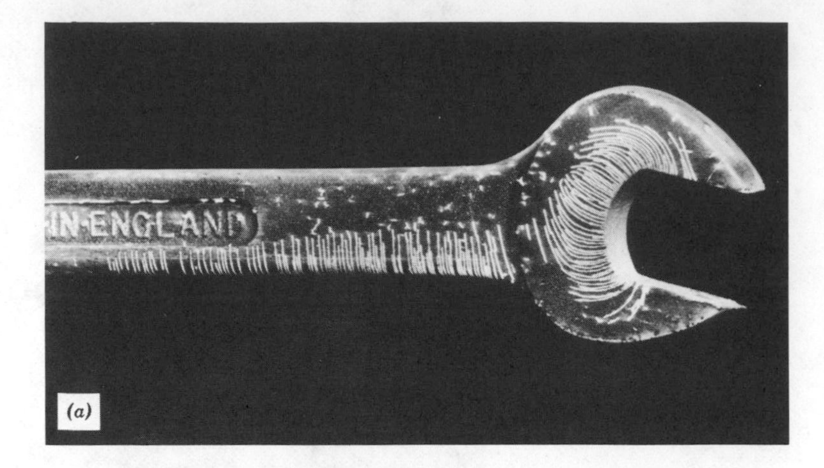

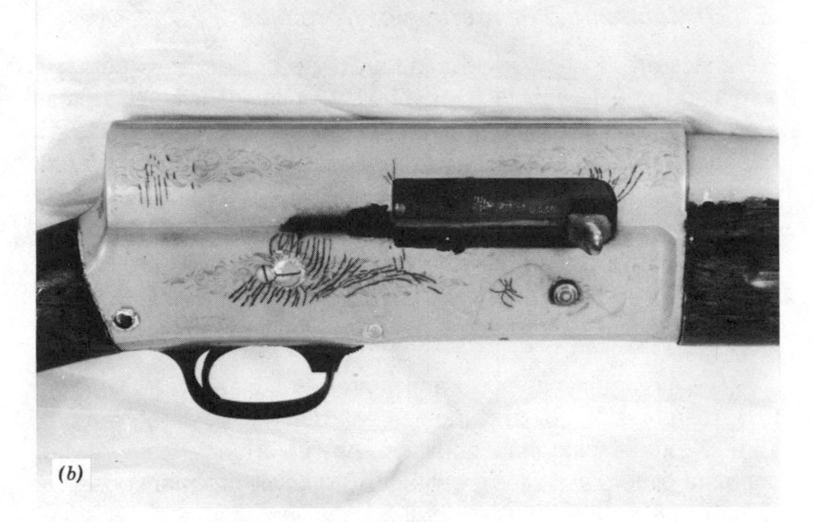

FIG. 5.33. Brittle coatings used in experimental stress analysis of an end wrench and shotgun. (Photographs courtesy of Magnaflux Corporation)

and hafnium carbide at 7520 F. The cooling of these ceramics induces thermal stresses that result in fine cracks. Compression loads push the cracks together, whereas tension loads separate and tend to pull the cracks apart. In tension, high-stress concentrations at the crack tips promote the growth of these cracks and fracture occurs. Some small glass fibers with essentially no cracks have been tested up to stresses of 10^7 psi before failure in tension occurred. Larger glass specimens with surface cracks failed around 10 000 psi.

The brittle cracking of ceramics can be used advantageously in mechanical testing of other materials. Brittle coatings such as the tradename "stresscoat" can be applied to structures and other products as shown in Figure 5.33. The coating cracks at strain levels much below the elastic limits of metals. The density of the cracks is indicative of the stress magnitude and the crack directions are perpendicular to the maximum tension stress.

Many ceramics experience some limited elastic action and therefore have a modulus of elasticity. Most glasses have a modulus of elasticity around $10(10)^6$ psi, which is less than steel, greater than wood and about equal to that of aluminum. Some mechanical properties of ceramics are given in Table D.2.

5.3.6/Mechanical Properties of Polymers

The behavior of polymers subjected to mechanical force depends on two important variables—time and temperature. These two variables are often appropriately neglected in the design of products made from metal materials. However, such an oversight in design that involves polymers could easily result in a failure. The deformation, tensile strength, and the modulus of elasticity of polymers depend significantly on the rate at which the force is applied. Furthermore, the deformation continues at a slow rate if the load is maintained.

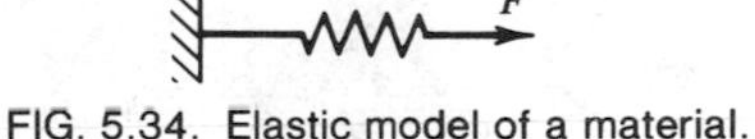

FIG. 5.34. Elastic model of a material.

The reaction of polymers to mechanical force is described on a microscopic level as the stretching and sliding of molecular chains. The stretching accounts for the elastic action, whereas the sliding of molecules is typical of fluids. On a macroscopic level, the elastic tendency of the material can be represented by a spring shown in Figure 5.34. The *spring* models the linear relationship between stress and strain, and implies the material will return to its original position when the load is released. As mentioned earlier, a simple spring is adequate in describing the behavior of metals when loaded below the elastic limit.

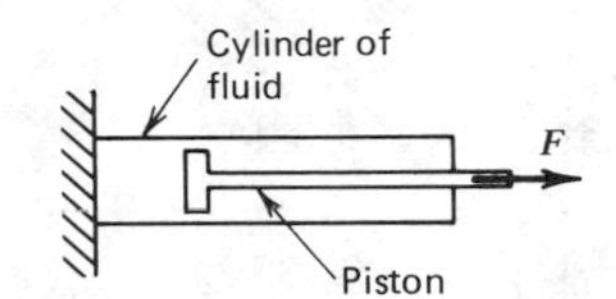

FIG. 5.35. Viscous model of a material.

The tendency of the polymer to flow as a liquid can be modeled by a *dashpot.* The dashpot shown in Figure 5.35 is a loosely fit piston inside a cylinder that contains a fluid. It takes more force to pull the piston rapidly. A slowly applied force allows the fluid to flow around the piston more readily. In other words, the displacement of the piston depends on the rate at which the force is applied in addition to the magnitude. Furthermore, the piston does not return to its original position after the load is released since the space behind the piston is filled with fluid.

The spring and dashpot represent the two material extremes: an elastic solid and a fluid. Polymer materials are somewhere in between and are said to be viscoelastic. There are several arrangements of springs and dashpots that could be used to model polymers. The model shown in Figure 5.36 is a typical viscoelastic model.

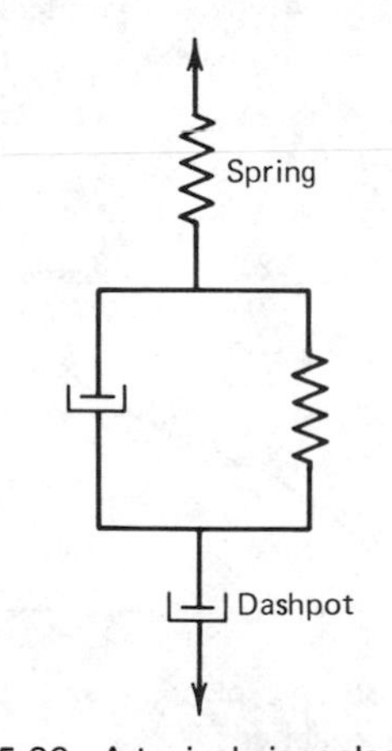

FIG. 5.36. A typical viscoelastic

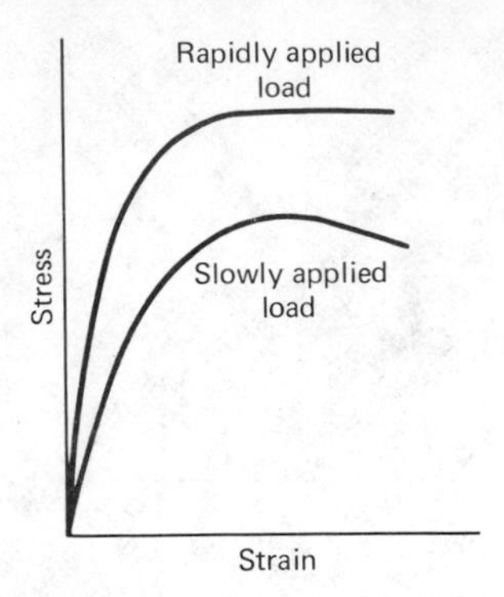

FIG. 5.37. In polymers, the stress-strain curve depends on the rate of the applied load.

In view of the dashpot, it is easy to understand why the deformation of polymers depends on the rate at which the load is applied. A load of 100-lb applied rapidly would cause less displacement than the same 100-lb load applied slowly. In Figure 5.37, a typical plastic specimen was subjected to a simple tension test. The lower curve is the stress-strain curve for a slowly applied load and the upper curve is a stress-strain curve for the same specimen subjected to a more rapidly applied load. The degree of dependence on the load rate is somewhat proportional to the dominance of the fluid action over the elastic action of the polymer. Some polymers that can be adequately modeled by a spring could be treated as an elastic solid with little dependency upon the rate of loading.

Not all polymers have a well-defined modulus of elasticity. As mentioned earlier, the stress-strain curve of most metals has an initial elastic range that is a straight line, and the modulus of elasticity is constant from zero to the elastic limit. Some polymers have these straight-line segments in their stress-strain curve. The stress-strain curve for polyethylene is shown in Figure 5.38. Note that the slope of the curve is not constant. Therefore, one value of *E* doesn't apply exactly to the complete curve. The stress-strain curve shown in Figure 5.39 is that of rubber polymer. Again, note that an *E* value would depend on the stress level. Some plastics such as polyvinyl chloride (PVC) do have a fairly well-defined elastic range as shown in Figure 5.40. Remember that the values for the modulus of elasticity given in standard tables might be highly dependent on the levels of stress.

There is another problem with the use of polymers that is related to time. If a weight is suspended on the end of the bar as shown in Figure 5.41, a polymer bar will deform initially and then continue to deform even though the applied load is constant. This

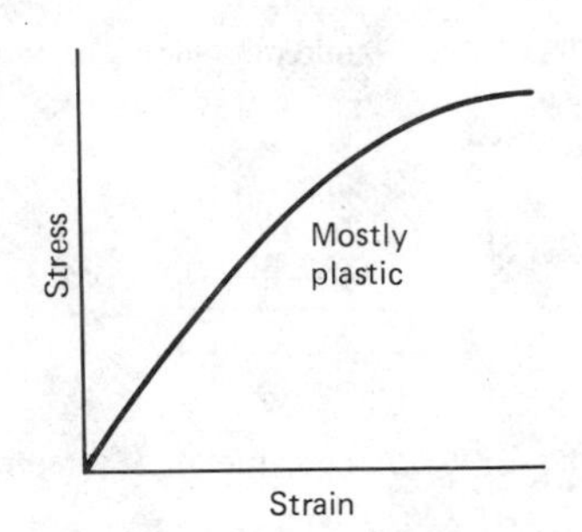

FIG. 5.38. A typical stress-strain curve for polyethylene.

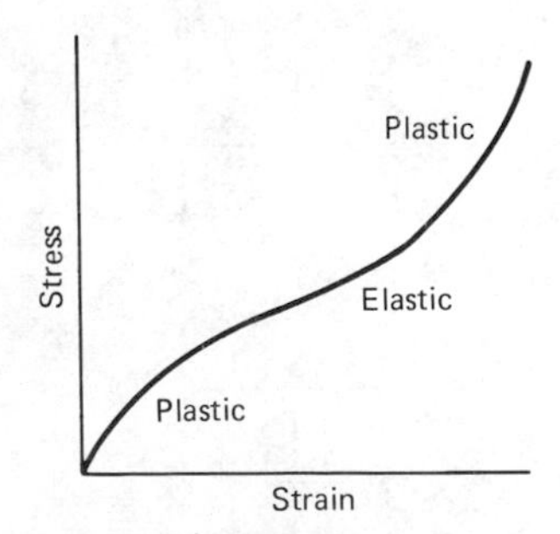

FIG. 5.39. A typical stress-strain curve for rubber.

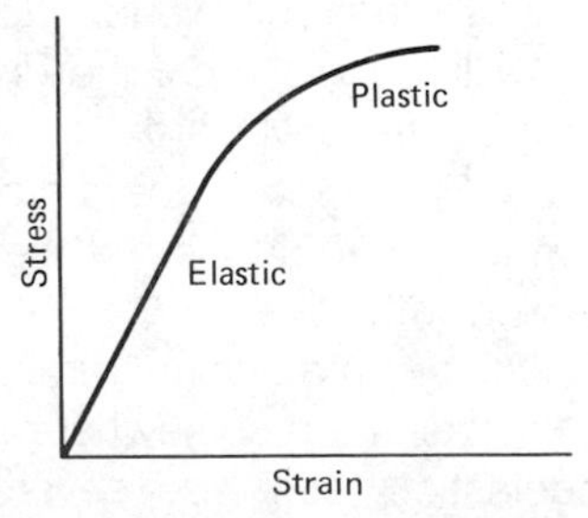

FIG. 5.40. A typical stress-strain curve for PVC.

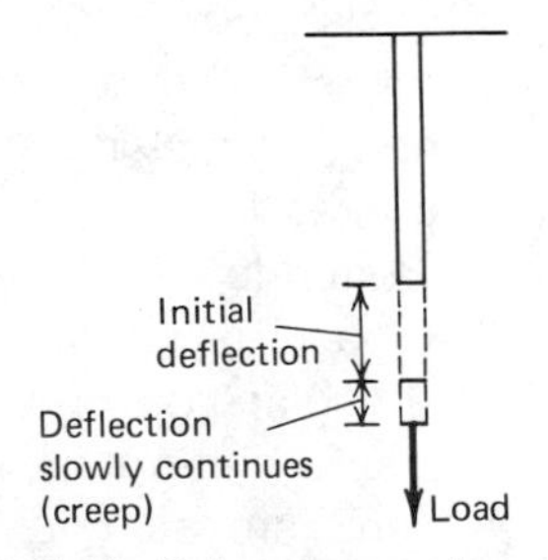

FIG. 5.41. Polymers experience creep.

continued deformation is called *creep*. Metals and ceramics experience little creep compared to polymers.

Intuitively, one would expect temperature to effect the stress-strain curve of polymers. Temperature increases will shift the behavior of a viscoelastic material towards the liquid extreme. The ductility of a polymer will increase as the temperature increases. In general, tensile strength will decrease as the temperature increases.

In view of load rates, deformation, creep, and temperature, how does one select an allowable design stress? Proper design using polymers in load-bearing situations requires a knowledge of both the loading and environmental conditions. The value for mechanical properties such as tensile strength and temperature expansion, should be related to loading rates and temperature conditions. If sufficient detail in relating to the exactness of the material properties is unavailable, then a significant safety factor might help to insure the safety of the design. Perhaps a better approach would be to contact the manufacturer for more complete data.

To demonstrate the selection of material properties, a typical test curve for polyvinyl chloride is given in Figure 5.42. The upper curve is a short time tensile strength. The lower curve is the stress-strain relationship when subjected to a slowly applied load. The effect of temperature is also taken into account. Although one of the two loading curves may not fit the exact loading conditions, the curve still assists in making a judgment. An allowable design stress that would be conservative might be selected as shown. Some mechanical properties of polymers are given in Table D.3.

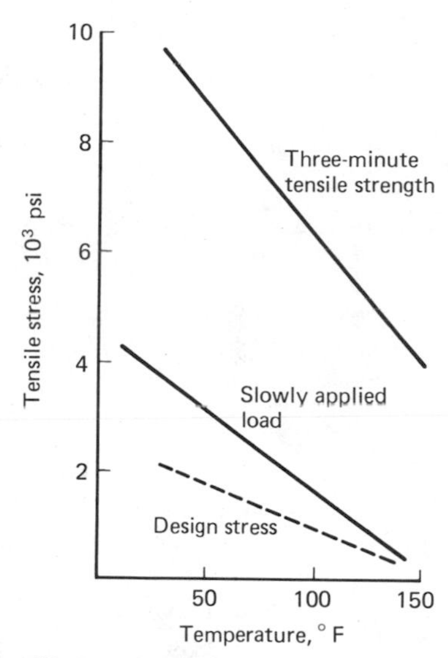

FIG. 5.42. A load curve for PVC.

EXAMPLE PROBLEM 5.1

A uniform round rod 0.75 in. in diameter and 3 ft long supports a weight of 1000 lb. The material is structural steel. Determine the total elongation at the end of the rod.

Solution

Since the rod is uniform, the stress is constant along the length and also across the cross section. The normal stress can be computed as follows:

$$\sigma = \frac{F}{A} = \frac{1000}{\pi(0.75)^2/4} = 2263 \text{ psi}$$

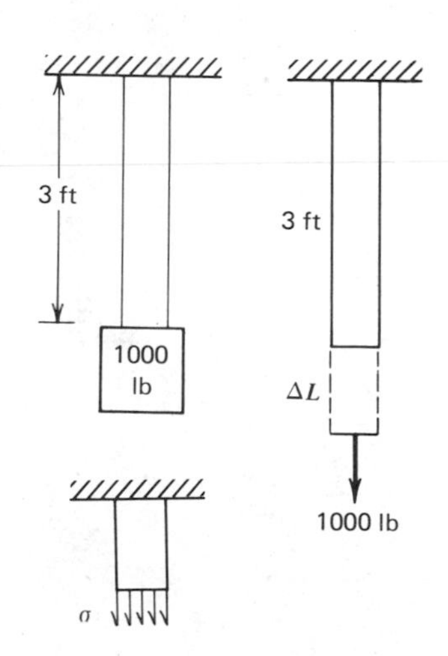

The strain can be determined by using Hooke's law. The modulus of elasticity is found from Table D.1.

$$\sigma = E\epsilon \qquad \text{or} \qquad \epsilon = \frac{\sigma}{E} = \frac{2263}{30(10)^6} = 75(10)^{-6} \text{ in./in.}$$

Since the strain is the elongation for only one unit, in this case an inch, the total elongation of the rod can be found as follows:

$$\epsilon = \frac{\Delta L}{L} \qquad \text{or} \qquad \Delta L = \epsilon L = 75(10)^{-6}(36) = 2.7(10)^{-3} \text{ in.}$$

EXAMPLE PROBLEM 5.2

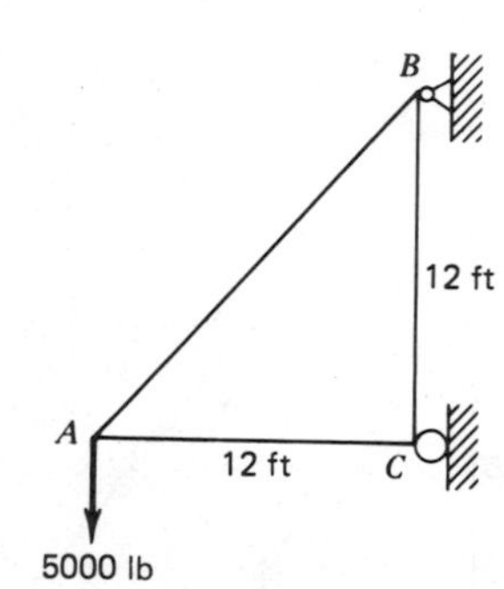

A pin-connected truss carries a load of 5000 lb and is to be made of aluminum (6061-T6). Determine the minimum cross-sectional area of the member AB such that the yield stress will not be exceeded.

Solution

The internal force carried by member AB can be computed as shown in Example Problem 3.12.

$$F_{AB} = 7072 \text{ lb tension}$$

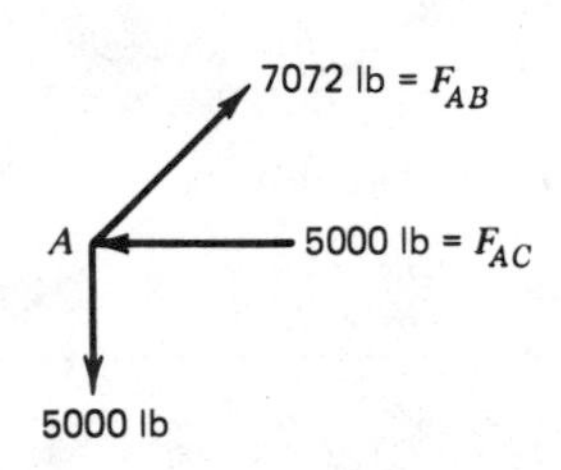

From Table D.1, the yield stress for aluminum (6061-T6) is 40 000 psi

$$\sigma = \frac{F}{A} \qquad \text{or} \qquad A = \frac{F}{\pi} = \frac{7072}{40\,000} = 0.176 \text{ in.}$$

If the member AB were round, the diameter of the aluminum rod would be

$$A = \frac{\pi d^2}{4} \qquad \text{or} \qquad d^2 = \frac{4A}{\pi}$$

$$d = 0.474 \text{ in.}$$

EXAMPLE PROBLEM 5.3

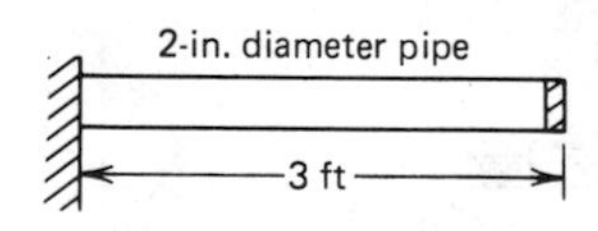

A nominal 2-in. diameter schedule 40 PVC pipe has a 0.308-in. wall thickness. A 3-ft length extends from a wall and is capped on the free end. The pressure of the fluid inside changes from 0 to 175 psi. Determine the longitudinal stress in the pipe wall and the total end deflection.

Solution

For equilibrium, the internal forces in the pipe shell must equal the total end force that results from the fluid pressure acting on the cross-sectional area.

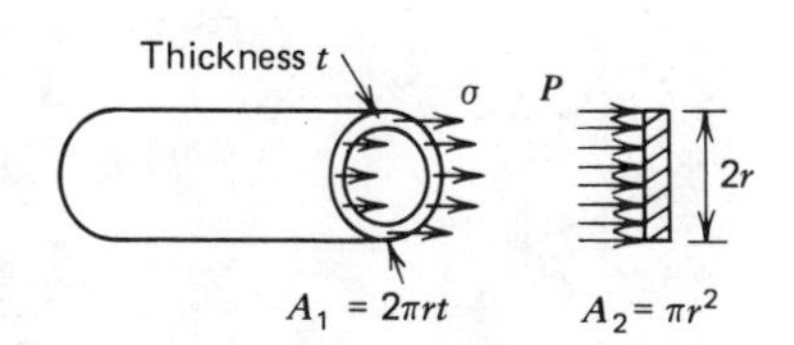

$$\sigma A_1 = PA_2$$

$$\sigma 2\pi rt = P\pi r^2$$

$$\sigma = \frac{Pr}{2t} = \frac{(175)(2)}{(2)(0.308)} = 568.2 \text{ psi}$$

The strain is determined from Hooke's law, where E is taken from Table D.3 as $0.2(10)^6$ psi.

$$\epsilon = \frac{\sigma}{E} = \frac{568.2}{0.2(10)^6} = 2.84(10)^{-3}\text{in./in.}$$

The end deflection δ follows as

$$\delta = \epsilon L = 2.84(10)^{-3}\,(3)(12) = 0.102 \text{ in.}$$

PROBLEMS

5.13. A $\frac{1}{4}$ in. $\times$ 2 in. flat bar of some unknown alloy elongates 0.7 in. in a length of 5 ft under an axial load of 13 000 lb. Determine the alloy's modulus of elasticity.

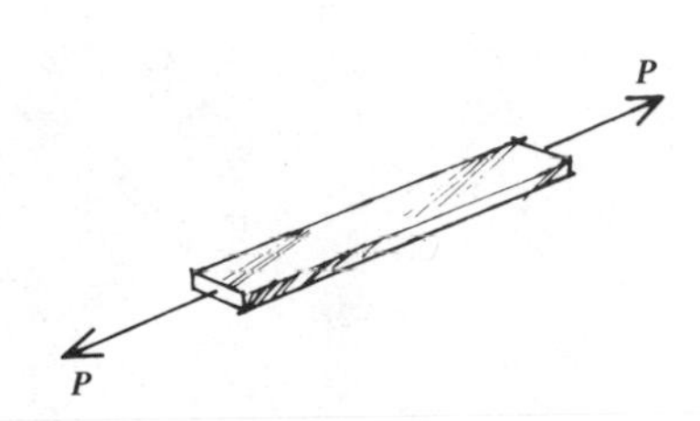

5.14. Determine the maximum allowable load P in Newtons that can be applied to a stainless steel (18-8 annealed) bar that is 4 cm wide and 1 cm thick. (1 psi = 6.895 Pa and 1 Pa = 1N/m^2)

5.15. A $\frac{1}{2}$-in. diameter brass (cold-rolled) rod is subjected to an axial load of 1500 lb. Determine the strain in the rod.

5.16. A pile driver is used to start a 6-in. diameter steel (0.4 percent C hot-rolled) rod into the ground. The impact of the driver on the rod is equivalent to a static force of 10 000 lb. Determine the total deformation of the 20-ft rod.

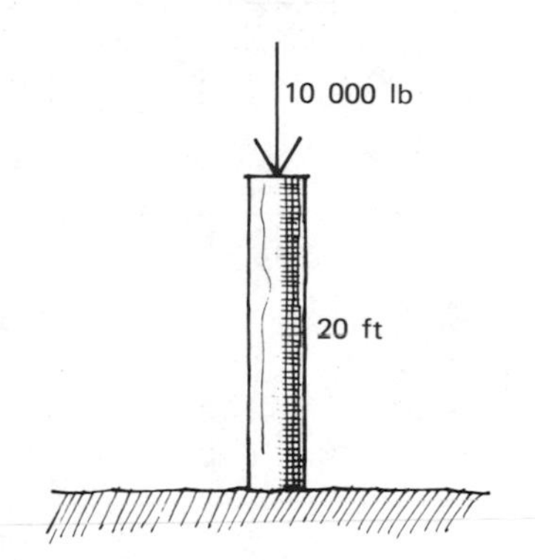

5.17. A well driller's bit becomes stuck in some hard rock. A large-tension load was applied to withdraw the pipe. A 12-in. gage length marked vertically on the pipe elongated to 12.0020 in. The top of the pipe elongated a total distance of 3 ft. Determine the depth at which the drill bit was stuck.

5.18. The truss is to be made from structural steel, and is loaded as shown. Determine the minimum cross-sectional area of member BC that will prevent yielding.

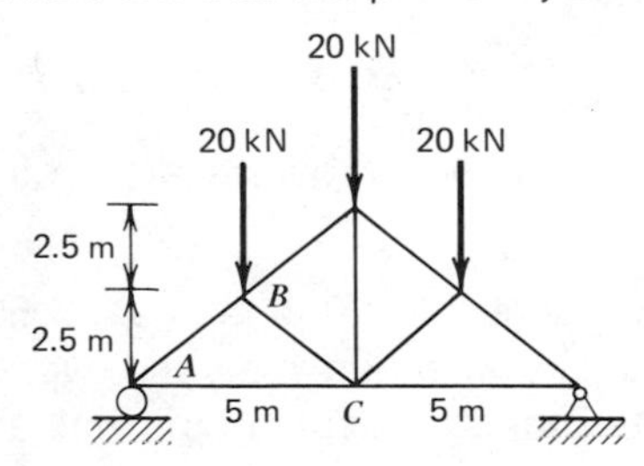

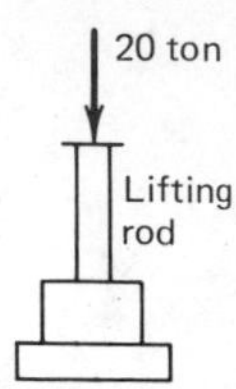

5.19. A hydraulic jack is to lift 20 tons. Determine the minimum diameter of the lifting rod if the material is stainless steel (18-8 annealed). Neglect buckling.

5.20. The truss is made from aluminum (6061-T6) bar 1.5 in. in diameter. Determine the maximum load *P* that can be applied without yielding. Neglect buckling.

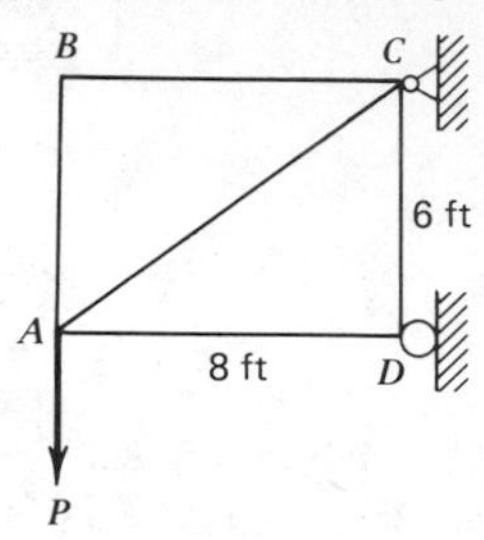

5.21. A chain on a wrecker is made from steel (0.8 percent C hot rolled) and must lift a 20-kN vehicle vertically. Estimate the minimum diameter of the links if a safety factor of 2 is required.

600 lb/ft

8 ft

Glass plate

10 ft

5.22. An architect wants to replace a brick-bearing wall with plate glass. The downward load on the wall from the roof is 600 lb/ft. Estimate the thickness of the plate glass required to carry the load in compression. Assume the compressive strength to be 90 000 psi. Note that this problem demonstrates the compressive strength of glass, but would not be a safe design due to other structural considerations.

5.23. A weight of 1000 lb must be suspended from a rod as shown. Determine the minimum diameter required if the rod material is glass.

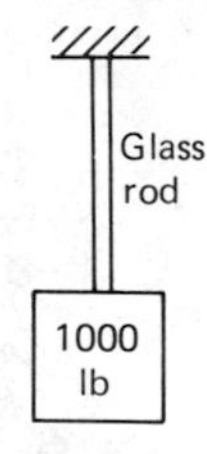

PVC pipe with end cap

4 ft

5.24. Explain why a nylon tow rope provides a smoother tow than does a steel chain.

5.25. A 4-in. diameter PVC schedule 40 pipe has a 4.500 O.D. and 4.026 I.D. A 4-ft length is suspended in air and capped as shown. Determine the longitudinal stress in the pipe wall and the end displacement if the pressure on the cap is 200 psi. Assume the base has zero deflection.

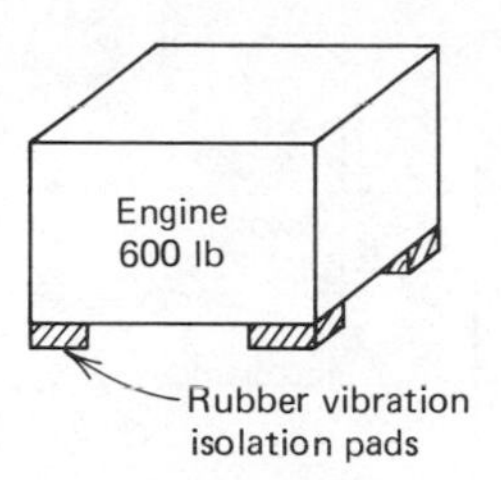

5.26. A 600-lb engine is placed on four vulcanized rubber mounts that are each 4 in. cubed. Estimate the downward deflection under the static weight of the engine.

5.27. Bone has a tensile strength of $12(10)^3$ psi and a modulus of elasticity of $2.5(10)^6$ psi. Determine the diameter of a polyethylene transplant with a tension strength equivalent to that of a natural bone 0.5-in. in diameter.

5.28. How would the total displacement of the polyethylene bone transplant in Problem 5.27 compare with the natural bone for a 6-in. segment?

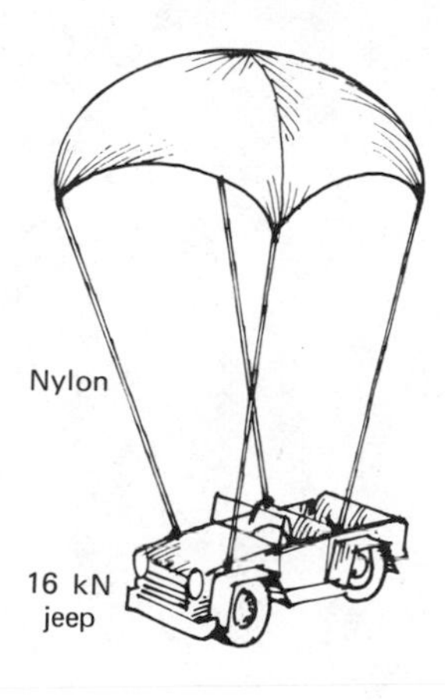

5.29. Four nylon cords attach a 16-kN Army jeep to a parachute. Estimate the minimum diameter of the cords that you would recommend.

5.30. Estimate the maximum slowly applied tension force allowable on a nylon shoestring that is 2 mm in diameter.

5.31. A power transmission cable is supported by a porcelain insulator as shown. The tension in the cable is 1500 lb. Determine the minimum cross-sectional area required for the insulator in compression.

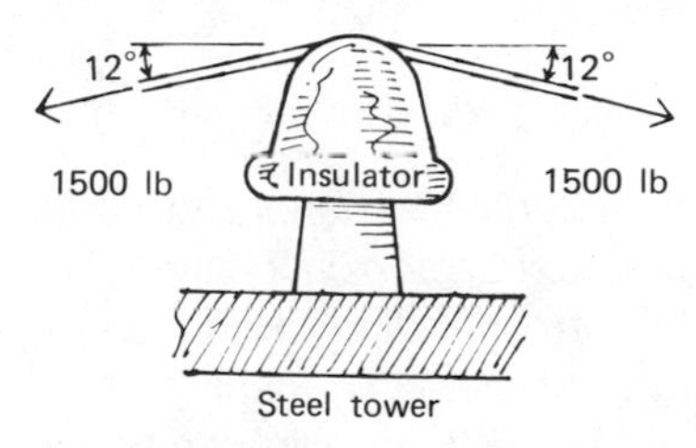

5.32. Determine the minimum cross-sectional area if the insulator in Problem 5.31 is turned over and acts in tension.

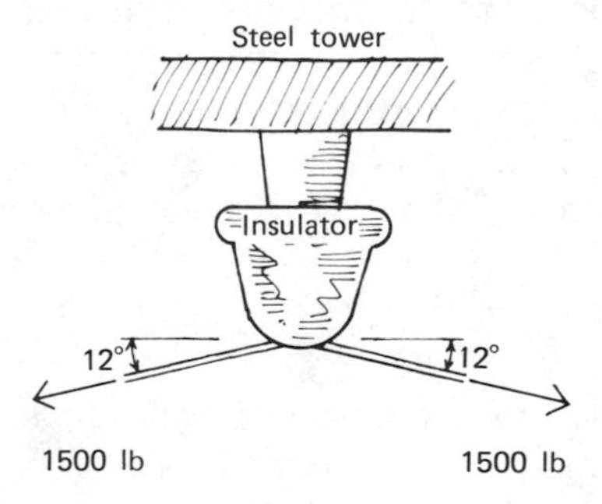

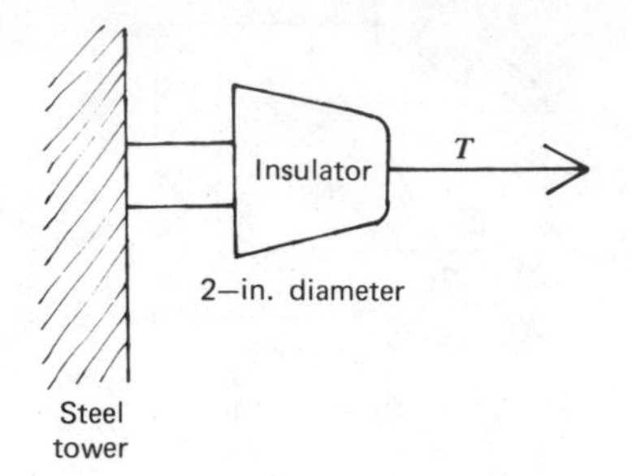

5.33. A polyethylene insulator in a power transmission line is used as a tension link between the metal tower and the conductor. If the minimum diameter of the insulator is 2 in., determine the allowable wire tension.

5.34. A 2.0-mm diameter copper wire is coated with a polyethylene insulation 0.6 mm thick. If the coated wire is subjected to a tension force of 500 N, estimate the stress in the polyethylene coating.

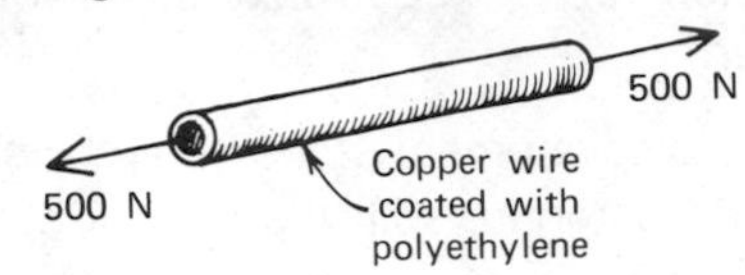

5.4/Thermal Properties of Materials

There are several thermal properties of materials that are of importance to engineers, such as heat capacity, conductivity, expansion and contraction, heat resistance, and thermal shock. These properties are determined by material tests and are available in several reference books. Most thermal characteristics of metals can be reasonably well understood by considering the atomic structure and the primary bonding of the electrons.

5.4.1/Thermal Expansion

Most everyone recognizes that materials expand and contract. The *coefficient of linear thermal expansion* is defined as the change in length per unit length corresponding to an increase in temperature of one degree. This coefficient is a function of material temperature; however, it can usually be considered as a constant. The units are typically inches per inch per degree Fahrenheit or perhaps meters per meter per degree Celsius. Thermal expansion is of primary importance to engineers that design structures. Steel buildings, bridges, and large pipelines must have expansion joints to allow for deformations induced by temperature changes. The thermal strain, ϵ_T, due to a temperature change of ΔT degrees in a material with a coefficient of linear expansion, α, is given as

$$\epsilon_T = \alpha \Delta T$$

This thermal strain is in addition to any other mechanical strain and can be added or "superimposed." If the member is free to move, then the thermal deformation induces no thermal stress. If the member is restrained while the temperature change takes place, then thermal stress is induced and can be computed using Hooke's law.

5.4.2/Melting and Boiling Points

Due to the crystalline nature of metals, they have a definite *melting point* at which the solid changes to a liquid. In contrast, amorphous solids such as some plastics soften gradually over an extended temperature range. Metal alloys usually have different melting points than pure metals. Most steels are an exception since they melt at approximately the same temperature as iron. Steel can be cut with the oxyacetylene torch whereas many of the other alloys cannot. Aluminum is often covered with a thin layer of aluminum oxide, which melts at around 3700 F. Aluminum melts at only 1200 F. If sufficient heat were applied to melt the oxide, then a hole would be burned through the aluminum. A flux is used to reduce the melting point of the oxide.

Boiling is a phase change from a liquid to a gas. The boiling point of materials is often important to engineers. For example, metals with low boiling points evaporate readily in a vacuum as found in outer space and the material experiences "outgassing." Nylon is seldom used in space missions due to outgassing characteristics. Some common adhesives also experience considerable deterioration in outer space.

5.4.3/Thermal Conductivity

The *thermal conductivity* of a material is a measure of the material's ability to transfer heat by conduction. As shown in Figure 5.43, heat flows from hot to cold temperatures through a material according to Fourier's heat conduction law as follows:

$$Q = \frac{KA(T_H - T_L)}{L}$$

where:

Q = heat conducted (Btu/hr)

K = thermal conductivity of the material (Btu/ft·h·F)

A = cross-sectional area of material (ft^2)

T_H = high surface temperature (F)

T_L = low surface temperature (F)

L = material thickness (ft)

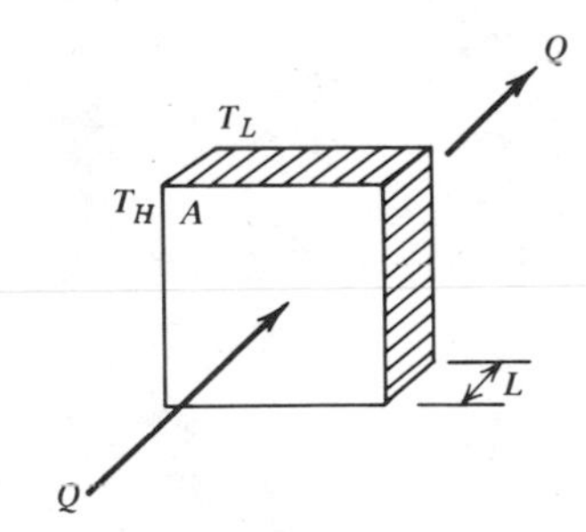

FIG. 5.43. The conduction of heat.

Conventional units are given in parenthesis. Other units could easily be used as long as the equation is dimensionally correct. The temperature difference divided by the thickness $(T_H - T_L)/L$ is frequently called the temperature gradient. Obviously, a high gradient tends to increase the heat flow.

Materials that readily conduct thermal energy are also good conductors of electricity. Poor thermal conductors or insulators are usually poor conductors of electricity. The Fourier conduction law applies to the transfer of heat and not the transfer of electrical energy. Nevertheless, a material microstructure that facilitates the conduction of heat usually allows the conduction of electrical charge.

5.4.4/Thermal Properties of Metals

All metals are relatively good conductors of thermal energy. The thermal energy is transmitted through lattice vibrations of the atoms and the motion of the free electrons involved in the metallic bonds. The thermal conductivity through the electron gas dominates in metals, whereas the lattice vibrations dominate in insulators or dielectrics. In alloys of metals, the thermal conductivity is decreased as compared to the pure metal. The addition of 1 percent manganese into pure copper reduces the thermal conductivity by 66 percent. Any impurities in pure metals also lower the conductivity.

The thermal conductivity in pure metals is essentially independent of moderate temperature changes. The thermal conductivity of alloys decreases as the temperature increases. Does a decrease in conductivity seem strange? After all, the increased temperature increases the velocity of the free electrons and should pass heat more quickly. Yes, indeed, but the lattice vibrations are also increased and increased movements in the lattice tend to block the flow of the electrons. A traffic jam of atom particles takes place.

What are some practical applications of thermal conductivity? Stainless steel cookware is lustrous and catches the eye of the housewife. Yet stainless is such a poor conductor of heat that the pan bottom is usually covered or sandwiched with copper or some thermally equivalent material to conduct heat more readily and uniformly. The thermal conductivity of stainless steel is only about 4 percent that of copper and about 30 percent that of a low-carbon steel. Thermal conductivity in metals also affects welding. Steel is easily welded, but, as the thermal conductivity increases, so does the difficulty in welding. Copper is such a good conductor that it is almost impossible to weld. The heat simply flows away from the spot to be welded. Copper and aluminum are often used for cooling fins to dissipate heat. Some space shuttle instruments use aluminum to conduct heat away from electronic devices and dissipate it into black space. Have you ever slept next to the metal zipper in a sleeping bag and found it to be cold?

5.4.5/Thermal Properties of Ceramics

The relatively strong atomic bonds in most ceramics can tolerate temperature induced vibrations better than metallic materials. These strong attraction forces keep electrons in their proper places and account for the high melting points, low thermal expansions and general thermal stability of ceramics.

Ceramics are nonconductors of thermal energy. The electrons that transmit these energies are more tightly bound to their atomic nuclei through ionic and covalent bonding in ceramics as compared to metals. The ceramic materials that consist only of covalent bonds are even better insulators or dielectrics than those with some ionic bonding.

It is good that nature did provide both conducting and insulating materials. Winter coats, thermos bottles, sleeping bags, houses, cooking handles, and ski boots are all devices that attempt to separate hot and cold temperatures. Without thermal and electrical conductors, it would be difficult to get energy into our homes, yet insulators help keep it there. Some of the common insulating ceramics are asbestos, fiberglass, porcelain, rock wool, glass wool, and earth. The thermal conductivities of some common materials are given in Table D.2.

5.4.6/Thermal Properties of Polymers

It would require a very extensive book to treat all the properties of over a million organic solids in detail. One of the advantages of polymers is the flexibility in material properties available through polymerization techniques. An engineer can usually find a commercial polymer with reasonable design requirements.

The thermal properties of polymers differ from metals and ceramics. Polymers are usually nonconductors. Their thermal conductivities are less than metals by a factor of roughly 10^3 as shown in Table D.3. Polymers usually have melting points below 500 F, whereas metal alloys such as steel melt between 2000 and 2800 F. In contrast to metals, the melting point of polymers is not sharply defined. A general softening of the polymer takes place at higher temperatures. The low melting point is an advantage when polymers are used as a sacrificial coating on heat shields for reentry of space vehicles into the atmosphere. When the ablative polymer on the nose cone of a spacecraft is subjected to high temperatures, it begins to vaporize and hot fragments fly off. The heat required to complete this vaporization is conducted away from the rocket material. The rocket is thereby cooled through the ablation of the polymer.

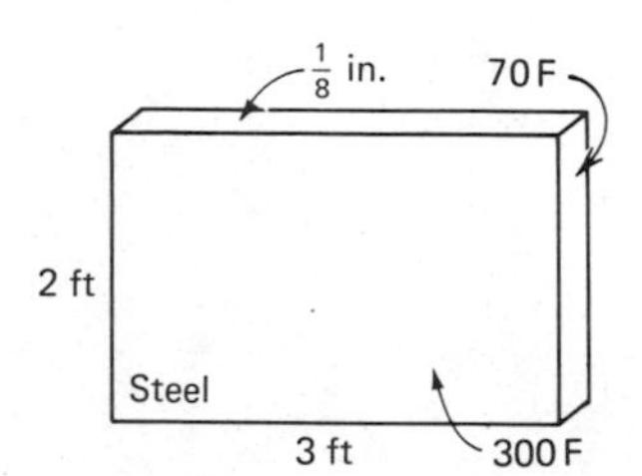

EXAMPLE PROBLEM 5.4

The metal side of a wood burning stove is 3 ft. long, 2 ft high, and $\frac{1}{8}$ in. thick. The inside surface temperature of the plate is 300 F and the outside surface temperature is 70 F. Assume the surface temperatures to be uniformly distributed. Calculate the heat flow through the plate if the material is structural steel.

Solution

The heat conduction can be calculated from Fourier's law

$$Q = \frac{KA(T_H - T_L)}{L}$$

where K = 29 Btu/h·ft· F from Table D.1. Upon substitution into Fourier's law:

$$Q = \left(\frac{29 \text{ Btu}}{\text{h·ft·F}}\right)\frac{(6 \text{ ft}^2)(230 \text{ F})}{(1/8)(1/12) \text{ ft}} = 3.8(10)^6 \frac{\text{Btu}}{\text{h}}$$

Note that if stove were made from aluminum (K = 97 Btu/h·ft·F) rather than steel (K = 29), that the heat conduction would be $\frac{97}{29}$ = 3.3 times greater.

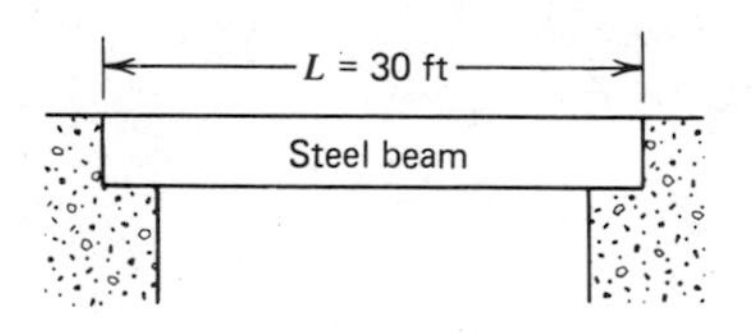

EXAMPLE PROBLEM 5.5

A structural steel I beam 30 ft long is fixed into concrete at both ends and spans a small creek. The cross-sectional area of the beam is 17.06 in.^{2} Determine the stress induced into the beam by a temperature drop of 80 F.

Solution

Since the beam ends are restrained, thermal stresses are induced into the beam when it wants to contract. One method of representing the stress problem is to assume one end unrestrained and compute the total thermal contraction along the beam as

$$\Delta L_T = \epsilon_T L$$

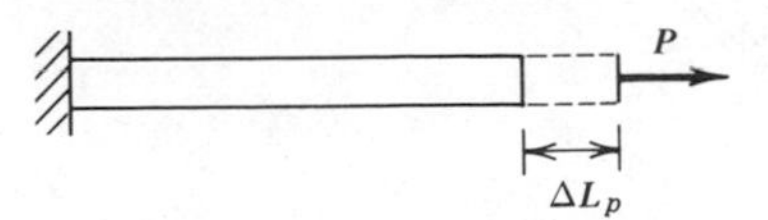

Then, a force P is applied that restores the beam to its original position by moving the beam end a distance ΔL_p. Since

$$\Delta L_p = \epsilon_p L = \frac{\sigma}{E} L$$

and $\Delta L_T = \Delta L_p$, it follows that

$$\epsilon_T L = \frac{\sigma}{E} L \quad \text{or} \quad \sigma = E\epsilon_T$$

which is Hooke's law as expected. Therefore,

$$\sigma = E\epsilon_T = E\alpha\Delta T = 30(10)^6\, 6.6(10)^{-6}\, (80)$$

$$\sigma = 15\,840 \text{ psi Tension}$$

EXAMPLE PROBLEM 5.6

The edge conditions on plate glass windows play an important role in determining the structural strength of the glass. In a particular case, assume that a narrow plate glass window is installed such that two opposite ends are fixed and the other two are essentially free to expand. The air temperature on the day of installation is 100 F. Determine the drop in temperature required to fail the glass.

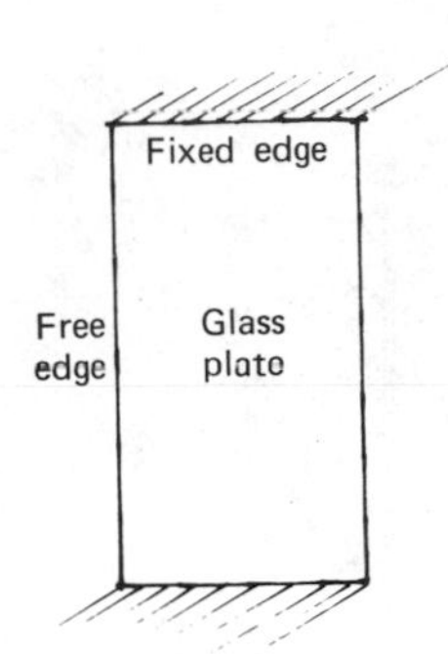

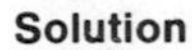

Solution

The glass plate will contract when the temperature decreases. Since two opposite sides are restrained and two free, assume a tension stress in only one direction. The strain can be determined as:

$$\epsilon_T = \alpha\Delta T$$

where $\alpha = 4.9(10)^{-6}$ in./in./F. The stress is related to the strain through Hooke's law

$$\sigma_T = E\epsilon_T$$

Since the allowable tension strength is 6000 psi, and $E = 10(10)^6$ psi for glass, it follows that

$$\sigma_T = E\,\alpha\Delta T$$

$$\Delta T = \frac{6000}{10(10)^6\, 4.9(10)^{-6}} = 122.4 \text{ F}$$

$$T = 100 - 122.4 = -22 \text{ F}$$

The glass plate would theoretically crack and fail at about 22° below zero.

Note that any edge movement that might be provided by some rubber caulking or similar material would eliminate the fixed restraint and allow a broader temperature change.

EXAMPLE PROBLEM 5.7

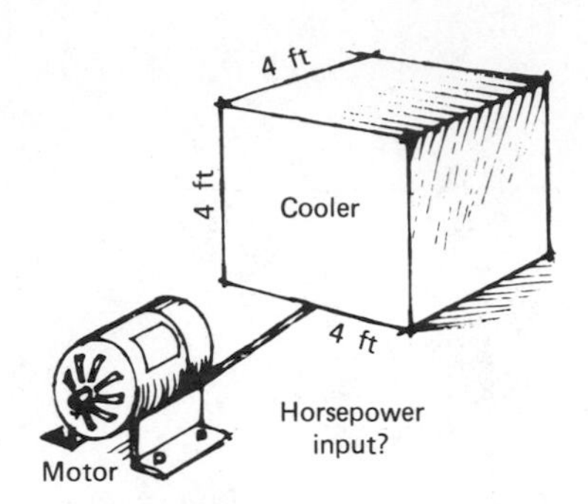

The inside of a box-type cooler is to be maintained at 15 F. The outside of the box is 4 ft $\times$ 4 ft $\times$ 4 ft. The six sides consist of a thin metal inside liner and 2 in. of mineral wool insulation. The outside surface temperature of the insulation is 70 F. Estimate the input motor horsepower required if the motor-refrigeration system is 50 percent efficient. In other words, 2 Btu/h are required to remove 1 Btu/h from the box.

Solution

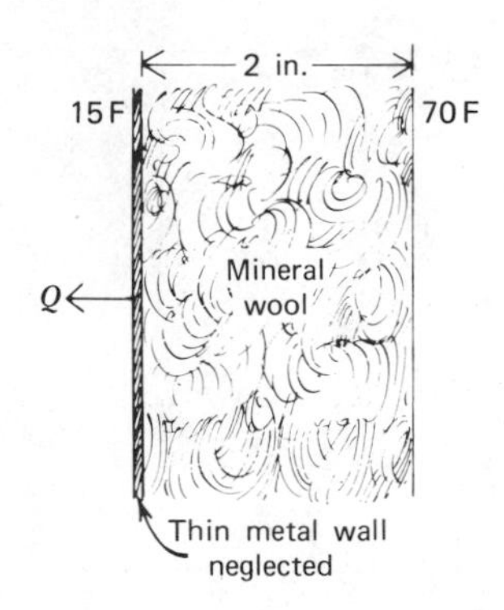

The heat conducted through one wall of the box can be computed using Fourier's law of heat conduction. Since the metal liner is thin and a good conductor, it can be assumed to have a constant temperature throughout of 15 F. Essentially, the metal liner is neglected since it provides negligible insulation compared to the mineral wool. The total heat flow through the six sides can then be estimated as six times the flow through a single wall. It follows that

$$Q = \frac{kA}{L}(T_h - T_L) = \frac{(0.025)(4 \times 4)}{(2)(1/12)}(6)(70 - 15)$$

$$Q = 792 \text{ Btu/h}$$

A more exact analysis would account for edge and corner effects that were neglected in this estimate. At steady-state conditions, 792 Btu of heat flow into the box each hour. Since the refrigeration system is only 50 percent efficient, 1584 Btu/h must be used to remove the heat and maintain a temperature of 15 F. Converting to horsepower, it follows that the input energy required is

$$1584 \frac{\text{Btu}}{\text{h}} \frac{\text{hp-h}}{2545 \text{ Btu}} = 0.62 \text{ hp}$$

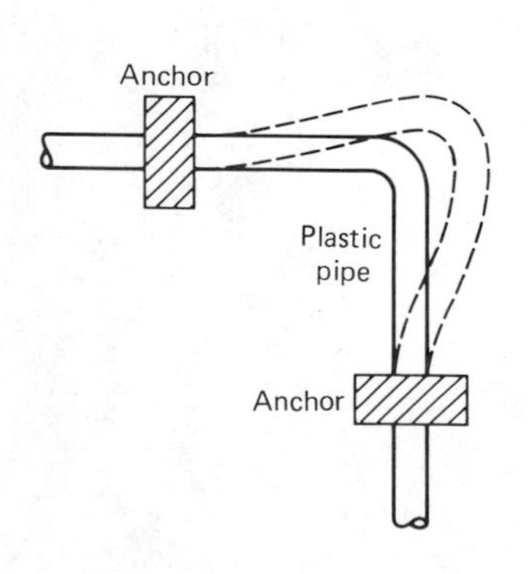

EXAMPLE PROBLEM 5.8

A polyvinylchloride pipe makes a 90° bend as shown. The end joints are anchored. High thermal stress levels can be initiated into the pipe at the anchors due to forced bending if sufficient length is not provided. Determine the total elongation of the pipe for a temperature increase of 60 F if the length between anchors is 4 ft.

Solution

The amount of thermal strain can be computed as

$$\epsilon_T = \alpha \Delta T$$

$$\epsilon_T = 55(10)^{-6}(60) = 3.3(10)^{-3} \text{ in./in.}$$

where α is found in Table D.3 as $55(10)^{-6}$/F. The total elongation ΔL follows as

$$\Delta L = \epsilon_T L$$

$$\Delta L = 3.3(10)^{-3}(4)(12) = 0.16 \text{ in.}$$

PROBLEMS

5.35. What thickness of concrete would be required to provide insulation equivalent to plate glass $\frac{1}{4}$ in. thick?

5.36. An appliance top is coated with a typical porcelain that has a compression strength of 49 100 psi and a tension strength of 2440 psi. Is the danger from sudden temperature changes at the surface (thermal shock) more severe for heating or cooling? Explain.

5.37. A lumber store sells both steel and aluminum casings for basement windows. They are identical, except for material, and must be placed in an 8-in. concrete wall. Compare the heat lost by conduction through the casings.

5.38. The bottom of a 14-in. diameter aluminum (6061-T6) frying pan is $\frac{1}{8}$ in. thick. Determine the thickness of a stainless steel frying pan equal in diameter if the heat conducted through the bottom is to be the same.

5.39. A stainless steel handle is connected to a stainless steel cooling pan by a $\frac{1}{4}$-in. steel bolt that is pretensioned to a stress of 1000 psi. When heated, will the bolt become tighter? Determine the bolt stress for a temperature increase of 110 F.

5.40. A high-pressure, high-temperature cast iron pipe line is connected at the flanges by bolts. Select a material for the flange bolts and justify your decision.

5.41. A schedule 40, 2 in. nominal diameter pipe has an outside diameter of 2.375 in. and an inside diameter of 2.067 in. The line is used to carry water at a temperature of 180 F and it passes through a room at 68 F. Estimate the heat lost by conduction per foot of pipe if the pipe material is cast iron.

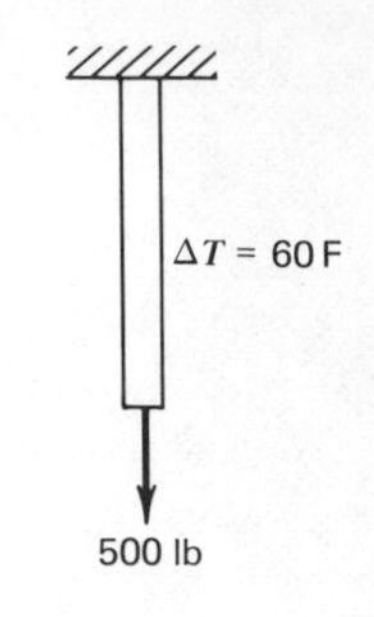

5.42. Determine the lineal expansion of a 10-m section of steel railroad track if the ends are unrestrained.

5.43. A $\frac{1}{2}$-in. diameter structural steel rod is 2 ft long. Determine the elongation if a weight of 500 lb is added to the end and the temperature is increased by 60 F.

5.44. A structural steel I beam has a cross-sectional area of 3.81 in^2. The beam is 12 ft long when under zero stress. The beam is positioned between end restraints with an initial stress of 500 psi tension. Determine the total stress in the beam if the temperature is increased by 60 F.

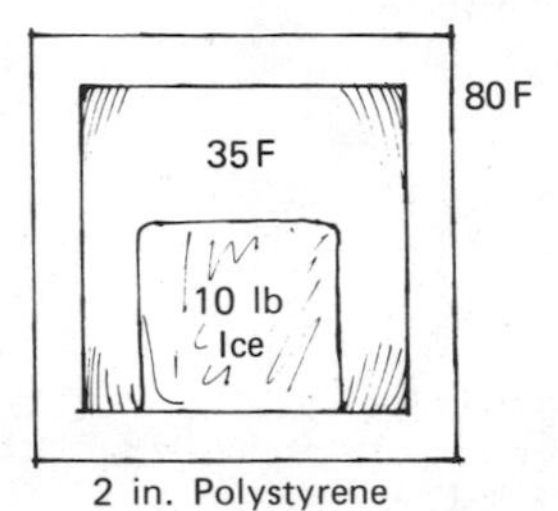

5.45. The total area of a polystyrene box-type cooler is 30 ft^2. The top, bottom, and sides are 2 in. thick. A 10-lb block of ice is inside. If 144 Btu are required to melt 1 lb of ice, estimate the time required to melt the ice if the outside surface temperature is 80 F. Assume the inside temperature to be constant at 35 F.

5.46. A plate glass window is $\frac{1}{2}$ in. thick. Estimate the percent reduction in heat flow if the plate glass is replaced with a polystyrene sheet 2 in. thick. Assume the same area and surface temperatures.

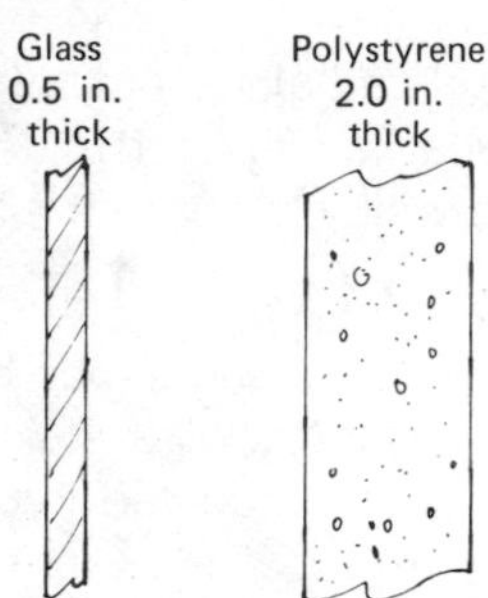

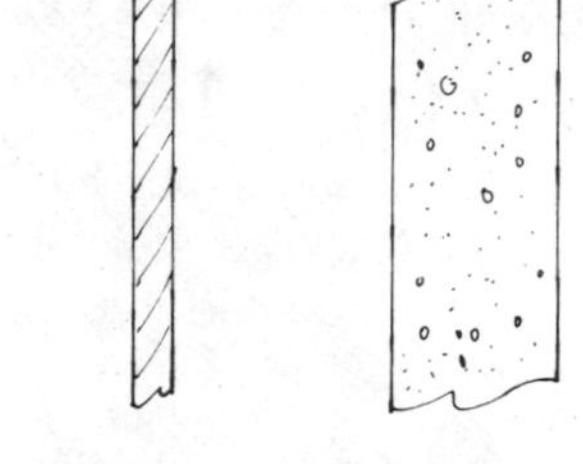

5.47. Explain how an ablative heat shield on the nose cone of a rocket dissipates heat.

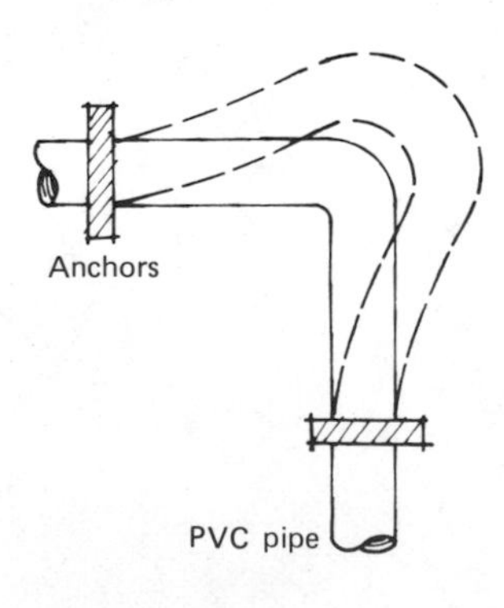

5.48. A 6-in. nominal diameter schedule 40 PVC pipe has a 6.625-in. O.D. and 6.065-in. I.D. The pipe bends 90° and is anchored with end connections as shown. Determine the total expansion of the pipe if the temperature is increased 70 F. The pipe length between anchors is 3 ft.

5.49. A solid aluminum (18-8 annealed) door on a bomb shelter is 80 in. high, 30 in. wide, and 2 in. thick. When closed, the door is surrounded by concrete. Determine the space between the door and the concrete at 68 F such that the door will just touch the concrete at a temperature of 150 F. Assume that the door only experiences a temperature change.

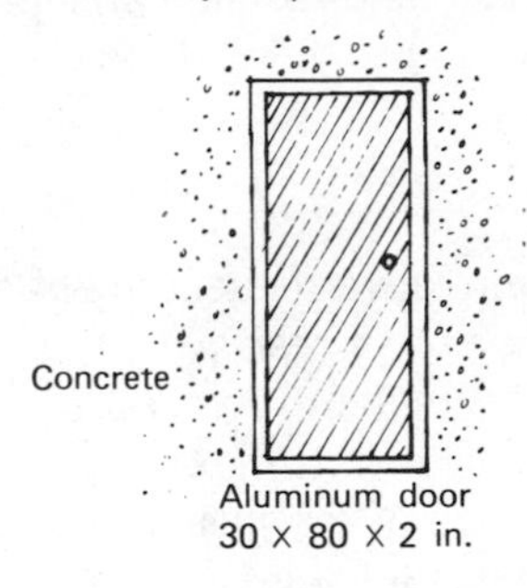

5.50. A porcelain insulator is restrained on both ends by mechanical fasteners. Estimate (a) the increase in temperature and (b) the decrease in temperature that will fail the insulator by exceeding its mechanical strength. Assume the modulus of elasticity to be $12.5\,(10)^6$ psi.

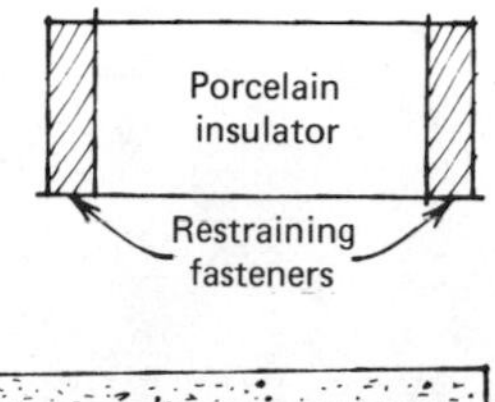

5.51. A 5-lb block of ice is placed in a picnic cooler. The cooler is 24 in. $\times$ 18 in. $\times$ 12 in. (outside dimensions) and is made from a fiberboard 2 in. thick with $k = 0.035$ Btu/h·ft·F. If 144 Btu are required to melt 1 lb of ice, estimate the time required to melt the 5-lb block of ice. The outside surface temperature of the cooler is 80 F and assume the inside temperature to be constant at 40 F.

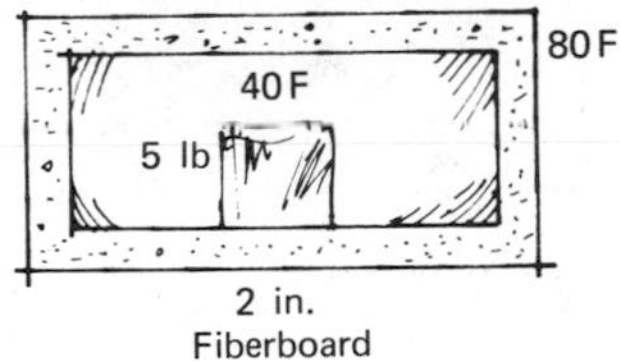

5.52. The thermal conductivity of foamglass insulation at 0°C is $1.3(10)^{-4}$ cal/cm s °C. Determine the thickness of a glass plate equivalent to 2 in. of foamglass insulation (1 Btu/h·ft·F = 0.4132 cal/s·m·°C).

5.53. What thickness of plate glass would provide insulation equivalent to 3 in. of glass wool?

5.5/Electrical Properties of Materials*

Some electrical properties of materials are somewhat analagous to thermal properties. The conduction of electricity is the result of a flow of electrons or ions through the crystalline lattice and is similar to the flow of thermal energy. Pure metals are the best conductors. Internal crystalline lattice vibrations, interstitial impur-

*Optional reading.

ities, lattice defects, alloying atoms, and other microstructural phenomena restrict the flow of electricity and result in an internal resistance.

5.5.1/Resistivity

The material property that provides a relative measurement of a material's opposition to current flow is called *resistivity.* Resistivity is expressed in terms such as ohm meters (ohm·m) or ohm centimeters (ohm·cm) and depends only on the material and not its dimensions. Materials with high values of resistivity are called *conductors.* An *insulator* or *dielectric* has a low resistivity.

The resistivity of copper is low and consequently copper is used extensively in electrical wires. Copper switches are popular since heat is dissipated easily and the terminals are not welded from an electric flash. Copper is about three times heavier than aluminum and is slightly heavier than iron, yet has a lower tensile strength. Copper has an electrical conductivity about eight times that of steel, yet has a tensile strength roughly five times less than low-carbon structural steel. To help overcome some of these related problems approximately 250 copper alloys are available. They are divided into the brass (copper + zinc) and bronze (copper + tin) groups.

The conductivity of aluminum is less than one-half that of copper, yet aluminum has much better structural properties and is only one-third as heavy. It is often used in large power transmission lines. A copper conductor usually needs an additional carrier cable such as steel for structural support, whereas aluminum can provide both.

The electrical resistivity of polyethylene is about 10^{20} times that of a metal conductor, and it is used as an insulation coating on some electrical wire. Proper insulation requires engineering. For example, the selection of materials to be used in the insulation of electricity is a highly specialized field and includes both printed circuit boards and wire. In the case of multiconductor cables as shown in Figure 5.44, the purpose of the dielectric or insulating polymer is to separate the conductors. The electrical signal flowing might be of low voltage such as found in communication cables, or it might be a high-voltage power transmission line. The electrical insulation must be designed in accordance to the signal strength and frequency, operating temperatures, magnetic fluxes, mechanical bending, and cost. The conduction and other electrical properties of polymers are not independent of the electrical signal being transmitted.

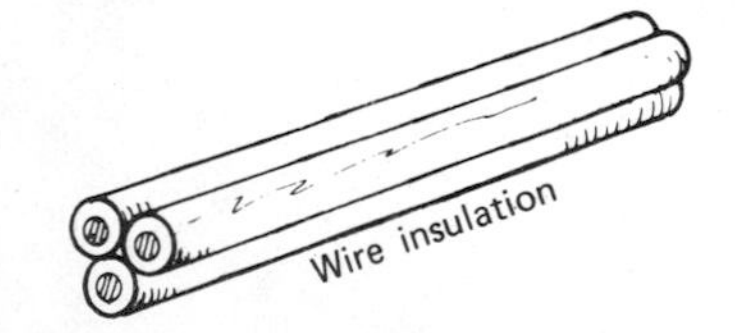

FIG. 5.44. Electrical conductors are coated with dielectric materials.

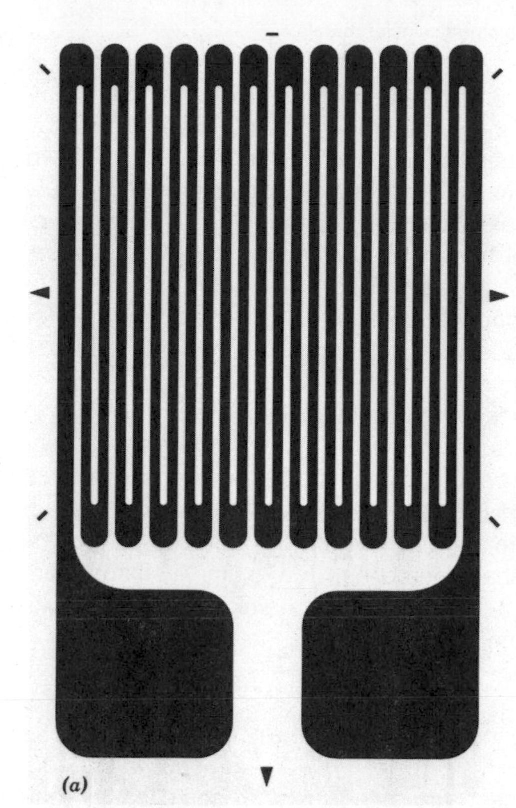

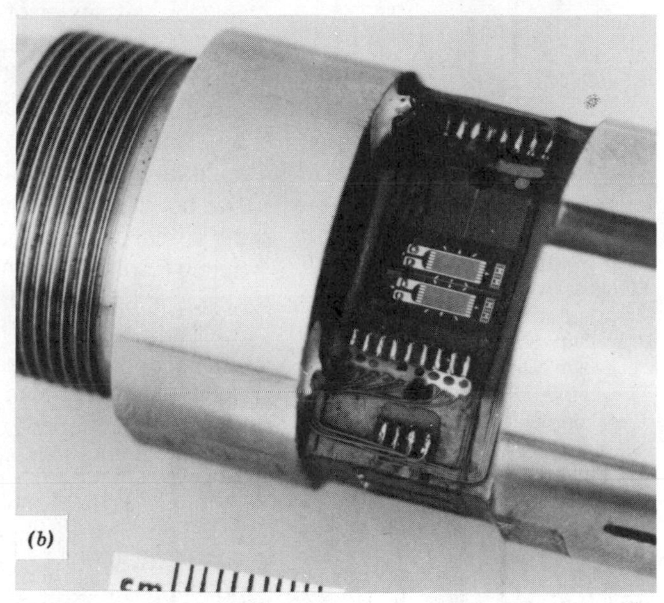

FIG. 5.45. A foil-type strain gage and a typical application. (Courtesy of Magnaflux Corporation)

An interesting application of the resistivity property is found in a foil strain gage as shown in Figure 5.45. A conducting grid is etched onto a small backing with terminals. The gage is then bonded to a material somewhat like a postage stamp. The bonded gage adheres to the material surface. As the material is stretched or compressed, the foil conductor decreases or increases in cross-sectional area due to Poisson's effect. This area change in addition to the change in the material's resistivity, results in a resistance change that can be measured as a small output voltage. Such a strain gage converts a minute mechanical displacement to an electrical voltage and is called a transducer. Strain gages may be used to measure strain, stress, force, displacement, and acceleration.

5.5.2/Corrosion

The corrosive nature of metals can also be discussed under the general area of electrical properties. *Corrosion* can be defined as

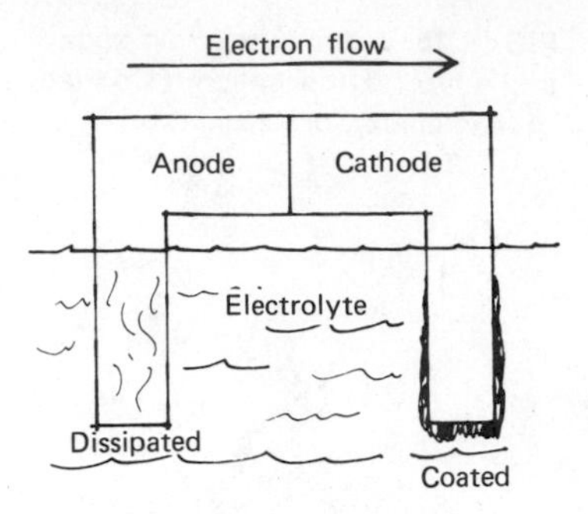

FIG. 5.46. Galvanic corrosion—dissimilar metals.

TABLE 5.3. Galvanic Series of Metals

Anodic end (Corroded)
Magnesium
Magnesium alloys
Zinc
Aluminum
Aluminum alloys
Low-carbon steels and iron
Cast iron
Stainless steel (active)
Lead
Tin
Nickel (active)
Brasses
Copper
Nickel alloys
Nickel (passive)
Stainless steel (passive)
Silver
Titanium
Gold
Cathodic end (Protected)

Note: In most cases, the passive metal should be selected if the electrolytic environment includes oxygen. In the absence of oxygen, choose the active metal.

the dissipation of a solid body through chemical or electrochemical action starting on the surface. Most metals are corroded by water and the atmosphere through electrochemical action. If two dissimilar metals are in proximity and are immersed in a liquid solution, a transfer of electrons may take place, as shown in Figure 5.46. The material that loses electrons is called the *anode,* and the material that gains electrons is the *cathode.* The conducting fluid is the *electrolyte.* During the electrochemical transfer of electrons, the anode material dissipates and the cathode material is coated. The behavior of materials under these corrosive conditions has been experimentally determined as is given in the so-called galvanic series shown in Table 5.3. The anodic tendency increases from bottom to top. In other words, the material above is anodic and dissipates; whereas, the lower material on the list is cathodic. The corrosive action will increase as the dissimilarity of the materials increases. For example, magnesium and copper will have a stronger electrochemical action than will iron and copper. In both cases, copper is the cathode.

As a practical example, aluminum nails on titanium sheets would eventually corrode away. The corrosion of the nails is excessive since the anodic area is small relative to the cathode as shown in Figure 5.47. Corrosion would also occur at the connection of copper pipes onto steel tanks. In some cases, galvanic-type corrosion can act on a single metal if cracks, dirt, or other barriers create concentration differentials, as shown in Figure 5.48.

Protection against corrosion in metals can be achieved by painting, vitreous enameling, or other coatings that isolate the anode from the cathode. In Figure 5.49 an underground steel pipe is protected by placing a sacrificial anode, such as magnesium, adjacent to the pipe. Any steel surfaces that might be exposed by scratches in the pipe coating would become the cathode. The magnesium anode would corrode as electrons passed through

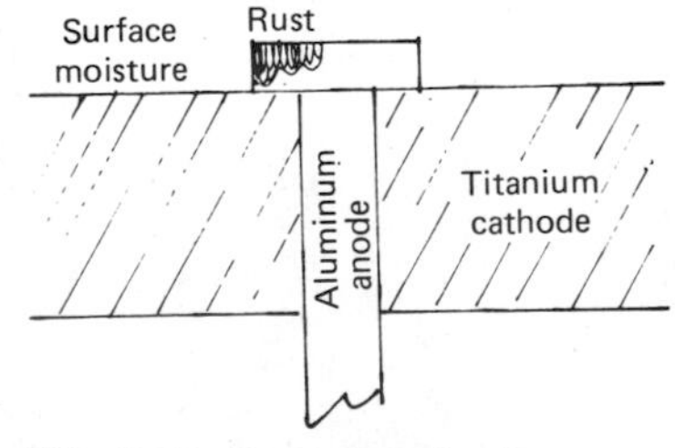

FIG. 5.47. Corrosion of nails.

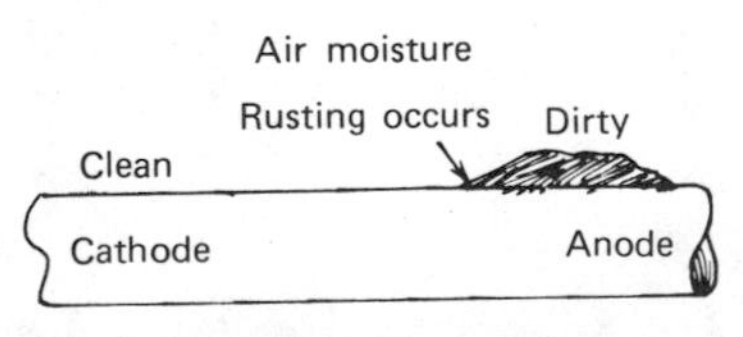

FIG. 5.48. Galvanic corrosion—surface differentials.

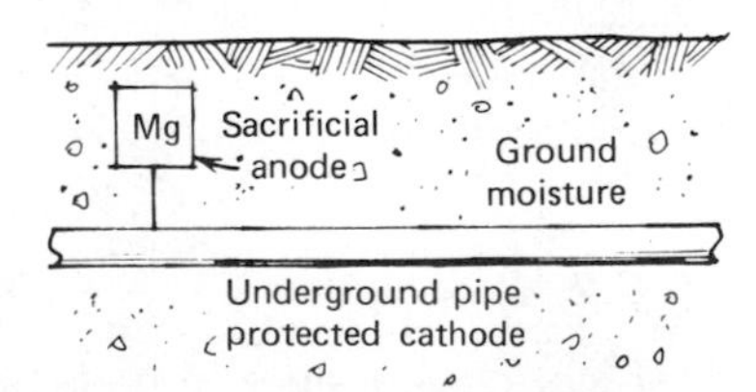

FIG. 5.49. Cathodic protection.

the underground moisture that serves as the electrolyte to the steel pipe.

There are several other electrical properties that assist engineers in selecting materials for light bulb filaments, heating coils, microelectronic circuits, photoconduction cells, and others. These characteristics will possibly be presented in future courses in engineering.

PROBLEMS

5.54. Aluminum sheets are nailed to a camper shell. Should you select lead or zinc-coated steel nails?

5.55. A bimetallic strip consisting of copper and magnesium are placed in water. Which of the two materials will oxidize (lose electrons) and therefore corrode away?

5.56. A sacrificial anode is to be attached inside a steel water tank to inhibit corrosion of the tank. Suggest a metal that might be used as the sacrificed anode.

5.57. Steel rivets are used to attach aluminum sheets. Water condenses and puddles around the rivets. Which of the two metals will corrode due to galvanic action?

5.6/Material Properties and Bioengineering*

In the field of bioengineering, materials have been used to design artificial organs to replace the heart, kidneys, joints, and lungs, for example. In fact, interdisciplinary teams of engineers and scientists are now technically capable of duplicating the function of most human organs, provided that there are no limitations such as size, maintenance, and cost. One of the primary challenges that separates success in the laboratory from successful "in vivo" results is that of materials. Epithelial tissue and the polymers of the human body seem to reject the presence of foreign transplants. The polymers of the body don't like synthetic materials. Interest in bioengineering has focused on the properties of materials that affect tissue compatibility rather than typical engineering properties.

Since the mid 1950s considerable progress has been made in using polymers and their acceptance by the human body. Many

*Optional reading.

of these improvements are the result of trial and error. In the early 1960s most of the pacemakers for the heart failed within 2 years—much to the dismay of some. Since that time, silicone-coated pacemakers with improved batteries last much longer. Cellophane has served continuously for over 30 years as a dialysis membrane to help remove the metabolic wastes normally removed by kidneys. Research efforts have attempted to increase cellophane's wet strength through reinforcing fibers, cross linking, and the addition of other organic materials. Silicone rubber has been used in blood flow, since it doesn't trigger natural clotting as readily as others. Silicone rubber has also been used in plastic surgery for such things as artificial chins. Polyvinylchloride is used to restore burned areas of skin. Teflon is used in artificial tendons to connect bone and muscle. The dental mercury-zinc and silver amalgams that accounted for roughly 80 percent of all fillings for teeth are being replaced by aromatic thermosetting acrylates that have compression strengths up to 50 000 psi—sufficient to chew both the meat and the bone of a beefsteak. What's more, they don't drop out after contracting from a cold drink. The natural fibers of silk and cotton are nonabsorbable when embedded in tissue, yet they lose their strength within six months. Nylon sutures will last much longer and have been shown to have 70 percent of their original strength after 11 years inside the body.

In view of the intensified research in the area of transplants, one might wonder if life could be extended indefinitely through the exchange of worn-out body parts. Perhaps so, yet there are some properties of living tissue that are still beyond our synthetic modeling capability. For example, bone is a natural composite that includes conductors of blood, organic collagen fibers for tensile strength, and inorganic mineral deposits that assist in compression loadings. In addition, however, human bone has the capability of self-regeneration and completely replaces itself several times during a normal lifetime—something our bridges, buildings, prostheses, and other structures are not yet capable of doing.

PART B/SELECTED TOPICS IN MATERIALS

5.7/Home Materials and Heat Transfer

The science that deals with the flow of energy that takes place as a result of a temperature difference is commonly referred to as *heat transfer.* Materials and heat transfer in the thermal design of homes is becoming more important. Mathematical models have been developed that predict the heat transfer rate as a function of temperature gradients, material constants, and dimensions. The three basic mechanisms by which heat is transferred are conduction, convection, and radiation.

Conduction heat transfer is the flow of energy through a material as a result of a temperature gradient. Heat flows from a high-temperature source to a low-temperature source somewhat similar to water flowing downhill—the steeper the gradient, the faster the flow. The conduction heat transfer equation was developed earlier, but the equation is repeated in Figure 5.50 where Q is the heat flow rate in British thermal units per hour (Btu/h), A is the cross-sectional area normal to the direction of the heat flow, K is the thermal conductivity of the material in Btu/h·ft·F, L is the material thickness in feet, T_H and T_L are the high and low temperatures, respectively, in degrees Fahrenheit. The thermal conductivities for some common materials are given in Table 5.4.

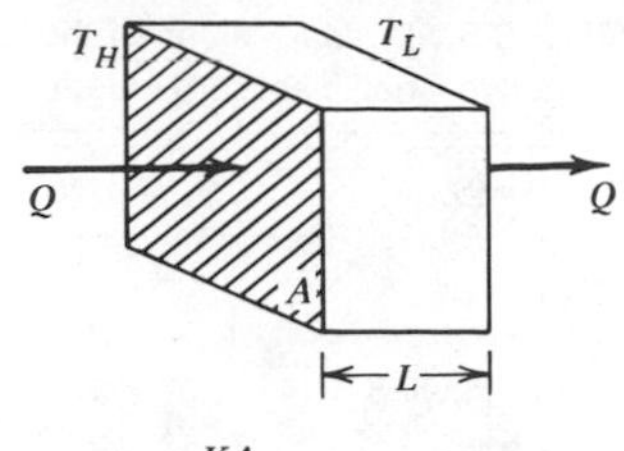

$$Q = \frac{KA}{L}(T_H - T_L)$$

FIG. 5.50. Conduction heat transfer.

TABLE 5.4. Thermal conductivity of Some Common Building Materials at 75 F

Material	K Btu/hr·ft·F
Copper	223
Steel	31
Concrete	0.80
Window glass	0.45
Brick	0.40
Fir	0.063
Sawdust	0.034
Glass wool	0.022
Cellulose	0.019
Air (dead)	0.015
Sheetrock	0.092
Vapor seal	0.029

Convection heat transfer occurs when a moving fluid, such as air or water, flows past a material and carries the heat away. If a hot metal plate were exposed to a moving air stream, it would cool more rapidly than if it were exposed to the same air under stagnant conditions. The end of your nose is much colder on a windy day as compared to a still day at the same temperature. One would expect the heat transferred through convection to be dependent upon the velocity of the fluid, the fluid thermal conductivity, density, and perhaps other things. The effect of all these variables is determined by experimentally measuring the heat convected under controlled conditions. The results can be generalized into an overall film conductance commonly referred to as a *convective coefficient.* The overall heat transfer rate can then be expressed in the form of Newton's law of cooling as shown in Figure 5.51, where Q is the heat transfer rate in British thermal units per hour, A the surface area in square feet, h the overall convective coefficient in Btu/h·ft²·F, T_H the surface temperature and T_L the fluid temperature in degrees Fahrenheit. Rather than measure the convective constant each time, one uses the results of other investigators. The values of convective coefficients are often available in various handbooks for particular problems. In the cases of houses, convective coefficients are given in Table 5.5. For example, the value of the convective constant, h, on an outside wall of a house is 6 Btu/h·ft²·F on a day when the wind is blowing at 15 mph.

Convection heat transfer

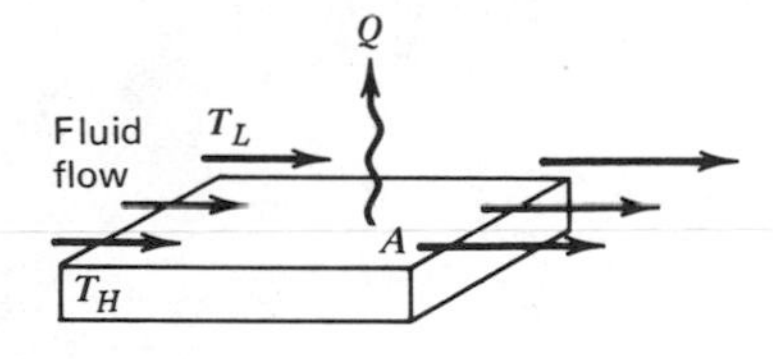

$$Q = hA(T_H - T_L)$$

FIG. 5.51. Convection heat transfer.

Radiation is the third heat transfer mechanism, and, unlike conduction and convection, no transporting medium is required.

TABLE 5.5. Typical Values of the Convective Coefficient for Houses

Surface	h $Btu/h \cdot ft^2 \cdot F$
Outside wall	
15 mph wind	6
7.5 mph wind	4
0 mph wind	2
Inside wall	1.4
Inside ceiling	1.6

Thermal radiation is part of the electromagnetic radiation spectrum that also includes radio waves, infrared light, visible light, ultraviolet light, and others. Electromagnetic waves are emitted from bodies in the form of thermal radiation as a result of the motion of the body's molecules and atoms. This emission of energy from a body is equivalent to the removal of heat. The rate of heat transfer is proportional to the fourth power of the absolute temperature. The convective coefficients given in Table 5.5 include corrections for thermal radiation effects.

The heat conducted through a single homogeneous body can be determined using Fourier's heat conduction law. The heat convected at the interface of a solid body and a flowing fluid can be determined using Newton's law of cooling. However, the prediction of the total heat flow through a *composite* body such as the wall of a house with different materials and different convective coefficients on both the inside and outside surfaces needs to be considered.

To simplify matters, define a term called *thermal resistance.* In the case of conduction, the temperature difference would represent the thermal potential or *driving force.* The heat flow is analagous to an electric current, and the thermal resistance R would be L/K as shown in Figure 5.52. For convection, the thermal potential would again be the temperature difference, but the thermal resistance would be $R = 1/h$ as shown in Figure 5.53.

Now an important point! In steady state, the heat that flows into the material must also flow through the material and out. Otherwise, the temperature would increase. Since the heat conducted through all sections must be the same, the total resistance to conduction through a composite wall can be determined by adding the individual resistances in series as shown in Figure 5.54. Once the total resistance has been found for the composite wall, then the total heat transfer can be determined using only one equivalent resistance.

In the case of a single-pane window, the heat flowing from the inside to the outside experiences three thermal resistances. As shown in Figure 5.55, convection takes place at the inside window surface where the temperature is T_H and the convective coefficient is h_1. The heat then conducts through the glass, which has a thermal conductivity of K and a thickness L. It is then convected away on the outside surface where the temperature is T_L and the convective coefficient h_2. The total resistance is the sum of all three resistors and includes both convection and conduction. Note that a large value of R indicates a high resistance to heat flow.

Thermal resistance—conduction

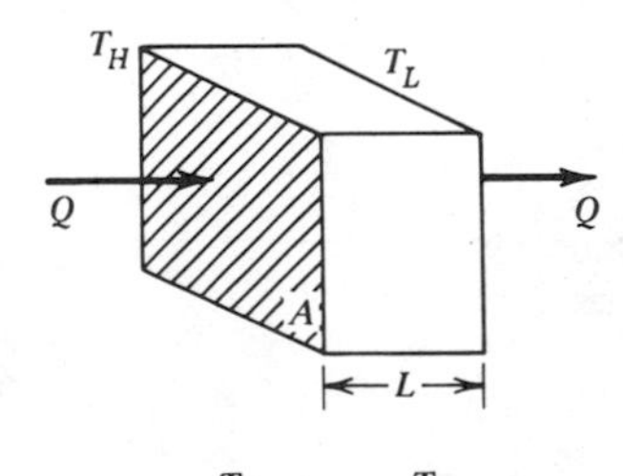

T_H R T_L

Q/A Q/A

$$\frac{Q}{A} = \frac{K}{L}(T_H - T_L) = \frac{1}{R}(T_H - T_L)$$

$$R = L/K$$

FIG. 5.52. Thermal resistance—conduction.

Thermal resistance—convection

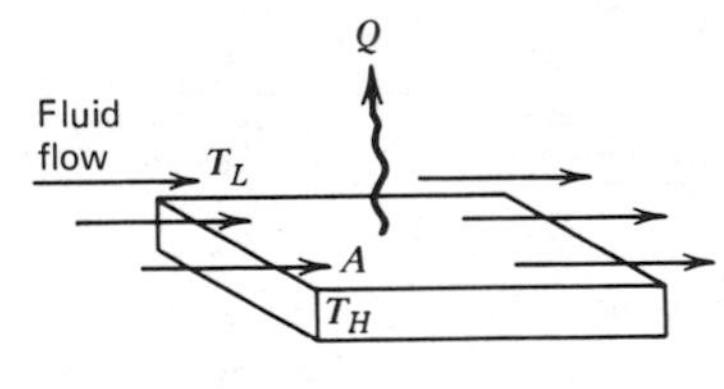

$$\frac{Q}{A} = h(T_H - T_L) = \frac{1}{R}(T_H - T_L)$$

$$R = 1/h$$

FIG. 5.53. Thermal resistance—convection.

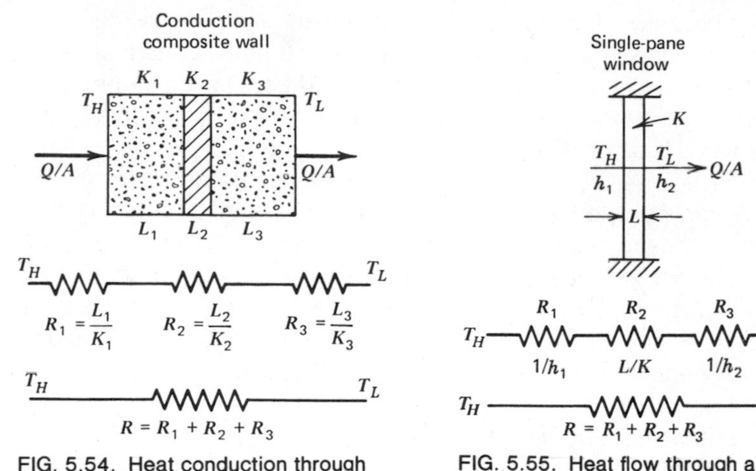

FIG. 5.54. Heat conduction through a composite wall.

FIG. 5.55. Heat flow through a window.

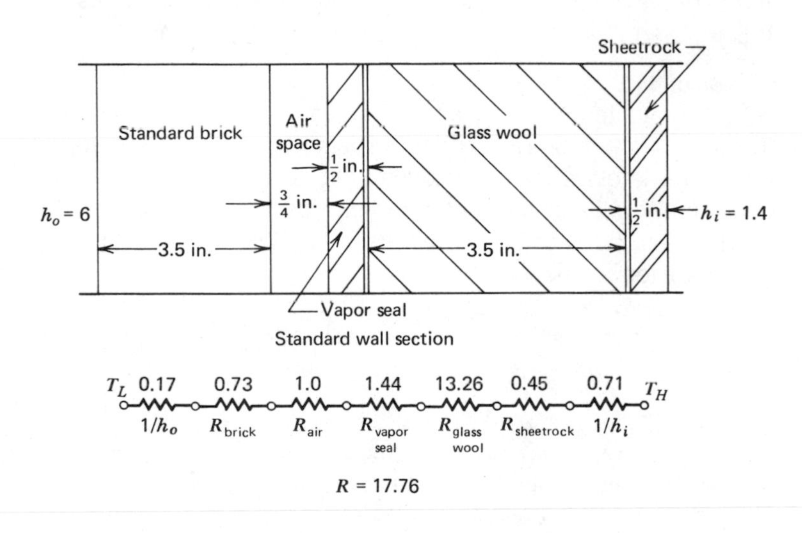

FIG. 5.56. Heat resistance for a standard wall of a house.

The total thermal resistance for a standard wall section of a house exposed to a 15 mph outside wind can be calculated as shown in Figure 5.56. The inside wall is made of Sheetrock and is nailed to 2 × 4 wooden studs; the space between studs, roughly 3.5 in. thick, is filled with glass wool for insulation. A vapor seal is nailed to the outside of the studs. An air space of roughly $\frac{3}{4}$ in. exists between the vapor seal and the outside brick. Note that the thermal conductivity of dead air given in Table 5.4 is very low. This low K value for dead air is achieved only if the air is completely stagnant. The air space between the vapor seal and the brick is not stagnant as a result of natural convection currents. Consequently, such entrapped air spaces have a thermal resistance that includes both conduction and convection. An equivalent R value for the air space was determined experimentally and a value is simply given as $R = 1.0$.

The relative values of thermal resistances given for a wall section indicate the value of insulation. Standard brick offers little resistance to heat flow. In fact, 1 in. of glass wool offers a thermal resistance equivalent to three inches of fir, 18 in. of brick, 20 in. of glass, or 36 in. of concrete. However, mass and specific heat are sometimes considered when temperature fluctuations between night and day are important.

EXAMPLE PROBLEM 5.9

Determine the heat lost per unit area through a double-pane window as shown. The inside temperature is 70 F and the outside temperature is 10 F. The glass is 0.16 in. thick and the panes are 0.5 in. apart. The outside wind is approximately 15 mph.

Solution

The total thermal resistance is determined using values from Tables 5.4 and 5.5. The thermal resistance of the air entrapped between the glass plates includes both conduction and convection. The R value of the entrapped air does not appear on standard tables, but can be estimated to be 0.80 h·ft²·F/Btu. The heat transfer can now be found as

$$\frac{Q}{A} = \frac{T_i - T_o}{R} = \frac{70 - 10}{1.71} = 35.08 \frac{\text{Btu}}{\text{h}\cdot\text{ft}^2}$$

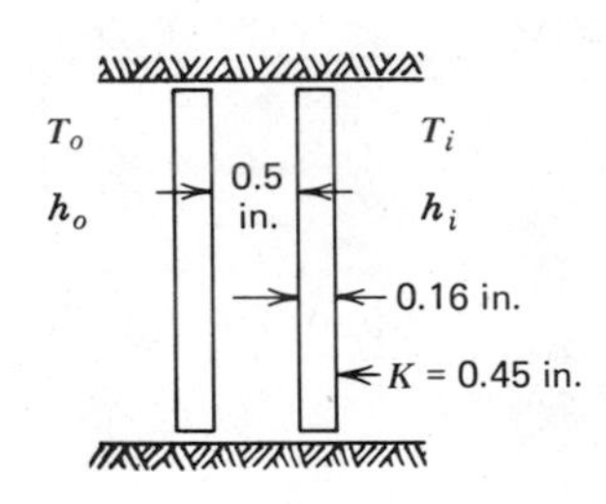

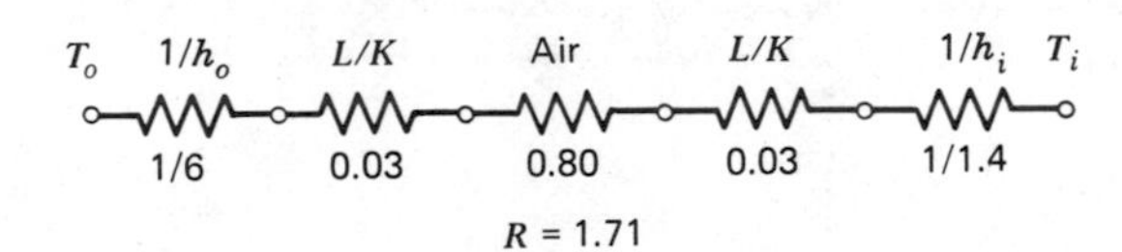

EXAMPLE PROBLEM 5.10

Determine the amount of energy lost in an 8-hour period through a wall 8 ft high × 10 ft long in a brick veneer home. Assume a standard wall section. The inside temperature is 70 F and the outside temperature is 10 F. Assume the wall has glass wool insulation between the 2 × 4 studs and is 3.5 in. thick. The outside wind blows at 15 mph.

Solution

The heat transfer rate is found from the equation:

$$\frac{Q}{A} = \frac{T_L - T_o}{R}$$

From Figure 5.55, the value of R is 17.76. The area is 80 ft^2. Upon substitution,

$$Q = \frac{(70 - 10)80}{17.76} = 270.3 \text{ Btu/h}$$

In an 8-hour period,

$$Q = (270.3)(8) = 2162 \text{ Btu}$$

PROBLEMS

5.58. Determine the percentage reduction in heat lost through a double-pane window as compared to a single-pane window. The double-pane window has two glass panes $\frac{1}{8}$ in. thick that are separated by an air space $\frac{1}{2}$ in. thick. The single pane glass is also $\frac{1}{8}$ in. thick. Assume that the outside wind is 7.5 mph and $R = 0.80$ h·ft^2·F/Btu for the air space.

5.59. Determine the rate of heat transfer per square foot through a typical noninsulated wall section in a brick veneer home when the inside temperature is 70 F and the outside temperature is −20 F. Approximate the thermal resistance R of the air space replacing the glass wool between the studs in Figure 5.56 to be 1.0 h·ft^2·F/Btu. Determine the percentage reduction in the heat transfer rate through the wall if 3.5 in. of glass wool insulation are added.

5.60. Compare the rate of heat transfer per square foot through a single-pane window $\frac{1}{4}$ in. thick with that through a standard wall with 3.5 in. of glass wool insulation. The inside and outside temperature is 70 F and 5 F, respectively. The outside wind blows at 15 mph.

5.61. Estimate the annual cost for heating a "typical" home in the Rocky Mountains that requires $100(10)^6$ Btu/yr and uses electricity.

5.62. Compare the heat lost per square foot through a 6-in. thick concrete wall with the heat lost through a typical wood frame home with glass wool insulation. Consider conduction only.

5.63. The ceilings of a particular house are made of $\frac{1}{2}$-in. Sheetrock. The house floor space is 1500 ft^2. Assume that $50(10)^6$ Btu/yr flow through the ceiling with no additional insulation. If coal costs \$55/ton and yields 12 000 Btu/lb, then determine the cost of the heat lost through the ceiling only.

5.64. If the house in Problem 5.63 were insulated with 12 in. of glass wool on top of the Sheetrock, *estimate* the savings in heat otherwise lost through the ceilings. Assume that the average temperature differential is the same.

5.65. A fire inside a log cabin burns 1.5 lb of coal in one hour. Assume that 10 percent of the heat generated flows through 4 wooden walls 8 ft $\times$ 10 ft and 1 ft thick. The temperature outside is -30 F. Determine the temperature on the inside of the wall. The inside temperature is uniform and the wind outside is about 8 mph.

5.66. Design a sleeping bag that will maintain an inside temperature of 75 F when the outside temperature is -5 F. Specify the material and thickness. Assume that a sleeping adult will generate approximately 90 W.

5.8/Materials and Structural Design

The challenge of an engineer engaged in the design of a structure is to create an assemblage of materials that will support a given external loading system as efficiently as possible. There are many alternative combinations of materials, configurations, sizes, and costs that might be selected. The engineer must be familiar with the different structural alternatives.

In general, complete structures or components of structures are identified by such basic terms as cables, beams, membranes, plates, and trusses. These terms describe the building blocks or structural alternatives that are available to the design engineer and are characterized by the internal force system that is developed in response to the applied external loads. One must understand the internal force system in order to determine the type and magnitude of material stresses that are developed. These material stresses must be less than the ultimate strength of the material in order to prevent failure. Considerable time and effort is expended by both the student and practicing engineer to understand the internal stresses that are developed in different types of structures. In the following paragraphs, some general discussion is presented to students that will introduce concepts and give the student some idea as to future challenges in preparing for professional engineering.

A *cable* is a very efficient-type structure in that the load-carrying capacity is high compared to the material required. In designing a cable, one assumes the internal force system to consist of tensile forces only. Both longitudinal and transverse external forces are supported by the axial internal tension force that is developed in the cable. By use of trigonometry and the equations of equilibrium, one can determine the magnitude of the internal force that is necessary. The internal force depends significantly on the angle of the applied force. A cable can carry a much larger force as a direct pull as compared to a force that acts transversely. If an element is removed from the cable as shown in Figure 5.57, the internal forces are in the direction of the cable and the normal stresses are distributed uniformly across the cross section. In most applications, there are no other internal stresses that need to be considered. In selecting a cable, the diameter required is usually of most concern. If the external force and the yield strength of the cable material are known, then the cross sectional area required is found by using the equation for normal stress. Obviously, the long cable has no moment rigidity. Transmission lines, guy wires, suspension bridges, and pulley systems are examples of cable applications.

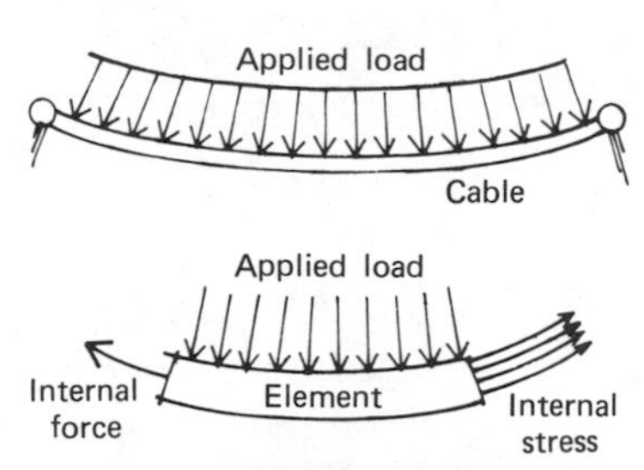

FIG. 5.57. Cable stress.

A beam, or flexural member, is another general type of structure that is commonly used in engineering. The term *beam* describes a member that is subjected to loads applied transverse to the long axis. A beam need not be horizontal. A long vertical member would be a beam if the applied loads were transverse and a column if the applied loads were in the axial direction of the member. Beams are usually given some special names that depend on the end conditions. For example, a beam that is supported on both ends by pins, rollers, or smooth surfaces is called a simple beam. A beam that is built into a wall or otherwise fixed on one end and free on the other end is called a cantilever beam.

In the case of a beam, the transverse loads tend to bend the beam. In contrast to the cable, the beam must offer some internal resistance to bending. An arbitrary section from a simple beam with internal forces and moments is shown in Figure 5.58. Internal moments of magnitude M_1 and M_2 along with vertical shear forces V_1 and V_2 that act on a small element of the beam are shown. Magnitudes M_1, M_2, V_1, and V_2 are unknown; however, they could be found using the equations of equilibrium. The important point to remember is that beams experience internal moments as well as internal forces.

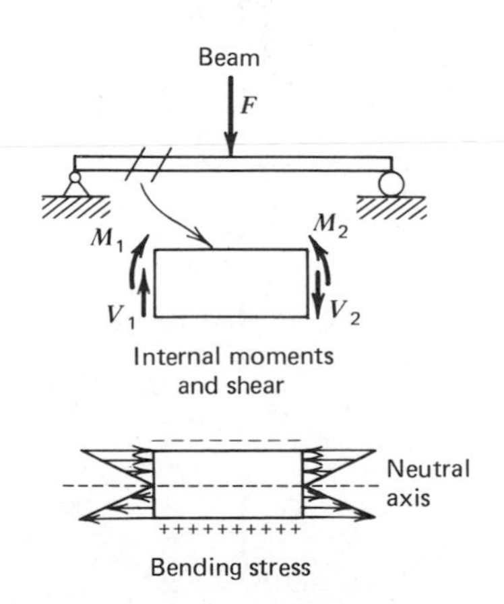

FIG. 5.58. Beam stress.

A thorough analysis of beam stresses is included in subsequent courses in engineering, and is beyond the scope of this introductory text. At this point, however, one can observe the bending action and get some understanding of the flexural stress developed. In Figure 5.58 the applied force would bend the beam such that the top fibers of the beam would be in compression and the bottom fibers would be in tension. Since the stress on the top is in a direction opposite to that on the bottom, the stress is obviously not uniform across the cross section. In fact, there must exist some point where the stress is zero. As one intuitively expects, it has been shown that the maximum stress appears on the outside surfaces. The center plane has zero stress and is referred to as the neutral axis. The magnitude of the maximum flexural stress depends on the applied loads, the beam length and the cross section of the beam. The determination of bending stresses in uniform beams is rather routine for most engineers; however, the stress analysis of nonuniform beams and beams of composite materials can become rather involved.

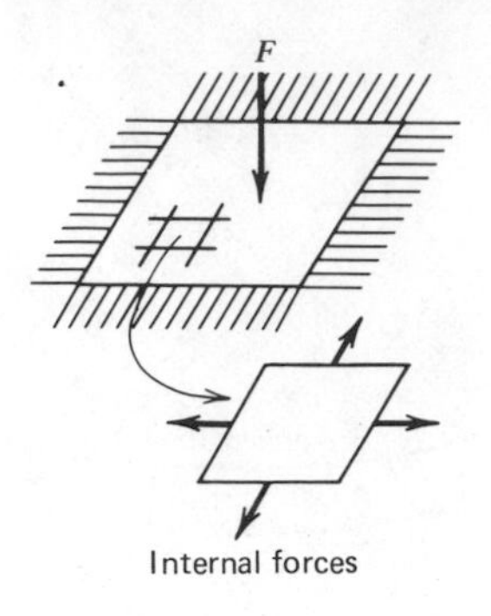

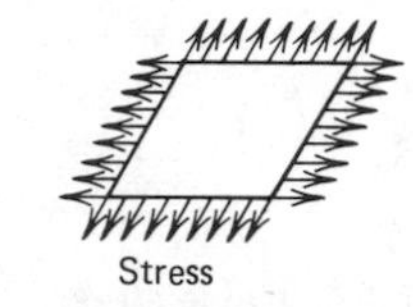

FIG. 5.59. Membrane stress.

The membrane is a thin-walled structure that is often used to carry transverse loads. In essence, a membrane is similar to a cable or a string in two dimensions. As shown in Figure 5.59 the membrane has internal tension only and offers no internal resistance to bending. The membrane forces are independent of bending and are determined by the equations of static equilibrium. A diaphragm is an example of a physical structure that would be analyzed as a membrane.

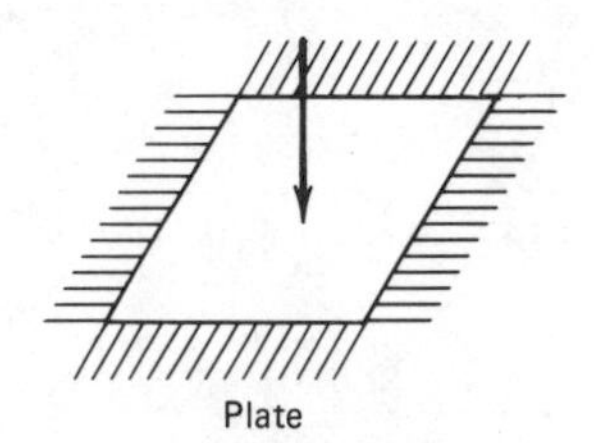

FIG. 5.60. Plate element.

A plate is another commonly used structure. Similar to the beam, a plate supports transverse loads through the action of internal moments and forces that resist bending. The internal stresses are a superposition of stresses that result from such things as internal bending moments, twisting moments and shear. A rectangular plate fixed on all sides and subjected to a concentrated load at the center is shown in Figure 5.60. The internal forces, moments and stresses are rather complicated and are not shown.

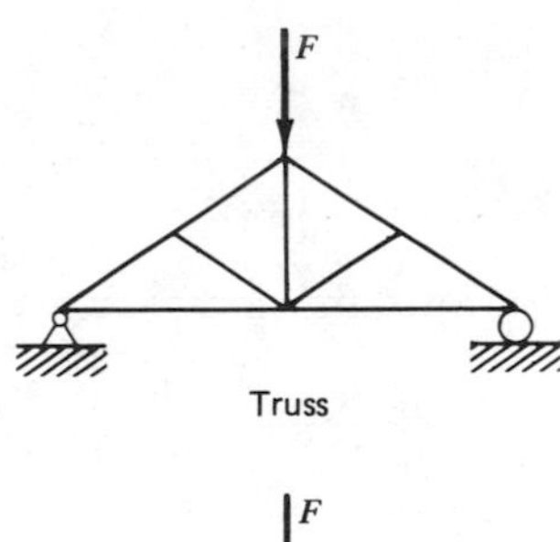

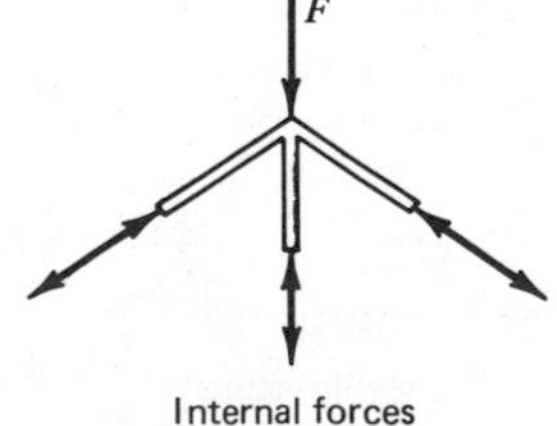

FIG. 5.61. Truss forces.

The truss is a basic type of structure that is used to support external loads. A truss consists of an assemblage of straight members that are pin connected at their extremities. External loads are applied only at the joints. It was shown in Chapter 4 that truss members are two force members and as a consequence experience no internal bending. Members of an ideal pin connected truss experience only axial tension and compression as shown in Figure 5.61. A truss is a very efficient use of materials and is commonly used in structural design.

The foregoing discussion of cables, beams, membranes, plates, and trusses provides an overview of some of the load-carrying characteristics of these structural types. The selection of the "best" structure for a particular situation depends on several constraints and decisions that must be evaluated at the beginning of the project. The design of a beam, membrane, and plate require relationships between external loads and internal stresses that have not been introduced. In the case of cables and some simple trusses, sufficient information has been given in Chapter 4 to provide students with the rudimentary design equations. In most truss design situations, the span and the external loads are known and the engineer must decide on the truss assemblage. This decision usually takes several trial-and-error attempts before an efficient assemblage is determined. Each time a different placement of truss members is considered, some calculations must be made to make certain that material yield stresses and so-called critical buckling loads have not been exceeded. The juggling of materials, member lengths, member cross sections, member placements, and so on, to obtain an efficient, dependable truss represents a challenge. In order to design a truss, the concept of buckling is introduced. A student is then prepared to complete a design of a simple house truss. An example problem is provided. Furthermore a student might be interested in using design concepts to actually build a model bridge.

The stress and strain formulas previously developed apply to members loaded in axial tension or compression and the stress magnitude in the material is independent of the member length. In the case of long members loaded in compression such as a column, a structural failure may occur before the compressive stress yield point is reached. For example, a long wire loaded in axial compression will deflect laterally into an unstable configuration before any stress failure is experienced. This instability due to lateral deflection from the equilibrium position is called *buckling* and is shown in Figure 5.62. The magnitude of the axial compressive load that is required to buckle a column is referred to as the critical buckling load, P_{cr}.

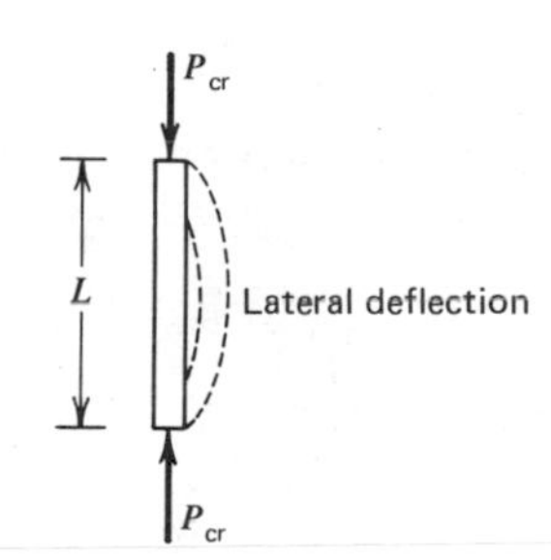

FIG. 5.62. Buckling.

The first theoretical equation to predict the critical buckling load in long, slender columns was published in 1757 by Leonard Euler, a Swiss mathematician. Since that time, several empirical formulas and empirical modifications of Euler's equation have been developed that include the effect of eccentric loading, different end conditions, and other departures from the ideal column. A thorough discussion of the development and application of these column formulas is usually contained in subsequent engineering courses. Some introductory experience in column design can be obtained, however, by attempting to understand

Euler's equation in a conceptual way. If one pushes on a flexible slender column, the buckling force can be roughly determined and mentally registered by the investigator. If the investigator holds the column at the midpoint, which reduces the unsupported length by a factor of $\frac{1}{2}$, the axial load required for buckling increases roughly by a factor of 4. After some thought, other parameters in addition to length that affect the buckling of the column include the cross section of the column and the type of material. Euler's equation specifies the relation of these parameters as:

$$P_{cr} = \frac{\pi^2 EI}{L^2}$$

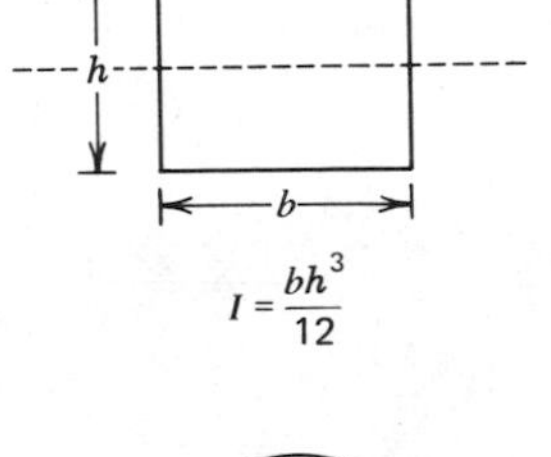

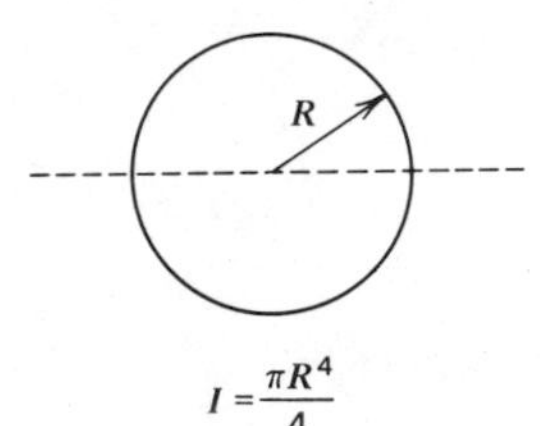

FIG. 5.63. Area moments of inertia.

where E is the material modulus of elasticity, L is the unsupported length, and I is the *area moment of inertia.* The area moment of inertia is a term that relates the effect of the cross section on the column rigidity. The moments of inertia for a rectangle and a circular cross section are given in Figure 5.63.

There are several structures that have obvious columns, such as buildings and bridges. There are several other structures, such as frames, trusses, and mechanisms that include long members in compression and require the column-design criteria. In the design of a truss for example, in addition to the stress, the engineer must also calculate the critical buckling load for long compressed members. Both the stress and the load required for buckling must be less than the yield stress and the applied load, respectively. A factor of safety is usually required.

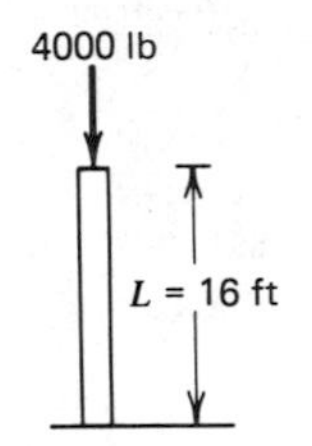

EXAMPLE PROBLEM 5.11.

A 4-in. square wood column 16 ft long is loaded in compression with a load of 4000 lb. Determine (a) the normal stress and (b) the maximum load before buckling would occur. Assume E $= 1.6(10)^6$ psi.

Solution

Determine the normal stress:

$$\sigma = \frac{F}{A} = \frac{4000}{4 \times 4} = 250 \text{ psi C}$$

Determine the critical buckling load:

$$P_{cr} = \frac{\pi^2 EI}{L^2} \qquad I = \frac{bh^3}{12} = \frac{(4)(4)^3}{12} = 21.3 \text{ in.}^4$$

$$= \frac{\pi^2(1.6)(10)^6(21.3)}{(16 \times 12)^2} = 9124 \text{ lb}$$

EXAMPLE PROBLEM 5.12

The truss shown is to be made from wooden 2 × 4's and will be used in the roof of a house. The external loads are given and the total span is 24 ft. Determine the internal force in each member. Determine the maximum stress and compare with the allowable stress for wood given as 1600 psi. Also check buckling. Assume $E = 1.6(10)^6$ psi.

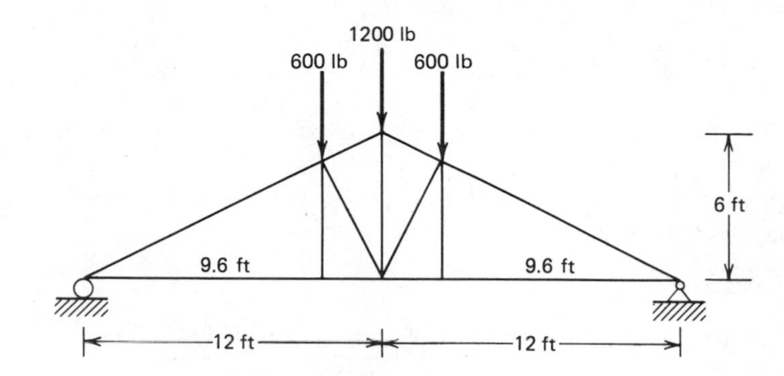

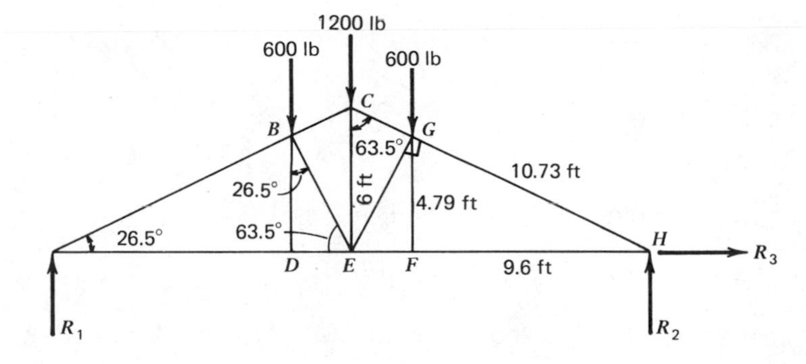

Solution

Draw a FBD of the complete structure and show forces, lengths and angles. Take advantage of symmetry, and recognize that joint B is identical to joint G, and so on. Apply the equilibrium equations to determine the truss reaction at A and H.

$$R_1 = R_2 = 1200 \text{ lb}$$

$$R_3 = 0$$

Draw a FBD of joint A and determine F_{AB} and F_{AD}. Assume F_{AB} compression and F_{AD} tension arbitrarily.

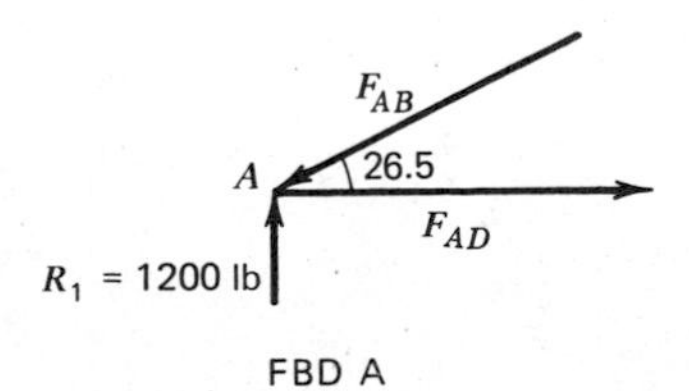

$$\Sigma F_v = 0$$

$F_{AB} \sin 26.5° - 1200 = 0 \qquad F_{AB} = 2689.3 \text{ lb C}$

$\Sigma F_H = 0$

$-F_{AB} \cos 26.5° + F_{AD} = 0 \qquad F_{AD} = 2406.7 \text{ lb T}$

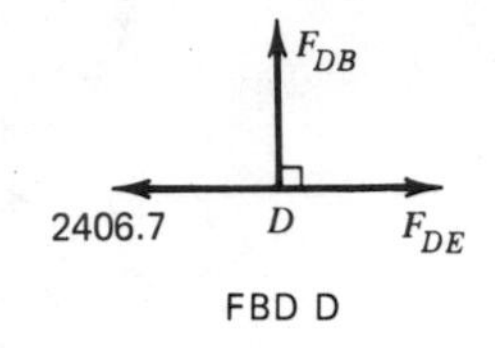

FBD D

Draw a FBD of joint D and determine F_{DB} and F_{DE}

$\Sigma F_v = 0$

$F_{BD} = 0$

$\Sigma F_H = 0$

$-2406.7 + F_{DE} = 0 \qquad F_{DE} = 2406.7 \text{ lb T}$

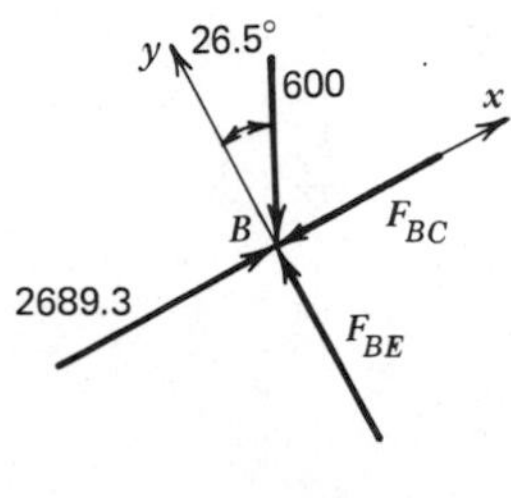

FBD B

Draw a FBD of joint B and determine F_{BC} and F_{BE}. Use an inclined axis for convenience.

$\Sigma F_y = 0$

$+F_{BE} - 600 \cos 26.5° = 0 \qquad F_{BE} = 537 \text{ lb C}$

$\Sigma F_x = 0$

$-F_{BC} - 600 \sin 26.5 + 2689.3 = 0 \qquad F_{BC} = 2421.6 \text{ lb C}$

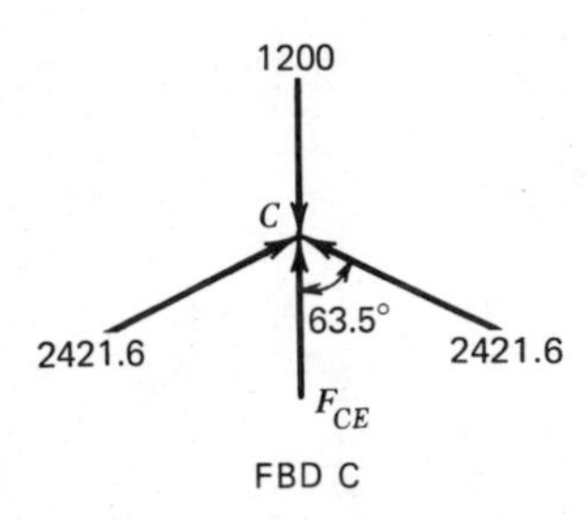

FBD C

Draw a FBD of joint C and determine F_{CE}

$\Sigma F_v = 0$

$(2)\ 2421.6 \cos 63.5 - 1200 + F_{CE} = 0$

$F_{CE} = -961 \qquad F_{CE} = 961 \text{ lb T}$

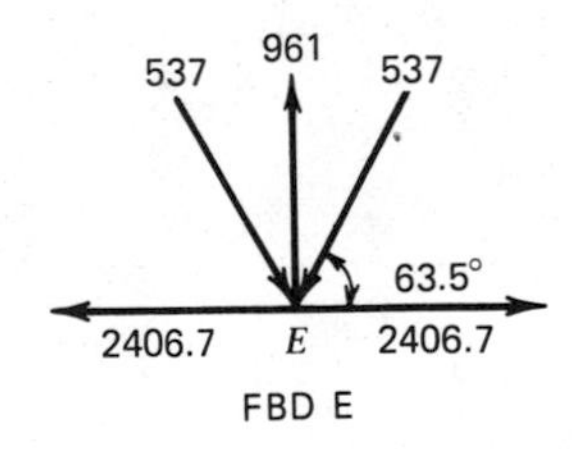

FBD E

Draw a FBD of joint E and check calculations

$\Sigma F_y = 0$

$-961 + (2)\ 537 \sin 63.5 \approx 0$

$\Sigma F_H = 0$

The internal forces are summarized as shown.

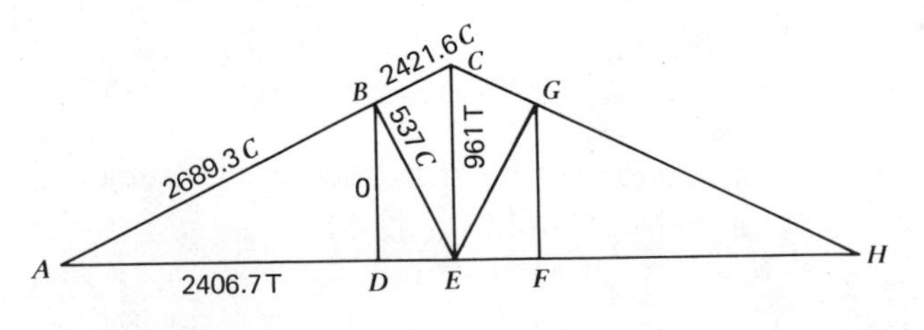

The maximum stress will occur in the member with maximum load, namely member *a*.

$$\sigma_a = \frac{F_a}{\text{Area}} = \frac{2689.3}{2 \times 4} = 336 \text{ psi}$$

$$336 < 1600 \text{ psi} \therefore \text{safe}$$

To check for buckling, consider member *a* since it has the maximum compression load and is also the longest.

$$P_{cr} = \frac{\pi^2 EI}{L^2}$$

To determine the area moment of inertia, recall the equation

$$I = \frac{bh^3}{12}$$

where b is the base (longest side) and h the height. For the 2 × 4, $b = 4$, and $h = 2$.

$$I = \frac{(4)(2)^3}{12} = 2.66$$

$$P_{cr} = \frac{\pi^2(1.6)(10)^6(2.66)}{[(10.73)(12)]^2} = 2533.6 \text{ lb}$$

Buckling will occur since $F_{AB} > P_{cr}$.

Note: Since member *AB* would buckle, a new assemblage would have to be considered. Perhaps the addition of a new member to shorten the unsupported length of member *AB* could be a solution.

PROBLEMS

5.67. Refer to Problem 4.35. Determine the stress and critical buckling load in member *AB* if wooden 2 × 4's are used.

5.68. Refer to Problem 4.37. Determine the stress and critical buckling load in member *AC* if the truss is made from round structural steel rods 25 mm in diameter.

5.69. In Problem 4.36, determine the minimum round cross-sectional area required for members *AB* and *CD* if the material is structural steel. The yield strength for structural steel is 30 000 psi.

5.70. The truss in Problem 4.33 is to be made from structural steel. All members are to be uniform in cross section. Select the minimum square cross-sectional area that could be used with a safety factor of 2.

5.71. A total volume of 3 m^3 is required for a sealed container in outer space. The internal pressure is 14.7 psi and the external pressure is zero. Describe the most efficient shape that will withstand the pressure differential with the least material.

5.72. Design a truss to be made of wooden 2 $\times$ 4's that will be used in the roof of a house. The span required is 30 ft. The local building code requires that the truss carry a minimum live load of 40 lb/ft^2 and a minimum dead load of 20 lb/ft^2. The trusses will be placed on 2-ft centers. Note: the total load of 60 lb/ft^2 will have to be approximated with concentrated loads placed at the joints. Assume that the yield strength for wood is 1600 psi and $E = 1.6(10)^6$ psi.

5.73. Design and construct a model bridge according to the specifications provided by your instructor.

5.9/Cement and Concrete

Concrete is a well-known composite material that consists primarily of aggregates of rock and sand bonded together by portland cement. The raw materials of concrete are usually categorized as ceramics; however, the nonhomogeneous mixture itself is not always associated with glass, vitrified silica, or other commonly accepted ceramics. The hardness, low ductility, insulating strength, and low-expansion characteristics of ceramics are certainly applicable to cured concrete.

On a cost per pounds basis, concrete is less expensive than brick, wood, steel, nylon, plastic or about any other structural material. It is used extensively in foundations, floors, walls, and roofs of buildings. Many miles of highways, pipelines, canals, ditches, and airfields are made from concrete. Dams, bridges, football stadiums, nuclear reactors, and vaults are other examples of concrete structures. In addition to being inexpensive, concrete is essentially fireproof, provides soundproofing, is easily maintained, the supply is essentially unlimited, and it is available in all parts of the world. One concern about concrete is the amount of energy required to process the clay and limestone to produce the portland cement. Approximately $(10)^6$ Btu of heat are required to produce one barrel of cement. This heat would be available from about 83 lb of coal.

Civil engineers frequently use the words *quality* and *workability* in describing concrete. The term *quality* refers to the strength of the concrete and its ability to withstand deteriorating forces such as freezing and thawing, wetting and drying, and chemical and environmental attacks. A quality concrete must also have dimensional stability. For example, a concrete that would shrink excessively while drying would be difficult to use in the construction of a foundation. A beam made from concrete that had randomly varying strength characteristics would certainly present problems.

The *workability* of concrete is a measure of its ability to be poured, moved about inside the forms and vibrated to completely fill the desired space without excessive voids or "honeycombs." A cement finisher who complains about a concrete mix being too "stiff" is complaining about the workabilty of the concrete. The slump test is a common method used to determine workability as shown in Figure 5.64. A truncated cone-type bucket is carefully filled with wet concrete mix. The container is slowly removed. The slump is then measured as the drop from the top of the original 12-in. truncated cone to the top of the concrete. A particular concrete might be referenced as one with a 3-in. or a 10-mm slump.

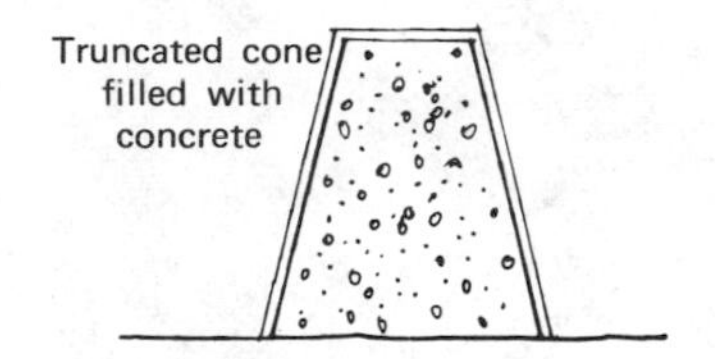

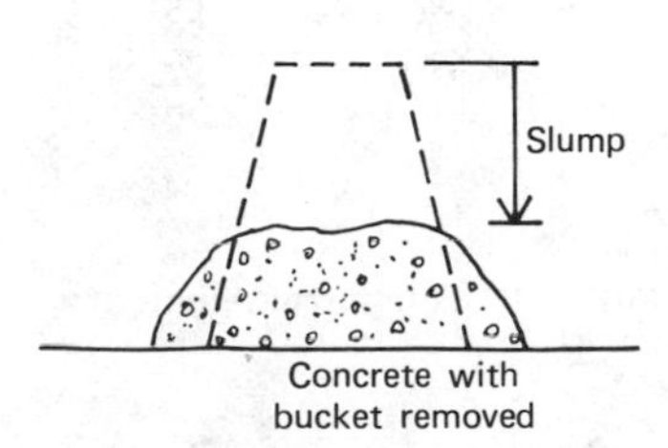

FIG. 5.64. Concrete workability

EXAMPLE PROBLEM 5.13

The initial mix proportions of a particular concrete mixture are given as shown. Assume that the strength versus w/c curve in Figure 5.68 applies to the mixture. Determine the compression strength of the concrete. Also, determine the expected strength if four more gallons of water per yard are added.

Mix Proportions	
Portland cement	530 lb/yd^3
Water	291 lb/yd^3
Crushed aggregate	3069 lb/yd^3
Total weight	3890 lb/cu. yd

Solution

The water-cement ratio in the initial mix is determined first:

$$w/c = \frac{291}{530} = 0.549 \qquad \text{or} \qquad 54.9 \text{ percent}$$

From Figure 5.68,

Strength = 4400 psi

If 4 gal of water are added and the water weighs 8.3 lb/gal, the new w/c ratio becomes

$$w/c = \frac{291 + 33.2}{530} = 0.61$$

The new deduced strength from Figure 5.68 is

Strength = 3600 psi

Both the quality and workability of concrete depend upon the individual characteristics and the relative amounts of the aggregate, portland cement, and water used in the mixture. In general, a relative increase in aggregate improves the quality and decreases the workability. A relative increase in cement improves the quality and the workability. An increase in water decreases the quality and increases the workability.

The properties of concrete are also influenced by the size and shape of aggregates. Too much cement reduces the plasticity of concrete and also causes increased drying shrinkage. Too much water dilutes the portland cement and reduces its strength. Also, an abundance of water increases the open void spaces and promotes deterioration of concrete exposed to freezing and thawing conditions. As shown in Figure 5.65, water penetrates these microsized voids either from the surface or through capillary action. Frequently, concrete workers will bring water to the surface in the form of a slurry by rubbing with a float or trowel. This dried slurry is attractive but often flakes off prematurely as a result of freezing and thawing.

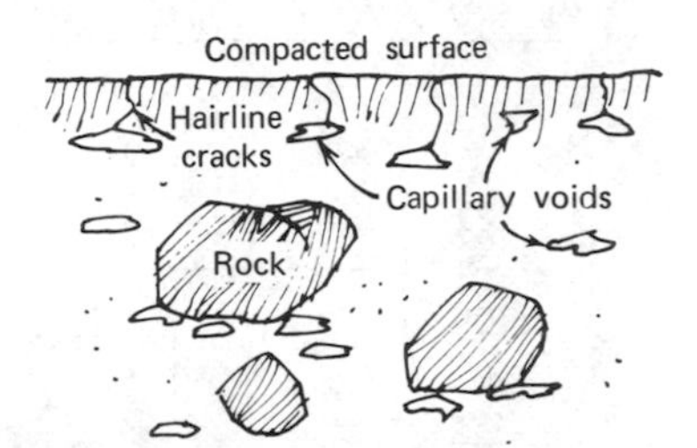

FIG. 5.65. Internal structure of concrete.

Portland cement is an adhesive material that absorbs water as it hardens. This so-called hydration process, along with related chemical reactions, continues for several days. The initial combination of cement and water takes place within a day and leaves the concrete in a hardened or solid condition. In the presence of moisture, the cement continues to harden with time. A typical compression strength curve of the concrete mix is shown in Figure 5.66. As evidenced by the strength-time curve, one should be careful in the loading of young concrete. Concrete walls have collapsed as the result of some over anxious dozer operator who backfilled and applied a load against freshly hardened concrete.

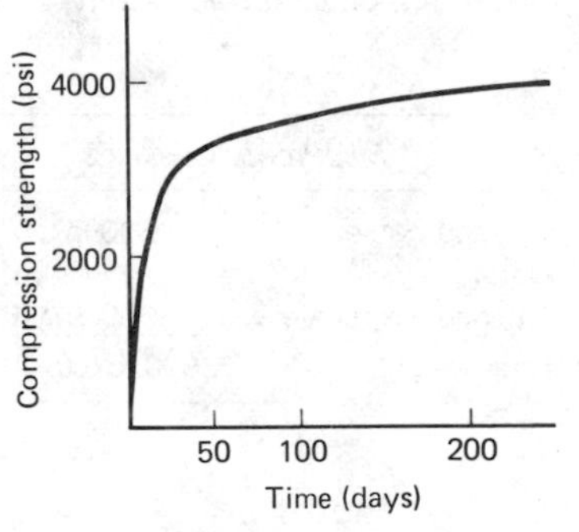

FIG. 5.66. The strength of concrete as a function of curing time.

Any additive to concrete other than aggregate, portland cement, and water is called an admixture. Lightweight additives are sometimes used in concrete to reduce the density and improve the thermal and accoustical insulation properties. Accelerating agents, such as calcium chloride, accelerate the hydration of the cement and reduce the hardening time. Pozzolans, such as fly ash, reduce the heat of hydration and the permeability of concrete which aids in waterproofing. They also reduce undesireable alkali-aggregate reactions. There are retarding agents, water-reducing agents, and agents that produce air pockets throughout the concrete.

The strength of concrete is affected by aggregate shape and size, curing time, mix proportions, admixtures, portland cements, and others. However, the water-cement ratio has the most profound effect. The addition of water reduces the strength of concrete. The compressive strengths of concrete based on tests of standard cylinders 6 in. in diameter and 12 in. long are given in Table 5.6. The photographs in Figure 5.67 show a concrete test cylinder as it fails in compression. The tensile strength is roughly one-tenth of the compressive strength; however it is often considered to be negligible. The modulus of elasticity is about 10^3 times the compression strength.

An engineer that anticipates using concrete as a material can control the mixture of crushed gravel, sand, water, and cement in

TABLE 5.6. Relationship Between Water-Cement Ratio and Compressive Strength of Non-Air-Entrained Concrete

Compressive Strength at 28 Days (psi)	*Water-Cement Ratio by Weight*
6500	0.38
5800	0.43
5000	0.48
4400	0.55
3600	0.62
2900	0.70
2200	0.80

FIG. 5.67. Concrete cylinder at the moment of failure in compression. (From W.A. Cordon, *Properties, Evaluation, and Control of Engineering Materials* Copyright© 1979 McGraw-Hill, Inc. Used with the permission of McGraw-Hill Book Company)

order to achieve the quality and workability desired. In the foundation of a house, strength is not critical in view of the anticipated loads. A concrete beam used in a bridge might require serious scrutiny of the concrete strength. A concrete collar poured around a pipe might require fluidity, whereas concrete dumped into a dam could be fairly stiff.

The proper proportioning of aggregate, sand, water, and cement to achieve the necessary slump, strength, cost, density, or whatever else can be accomplished through trial and error. A scaled-down mixture, perhaps what a wheelbarrow would contain, is mixed as a trial batch. The weight of the aggregate, water, sand, and cement are predetermined and then mixed, poured, and appropriately tested. These tests usually include a slump and a strength test. The proportions of the mixture can then be changed in an attempt to achieve the desired properties. The relationship between compressive strength and the water-cement ratio for a particular batch could be determined by testing three different batches of each and plotting as shown in Figure 5.68. Although the relationship between strength and w/c ratio is nonlinear, it can be approximated as a straight line within a narrow range.

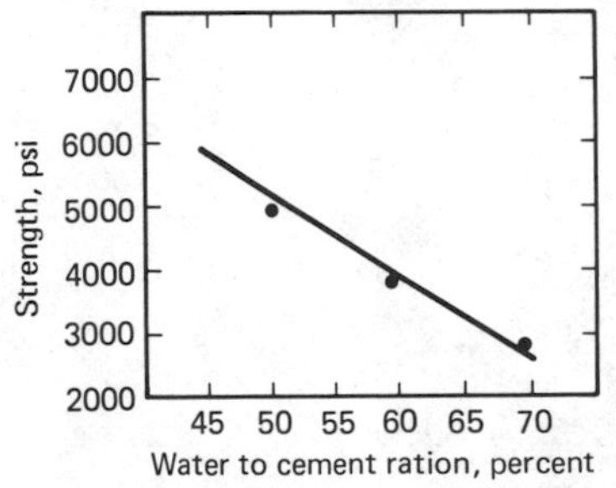

FIG. 5.68. Concrete strength versus water-cement ratio.

PROBLEMS

5.74. Identify some reasons as to why concrete is used extensively in structures.

5.75. How could the thermal conductivity of concrete be reduced?

5.76. How does the strength of concrete vary with curing time?

5.77. Estimate the allowable weight per lineal foot that could be carried by a concrete foundation 8 in. thick. Assume a w/c ratio of 0.53.

5.78. Name two reasons for placing steel rebar in concrete.

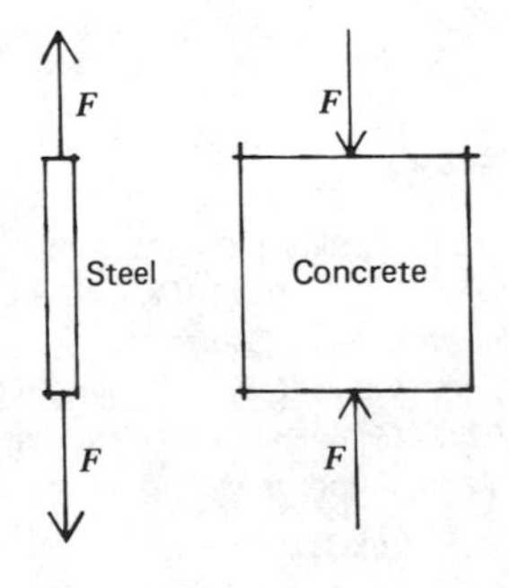

5.79. A short pedestal of concrete is 1 foot square. Determine the diameter of a low-carbon steel rod that would carry a tension load equivalent to the load allowable on the concrete in compression. Assume that $w/c = 0.48$ for the concrete.

5.80. Determine the heat conducted through a concrete wall that is 8 in thick, 8 ft high, and 40 ft long. The inside surface temperature is 70 F and the outside is 10 F. Assume that there are no admixtures in the cement.

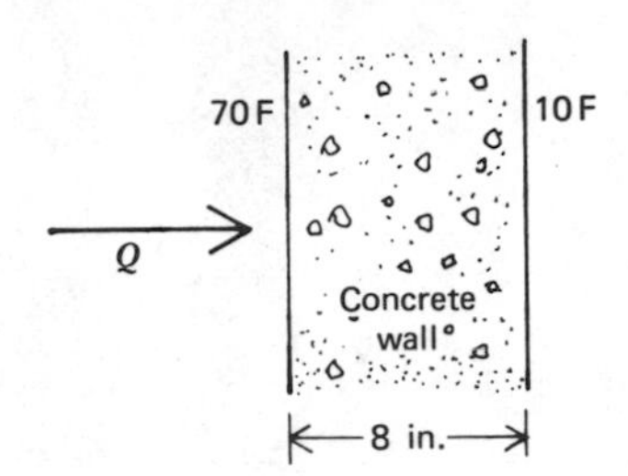

5.81. Assume that you are a civil engineer and are responsible for pouring a concrete retaining wall. A truck loaded with 7 yd^3 of concrete begins to unload. The workers claim the mix is too stiff and desire more water. If the mix proportions are as given, determine the percent reduction in strength if 12 gal of water are added to the contents of the truck.

Mix Proportions	
Cement	500 lb/yd^3
Water	200 lb/yd^3
Concrete aggregate	3200 lb/yd^3

5.82. The mixture of a concrete is given as 359 kg/m^3 of cement, 161 kg/m^3 of water, and 902 kg/m^3 of rounded aggregate. Estimate the strength of the concrete.

5.83. Assume that a typical bag of portland cement weighs 94 lb. Compute the new strength of the concrete if one bag of cement is added to the mixture given in Example Problem 5.13.

5.84. A C-5A aircraft is loaded and weighs 580 000 lb and rests on a concrete pavement. It is suspended equally on 24 wheels. Determine the minimum contact area (footprint) required on each tire if the compressive strength of the concrete is not to be exceeded. Assume a safety factor of 2.

5.85. A 4 in. × 4 in. nonreinforced concrete beam 12 ft long is fixed on both ends and carries a load at midspan. Where and how will the beam fail? How could the strength of the beam be increased without changing the beam size?

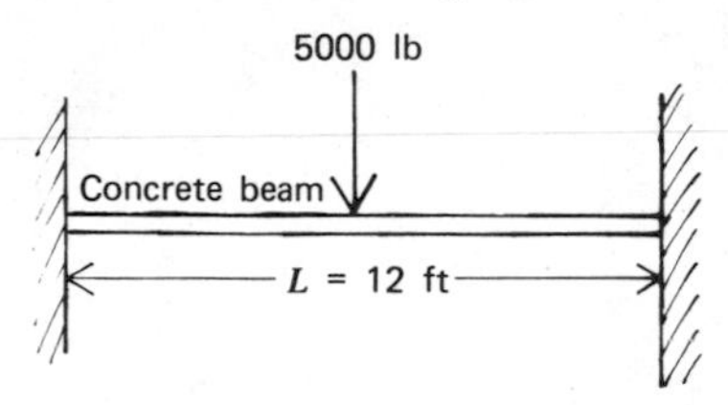

CHAPTER SIX

ELECTRICAL CIRCUITS

6.1/Introduction

In engineering, the term *circuit* refers to a path. There are several different types. A network of underground culinary water pipes and valves provides several circuits or paths along which water may flow. Along these circuits, the fluid may change directions, the volume of flow may increase or decrease, and the pressure may vary. In the nervous system of our bodies, a signal travels in a circuit from our preceptors at the ends of our fingers that might be placed on a hot stove, to the control center in the brain and back to the appropriate muscles with a message to move. Some physiological circuits control involuntary muscles without even notifying our consciousness each time, such as the stimulator of our heartbeat. Our highways, railways, and airways are transportation circuits. The hydraulic hoses on a backhoe provide circuits that enable fluids to perform work. Heat conductors provide a path for the transfer of heat. Telemetry signals follow wireless paths to control logic circuits housed in the computers of the space shuttle.

In this chapter, attention is focused on electrical circuits, shown in Figure 6.1. They provide a path for charged electrons to energize light bulbs, drive motors, communicate words, compute numbers, control valves, extract aluminum from clay, heat and cool homes, toast bread and replay football games in the family room. The student will learn to analyze and design both resistance and logic circuits.

FIG. 6.1. Electrical circuits have multiple applications.

6.2/Resistance Circuits

An electric circuit that consists of power supplies and resistances is called a resistance circuit. The power supply might be a battery, generator, or a similar device. The resistances would represent the toasters, television sets, heaters, and other appliances that convert electrical energy to useful purposes. Understanding resistance-type circuits is a first step toward more sophisticated circuit analysis.

6.2.1/Circuit Components

Electric *current* is the flow of electric charge. The actual transfer of charge from atom to atom along a conductor is rather complicated. In the case of metals, the mobile cloud of electrons associated with the metallic bonding of atoms provides the mechanism for the energy transfer. The charge is carried by the electrons. Since the magnitude of charge on a single electron is so small, someone decided to define a *coulomb* as the charge on $6.24(10)^{18}$ electrons. A flow of one coulomb per second is defined as an *ampere*. Electric current is conventionally measured in amperes.

The flow of charge or current through a conductor can take place in any direction. Sometimes, there is no flow at all. Current flows only if there is an energy gradient. Just as water runs downhill, heat transfers from hot to colder temperatures and winds blow from high to lower pressures, so does current flow in response to differences in energy levels.

In an electric circuit, electrons can be forced to flow by pumping energy into the electrons in a manner similar to mechanical pumps that either push or pull water particles over hills. The electron pumps in an electric circuit are more appropriately called electromotive forces (emf). A battery produces energy through electrochemical action of metals in an electrolyte and converts it to the electron flow in the circuit. The difference in energy levels in a circuit, whether caused by electromotive forces such as batteries and generators, or perhaps by a reduction of energy at a neighboring station, is commonly referred to as a *potential difference* and is measured in volts. The term *volt* refers to energy per unit charge.

The opposition to current flow is called *resistance* and is measured in ohms (Ω). The total resistance R of a wire shaped material can be expressed as

$$R = \frac{\rho L}{A}$$

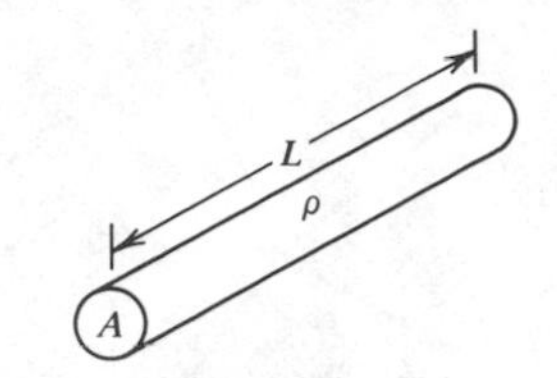

FIG. 6.2. Resistance parameters.

where ρ is the resistivity of the material ($\Omega \cdot$ in.), L is the length (in.), and A is the cross-sectional area (in.^{2}) as shown in Figure 6.2. The total resistance of a typical wire is proportional to its length and inversely proportional to its cross-sectional area.

Engineers have adopted some models to represent batteries, current, and resistors in a circuit. A battery and a resistor are shown in Figure 6.3. The plus and minus symbols on the battery indicate the direction of current flow. At the present time, electrical engineers use the convention that current flows from positive to negative, although in reality the electrons are thought to drift the other way.

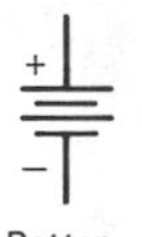

FIG. 6.3. Battery and resistor symbols.

The current that will flow through a conductor depends upon both the applied voltage and the resistance. A German physicist, George Ohm, is credited with the equation

$$V = IR$$

where V is the voltage, I is the current in amperes, and R is the resistance in ohms. Since voltage is energy per unit charge, Ohm's law relates the electrical energy per unit charge across a resistance to the product of current and resistance. This energy is withdrawn from the electrical circuit to perform useful work or is dissipated in the form of heat. Hence, one commonly refers to voltage drop across a resistance.

There are other circuit components that are frequently used by electrical engineers, such as capacitors and coils. The voltages across such components are proportional to the rate of change in current flow and require some knowledge of calculus. These more advanced electrical components will not be treated in this introduction.

6.2.2/Circuit Fundamentals

A simple series circuit can be made by connecting the terminals of a battery to a single resistor, R_1, as shown in Figure 6.4. The single resistor includes the wire resistance and other load resistances, such as light bulbs, all lumped together. Such a system is called a closed-loop circuit. If one assumes that the energy input to the circuit from the battery is constant, then the same voltage must be dissipated across the resistance. Where else would it go? Furthermore, if the resistance were constant, then the current flow around the circuit would also be constant.

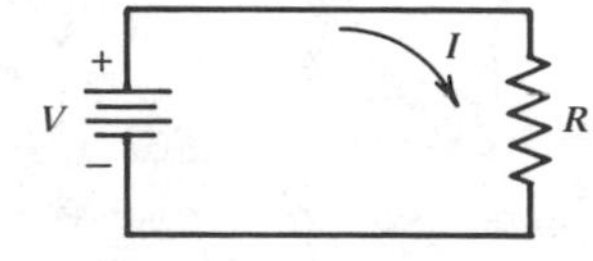

FIG. 6.4.

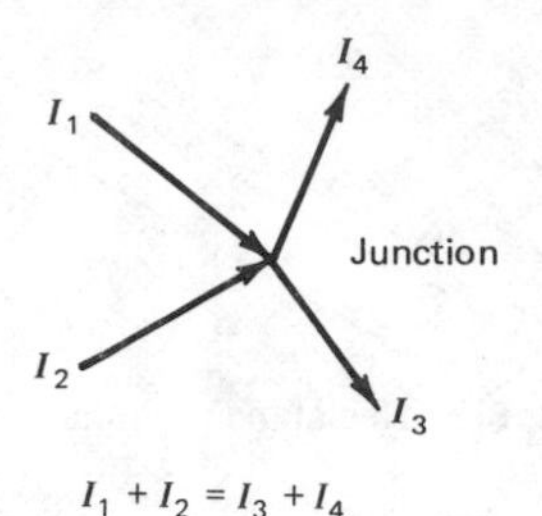

FIG. 6.5. Kirchhoff's current law.

Over 200 years ago, Gustav Kirchhoff established two principles through experimental evidence that apply to the analysis of circuits. *Kirchhoff's current law* states that:

> *The algebraic sum of the currents into a junction at any instant is zero.*

In other words, the total current flowing into a junction must also flow out as shown in Figure 6.5. If the flow of charge were compared to a stream of water, it's intuitively acceptable that all the water flowing into a joint must leave unless the pipe joint expands considerably.

Kirchhoff's voltage law states that:

> *The algebraic sum of all the voltages around any closed loop is zero at any instant.*

This voltage law is equivalent to the principle of conservation of energy. One difficulty that often appears is related to the sign on the voltage. Is the voltage positive or negative? A convention can arbitrarily be adopted as follows. As one moves around a closed loop in the direction of the assumed current flow, the voltage across a resistance is negative. The voltage across a battery is taken to be positive if the current flows through from minus to plus. The voltage across a resistor is positive if one moves around the loop opposite to the current flow. From Figure 6.4, a clockwise application of the sign convention says that

$$V - IR_1 = 0$$

whereas a counterclockwise application says that

$$-V + IR = 0$$

The equations are identical if either is multiplied by -1.

If a second resistor, R_2, were added to the simple circuit shown in Figure 6.4, there are two different ways in which it could be attached. If attached as shown in Figure 6.6, the resistors are in series and the same current flows through both. A single resistor, R_{eq}, can be found that is equivalent to both R_1 and R_2. According to Kirchhoff's voltage law,

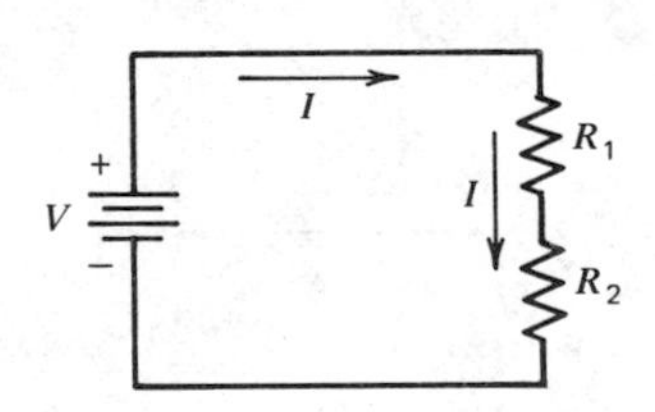

FIG. 6.6. Resistances in series.

$$V = IR_1 + IR_2 = I(R_1 + R_2) = IR_{eq}$$

where $R_{eq} = R_1 + R_2$. In general, a single equivalent resistor can be found by adding resistors in series, that is,

$$R_{eq} = R_1 + R_2 + R_3 + \cdots R_n$$

A second resistor, R_2, could also be placed in parallel with R_1

as shown in Figure 6.7. Note that the voltage across both R_1 and R_2 is identical, yet the currents I_1 and I_2 are not equal. Using Kirchhoff's current law at the junction, it follows that $I = I_1 + I_2$. Since $I_1 = V/R_1$ and $I_2 = V/R_2$, it follows upon substitution that

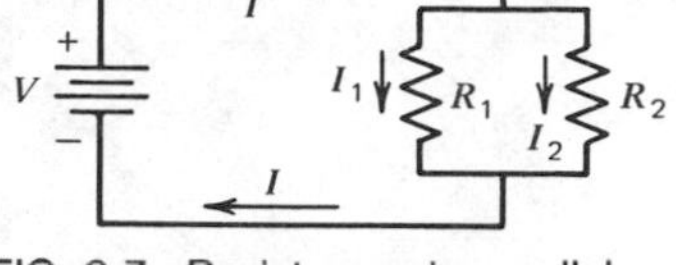

FIG. 6.7. Resistances in parallel.

$$I = \frac{V}{R_{eq}} = \frac{V}{R_1} + \frac{V}{R_2}$$

Since the voltage V is the same, it can be canceled from both sides of the equation. The equation for two resistors in parallel can then be written as

$$\frac{1}{R_{eq}} = \frac{1}{R_1} + \frac{1}{R_2}$$

In the case of several resistances in parallel, an equivalent resistance can be written as

$$\frac{1}{R_{eq}} = \frac{1}{R_1} + \frac{1}{R_2} + \cdots \frac{1}{R_n}$$

6.2.3/Electrical Power

Power is defined as the rate at which energy is expended. Power is the ratio of energy over time. In electrical terms, V is the energy per unit charge and I is the amount of charge flowing per unit time. The total electrical energy available in some time t may then be written as the product VIt. Dividing by time, one obtains the power equation as

$$P = \frac{VIt}{t} = VI = I^2R$$

where the extreme right-hand side of the equation follows from substitution of Ohm's law. Power is measured in watts or joules per second. Remember, a watt is a unit of power, not energy.

EXAMPLE PROBLEM 6.1

For the series-parallel circuit shown, determine an equivalent circuit with only one resistor.

Solution

The parallel resistors R_2 and R_3 can be replaced by an equivalent resistor R_4.

$$\frac{1}{R_4} = \frac{1}{R_2} + \frac{1}{R_3}$$

$$R_4 = \frac{R_2R_3}{R_2 + R_3}$$

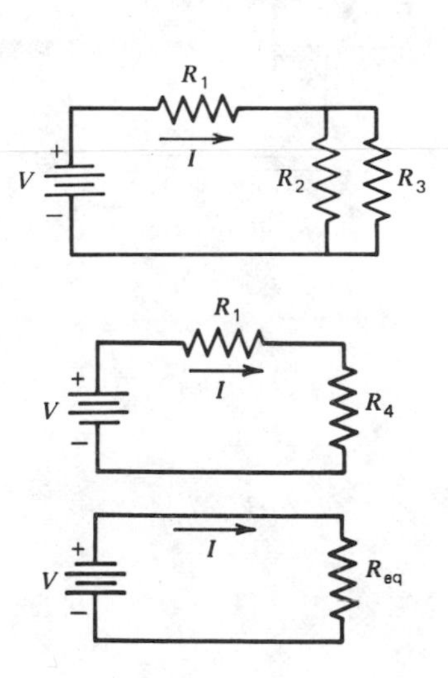

The series circuit can now be reduced to include only one equivalent resistor.

$$R_{eq} = R_1 + R_4$$

$$R_{eq} = R_1 + \frac{R_2 R_3}{R_2 + R_3}$$

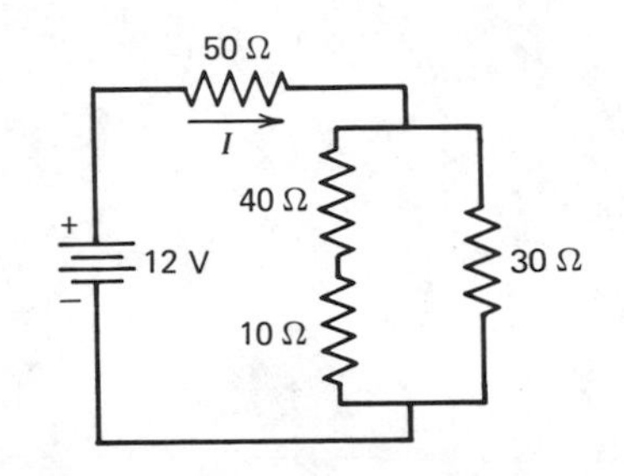

EXAMPLE PROBLEM 6.2

Determine the current through the 30-Ω resistor in the circuit given as shown.

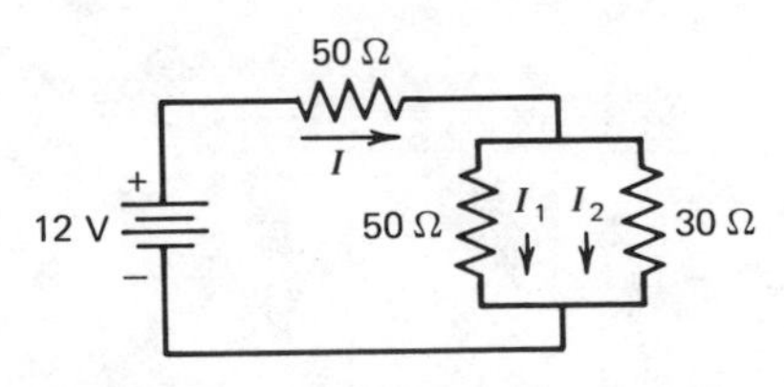

Solution

Apply Kirchhoff's voltage law to the outside loop to obtain

$$12 = 50I + 30I_2 \tag{1}$$

Apply Kirchhoff's voltage law to the inside loop and obtain

$$12 = 50I + 50I_1 \tag{2}$$

Substitute $I_1 + I_2 = I$ into Equation 2 and write

$$12 = 50I + 50(I - I_2) \tag{3}$$

Multiply (1) by 2 and subtract from (3) to determine that

$$I_2 = 0.109 \text{ A}$$

which is the current that flows through the 30-Ω resistor.

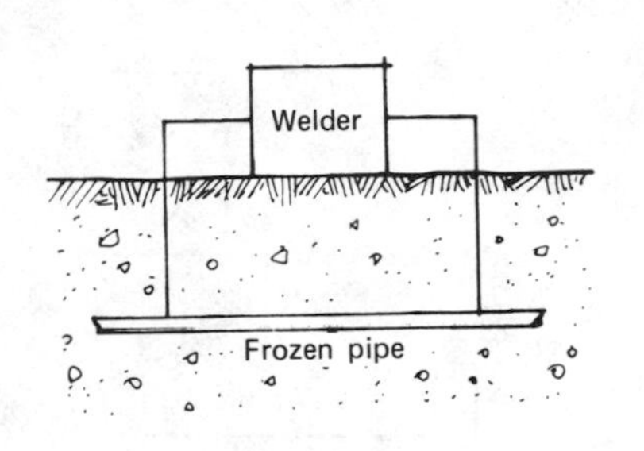

EXAMPLE PROBLEM 6.3

A standard 2.5-in. nominal diameter aluminum pipe has a cross-sectional area of 1.704 in. The pipe is used as a waterline and is frozen. It can be thawed by running a current from a welder through it. The pipe acts as a resistor and dissipates heat, which melts the ice. Determine the thermal energy available each hour (power) per lineal foot of pipe if the welder provides a current of 100 A.

Solution

Determine the resistance of the pipe as follows, where $\rho = 3.5(10)^{-6}\ \Omega\cdot\text{in.}$ from Table C-1.

$$R = \frac{\rho L}{A} = \frac{3.5(10)^{-6}\ \Omega\cdot\text{in. } 12\text{ in.}}{1.704\text{ in}^2\text{ ft}}$$

$$R = 2.46(10)^{-5}\ \Omega/\text{ft}$$

The electrical power dissipated in each lineal foot can be determined as

$$\text{Power} = I^2R = (100)^2(2.46)^{-5} = 0.246\ \text{W/ft}$$

Since 3.413 Btu/h = 1 W

$$\text{Power} = 0.246\ \text{W/ft}\ (3.413\ \text{Btu/h}\cdot\text{W})$$

$$\text{Power} = 83.95\ \text{Btu/h}\cdot\text{ft}$$

PROBLEMS

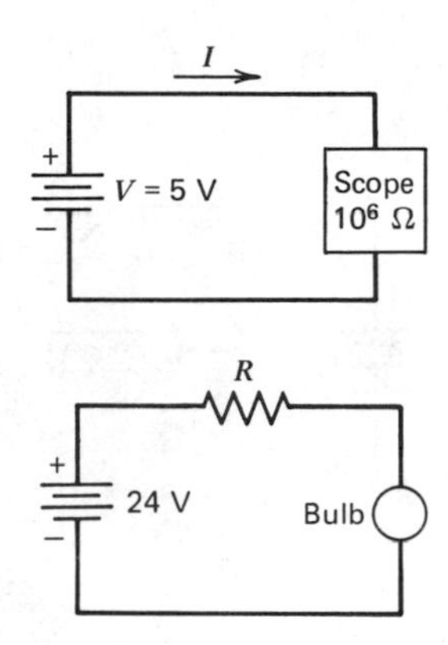

6.1. Determine the current that flows through the oscilloscope with a resistance of $10^6\ \Omega$. The DC power supply is set at 5 V.

6.2. Determine the value of a resistor R placed in series with an 8-Ω light bulb such that the voltage drop across the bulb will be 6 V.

6.3. A 12-V tractor battery is used to light a 50-W bulb in a camping trailer. Determine the current flow in the circuit.

6.4. The resistivity of a particular copper wire is $1.7(10)^{-6}$ $\Omega\cdot$cm. Determine the total resistance offered by 10 m of wire 25 mm in diameter.

6.5. Determine the value of a single equivalent resistor.

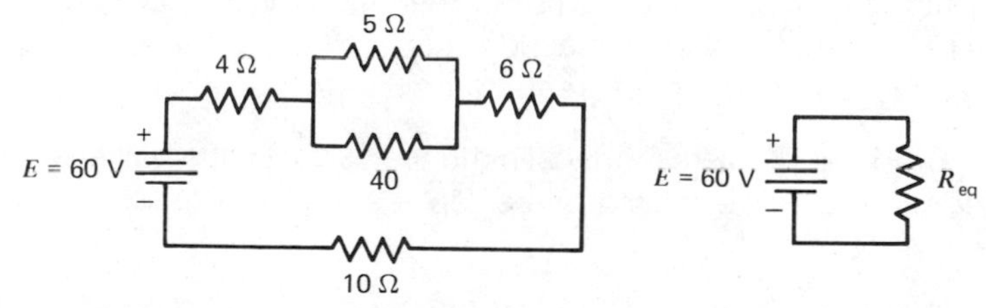

6.6. Determine the voltage drop across each resistor.

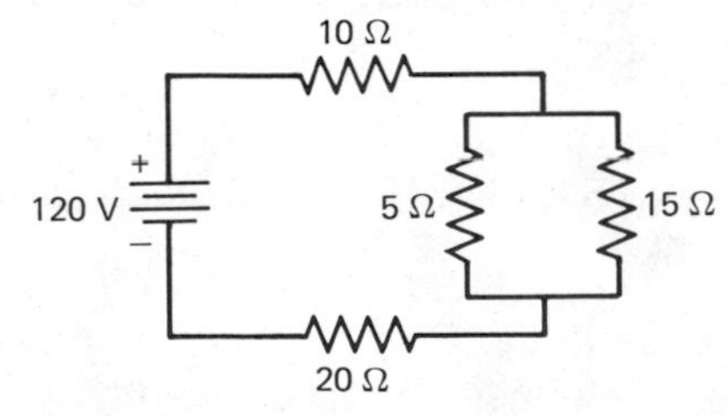

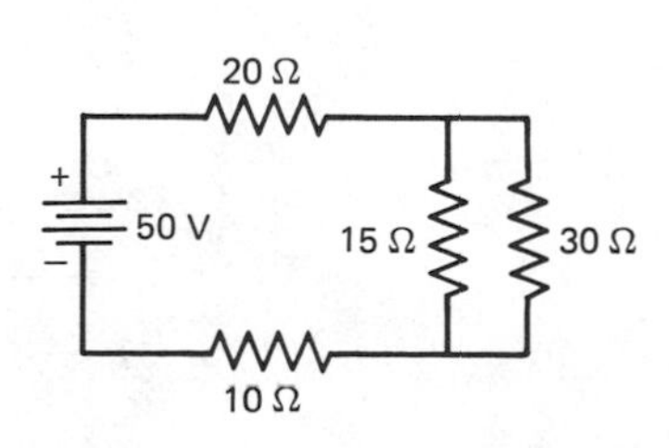

6.7. Determine the current that flows through each resistor.

6.8. Assume that electrical power costs \$0.04/kW·h. Determine the cost of a 100-W bulb burning for 24 h.

6.9. If electrical power costs $0.04/kW·h, determine the cost of operating a 50-hp irrigation pump continuously for 30 days.

6.10. A standard 4 in. nominal diameter steel pipe has a cross-sectional area of 3.17 in.^{2} and an inside diameter of 4.026 in. The resistivity of steel is $6.65(10)^{-6}$ Ω·in. The pipe is used as a water line and is frozen. A welder is attached to a 100-ft section of the pipe and a current of 300 A is provided continuously. If 144 Btu are needed to melt 1 lb of ice, *estimate* the time required to thaw the ice completely. Assume that the water is stagnant.

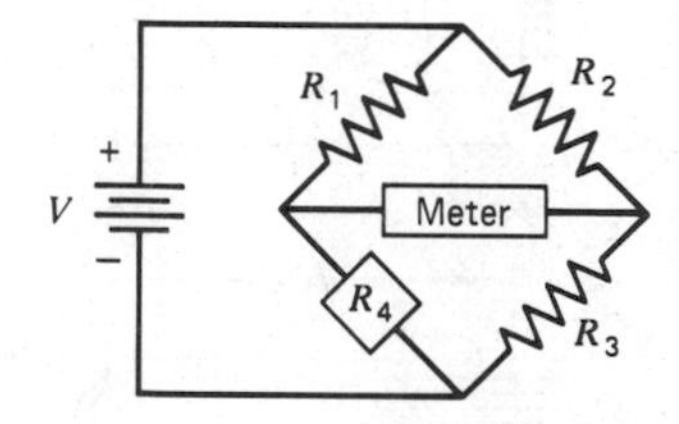

6.11. A Wheatstone bridge has arms of $R_1 = 150$ Ω, $R_2 = 250$ Ω, and $R_3 = 300$ Ω. The bridge excitation voltage is 12 V. Determine the nominal value of the internal resistance of an accelerometer, R_4, such that zero current will flow through the meter.

6.12. Assume that you have a 12-V battery, all kinds of resistors, and an appliance that has an internal resistance of only 6 Ω and a maximum current rating of 50 mA. Design a circuit that would allow you to operate the appliance without exceeding the current limit.

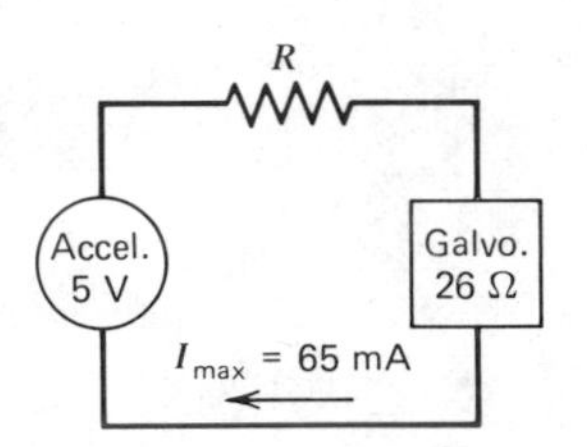

6.13. A galvanometer has a maximum current rating of 65 mA. The internal resistance of the instrument is 26 Ω. The input signal from a piezoelectric type accelerometer is 5 V. Determine the resistor that should be placed in series with the galvanometer to protect it from currents greater than 65 mA.

6.14. In Problem 6.13, estimate the current if the galvanometer and protective resistor were replaced with an oscilloscope ($R = 10^6$ Ω).

6.15. A 115-V house circuit is fused with a 15-A fuse. Determine the maximum number of 100 W bulbs that may burn simultaneously if connected in series.

6.16. Determine the heat per hour provided by a 110 V resistance heater made from 75 ft of gage No. 18 stainless steel wire (diameter = 1.024 mm).

6.17. A particular house in the Rocky Mountains requires $2.14(10)^4$ Btu/h for heating. Determine the total length of No. 18 stainless steel wire (diameter = 1.024 mm) required if the system is 110 V and the complete house must be heated. The current is 15 A.

6.3/Computer Circuits

There are two basic types of electronic computers—analog and digital. The analog computers utilize electric circuits that consist primarily of resistors, capacitors, and high-gain amplifiers. Combinations of these circuit components can be made to perform such mathematical operations as addition, subtraction, multiplication, and integration. The analog computer solves a problem by simulating the mathematical manipulations electronically. The voltages represent the magnitude of such physical quantities as displacement, velocity, acceleration, force, and temperature. A variable in the analog system is similar to a variable in an equation and can take on several different values. The variables are represented by continuous signals that can be graphically represented as a line or a curve. The continuous output voltage of an analog computer in a vibration problem might look like a sine wave.

A digital computer is an electronic system that contains basically an input/output unit, memory unit, arithmetic unit, and control unit. The input/output unit provides an interface between the user and the computer. The memory is a storage place for data and instructions. The arithmetic and control units solve problems through the use of logic circuits. In contrast to the analog system, the logic circuit signals are constrained to take on only discrete values. In most digital computers, the signals are limited to only two values and the system is said to be binary. In a binary system, the signal must be either one or zero. This may be interpreted as either off or on, yes or no, asserted or nonasserted, or perhaps zero or nonzero. These values are achieved electronically as high and low voltages.

Since digital computers utilize components that are binary in nature, it is important that one understands the binary number system. For example, how would numbers in our common decimal system, such as 8, be written if only 0 and 1 could be used? Binary arithmetic, switching algebra, and some logic circuits are discussed in this section.

6.3.1/Binary Arithmetic

When counting in the decimal system, one starts with zero in the units position and continues until the number 9 is reached. Since the number 9 is the highest one can go in the units column, the next count starts with 1 in the tens column and the unit column returns to zero. The count continues until 9 is reached again in the

TABLE 6.1. Binary Numbers from 0 to 12

Decimal	*Binary*
0	0
1	1
2	10
3	11
4	100
5	101
6	110
7	111
8	1000
9	1001
10	1010
11	1011
12	1100

unit column and then the next count advances the 1 in the tens column to a 2 and the unit column again starts at zero. This process continues into columns of hundreds, thousands, and so on.

When counting in the binary system, one carries in the same manner as in the decimal system. The binary numbers from 0 to 12 are shown in Table 6.1.

Numbers in the binary system can be converted to equivalent numbers in the decimal system. First, note that the number 452 in the decimal system could be written in terms of the base ten as

$$4(10)^2 + 5(10)^1 + 2(10)^0 = 452$$

Using the same concept, the number 10111 in the binary system may be converted to the decimal system as follows:

$$1(2)^4 + 0(2)^3 + 1(2)^2 + 1(2)^1 + 1(2)^0 = 23$$

A decimal number can be converted to a binary number by dividing repeatedly by 2 and tabulating the remainder. The method is demonstrated by converting 23 to a binary equivalent:

$$2\overline{)23} = 11 \text{ remainder of } 1$$

$$2\overline{)11} = 5 \text{ remainder of } 1$$

$$2\overline{)5} = 2 \text{ remainder of } 1$$

$$2\overline{)2} = 1 \text{ remainder of } 0$$

$$2\overline{)1} = 0 \text{ remainder of } 1$$

The binary number 10111 is obtained by reading the remainders from bottom to top. The binary digits are called bits in computer jargon.

TABLE 6.2. Binary Addition

0 + 0 = 0
0 + 1 = 1
1 + 0 = 1
1 + 1 = 10

Rules for addition are given in Table 6.2. Examples of binary addition are given as follows:

$$\begin{array}{r} 1010 \\ +\,1101 \\ \hline 10111 \end{array} \qquad \begin{array}{r} 11101 \\ +\,1111 \\ \hline 101100 \end{array}$$

Note that $1 + 1 + 1 = 10 + 1 = 11$.

TABLE 6.3. Binary Subtraction

1 − 0 = 1
1 − 1 = 0
0 − 0 = 0
10 − 1 = 1

The rules for subtraction in binary are summarized in Table 6.3. In order to subtract a binary 1 from a 0 in any given column, a 1 must be borrowed from the next larger column. Examples of subtraction are given as follows:

$$\begin{array}{r} 1100 \\ -\,1010 \\ \hline 0010 \end{array} \qquad \begin{array}{r} 1001 \\ -\,111 \\ \hline 010 \end{array}$$

In the second example, the borrow required in the second column from the left must come from the fourth column from the left. In the process, the third column zero goes to 1 since 10 − 1 = 1. The subtraction result can be checked by the usual addition process.

Binary multiplication is summarized in Table 6.4 and is rather simple. The addition of partial products must be done with care since many carries are often generated. Two examples are given as follows:

TABLE 6.4. Binary Multiplication

$0 \times 0 = 0$
$0 \times 1 = 0$
$1 \times 0 = 0$
$1 \times 1 = 1$

```
    1010        110
   ×101        ×11
    1010        110
   0000        110
  1010        10010
  110010
```

Binary division is similar to division in the decimal system and requires some subtraction. An example of binary division is given as follows:

```
          11
  101)1111
      101
       101
       101
         0
```

6.3.2/Switching Algebra

Addition, subtraction, multiplication, and division of binary numbers are quite similar to arithmetic operations in the common decimal system. However, a new algebra for digital systems that use the binary system must be introduced. In ordinary algebra, the variables may take on many values and are usually continuous functions. Binary digital systems require an algebra where the variable represents only two discrete values. A digital variable in the binary system is either zero or nonzero. There are no other "intermediate" values possible as there are in analog systems.

The mathematical formulation of digital electronics is based on Boolean algebra, which was first introduced in 1849 to describe logical thought and reason. Boolean algebra is a closed algebraic system containing a set of two or more elements and is based on some fundamental postulates and theorems. In the discussions that follow, the elements of the set will be specified as zero or one and only a few of the basic postulates will be

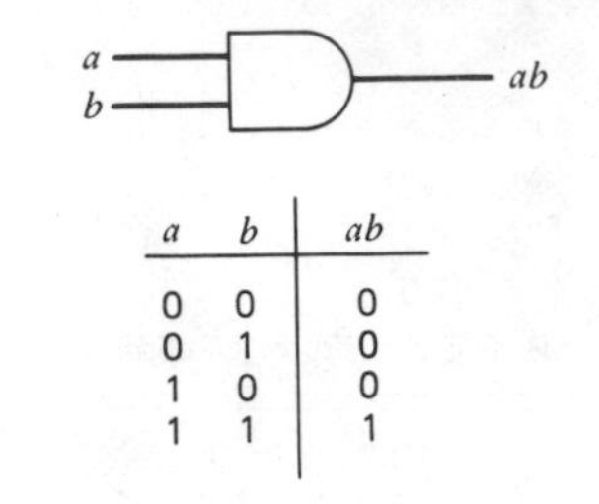

a	b	ab
0	0	0
0	1	0
1	0	0
1	1	1

FIG. 6.8. AND operator.

required. This formulation is commonly referred to as the algebra of switching functions.

There are three basic Boolean operations that provide the building blocks for all switching in logic circuits used in digital systems. These are the AND, OR, and NOT operators. These operators are sometimes called gates. The symbol for the AND operator is shown in Figure 6.8. The logic of the AND operator is summarized in the so-called truth table. The truth table shows the output of the gate as related to all the possible combinations of the inputs. The number 1 represents a true or an asserted statement and is electronically achieved by a relatively high voltage. The number 0 represents a false or a nonasserted statement and is a low or zero voltage electronically. Note that the output of the AND gate is 1 if and only if the input values are 1 simultaneously. An electric circuit representing the AND operator is also shown in Figure 6.8 where the light would go on only if the two switches were closed simultaneously.

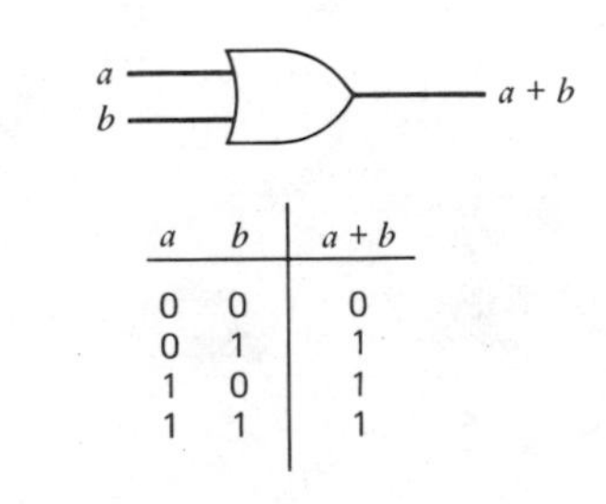

a	b	$a + b$
0	0	0
0	1	1
1	0	1
1	1	1

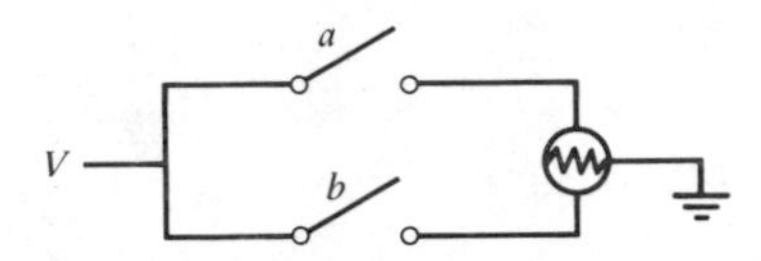

FIG. 6.9. OR operator.

The symbol for the OR operator is shown in Figure 6.9. The truth table indicates the logic associated with the OR operator. The output of the OR gate is 0 if and only if both inputs are zero. An electric circuit with switches in parallel represents the OR gate.

The NOT gate is shown in Figure 6.10. There is only one input designated by the letter *a*, and the output is opposite the input as designated by placing the bar over the input. The NOT gate is an inversion operator and is equivalent to the complement concept in Boolean algebra.

There are several possible combinations of the basic AND, OR, and NOT operators and many switching networks may be constructed. Three common combinations are worth mentioning, since they are also often considered as basic building blocks in a switching circuit. The NAND gate is a combination of the AND followed by a NOT gate and is shown in Figure 6.11. The NOR gate consists of an OR operator followed by a NOT operator as shown in Figure 6.12.

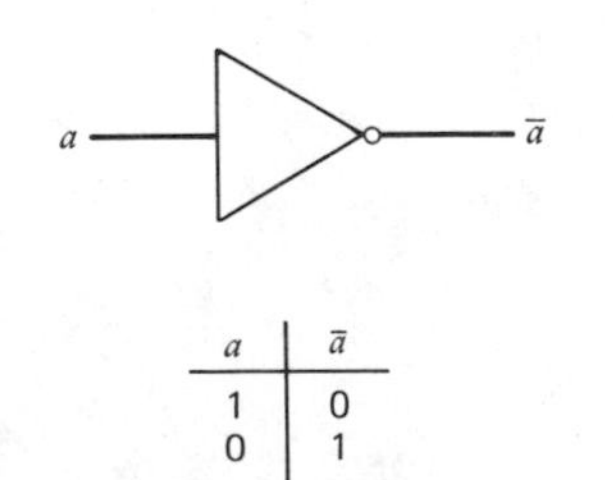

a	$\bar{a}$
1	0
0	1

FIG. 6.10. NOT operator.

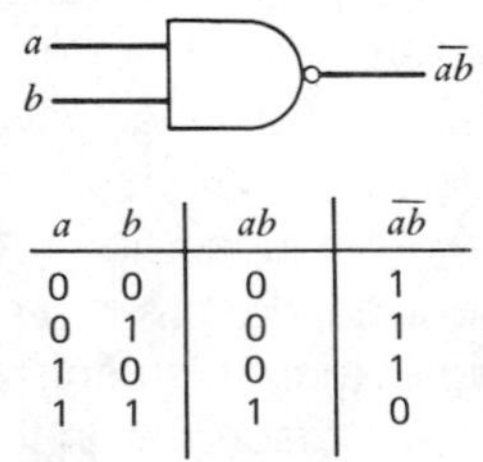

a	b	ab	$\overline{ab}$
0	0	0	1
0	1	0	1
1	0	0	1
1	1	1	0

FIG. 6.11. NAND operator.

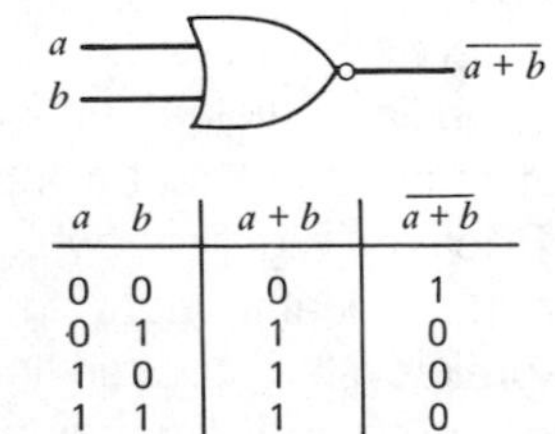

a	b	$a + b$	$\overline{a + b}$
0	0	0	1
0	1	1	0
1	0	1	0
1	1	1	0

FIG. 6.12. NOR operator.

6.3.3/Logic Circuits

In the design of a switching network that will perform some particular task, it is convenient to begin with a truth table. A truth table specifies the possible combinations of the inputs and the corresponding outputs. The individual logic statements are identified and tabulated. Boolean expressions can then be written from the truth table in terms of AND, OR, and NOT operations. A switching network using symbols for the AND, OR, and NOT gates can then be drawn in accordance with the Boolean algebraic expressions. Since the construction of a digital network from the Boolean algebraic expression might cause some difficulty, some typical logic circuits and the corresponding algebraic equations are shown in Figure 6.13. The output of the network is the algebraic expression and the inputs are the elements of the expression.

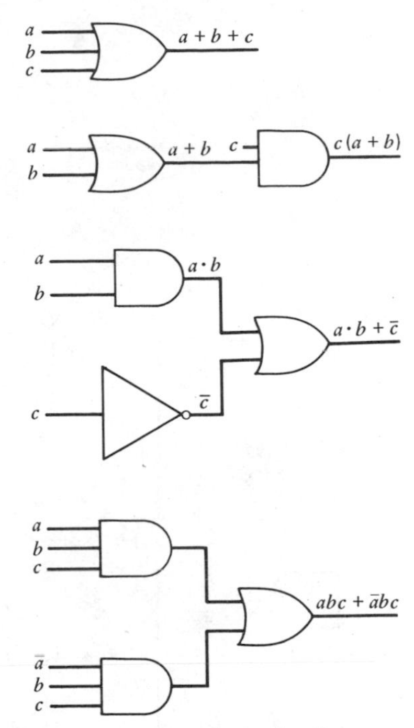

FIG. 6.13. Some logic circuits.

As a simple example of the design of a logic circuit, consider a bell that is controlled by two different switches. It is desired that the bell will ring when either switch is actuated. Initially, the bell is off ($B = 0$) when both switches are down ($a = b = 0$). The truth table is constructed as shown in Figure 6.14. When switch one is down ($a = 0$) and switch two is up ($b = 1$), the bell rings ($B = 1$) as shown in row 2 of the truth table. When switch 1 is up and switch 2 is down, the bell rings. When both switches are up, the bell does not ring. The truth table indicates that the bell rings only on rows 2 and 3, therefore, the Boolean expression could be written as:

$$B = \bar{a}b + a\bar{b}$$

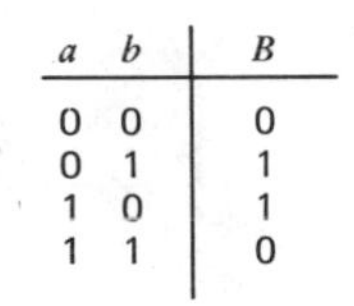

a	b	B
0	0	0
0	1	1
1	0	1
1	1	0

Note that only the rows from the truth table that have 1 in the output column need to be considered in the Boolean expression. Also, the symbol $\bar{a}$ represents $a = 0$ and a represents $a = 1$. The logic circuit could then be drawn as shown in Figure 6.14.

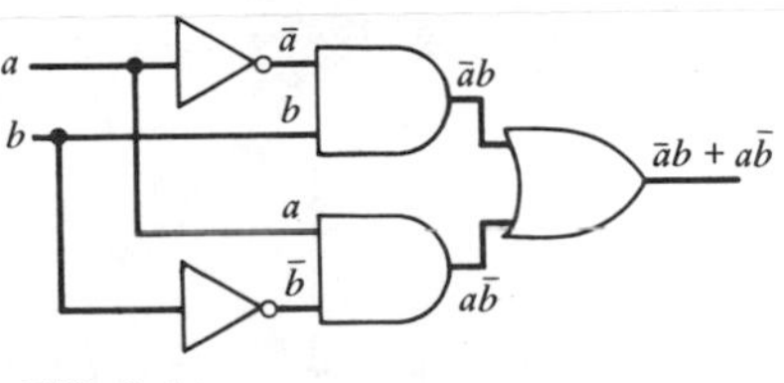

FIG. 6.14.

A second example of a logic circuit is the binary adder, which is found in the arithmetic units of a digital computer. The binary numbers to be added are stored in registers in the computer. A 16-bit register could contain 16 spaces and each space would be filled with either 1 or 0. The binary number 111 would have 13 leading zeroes in a 16-bit register as shown in Figure 6.15. Since the addition of two 16 digit binary numbers could result in a 17-digit sum, the 16-bit register would be exceeded and the computer would indicate an overflow.

0	0	0	0	0	0	0	0	0	0	0	0	0	1	1	1

FIG. 6.15.

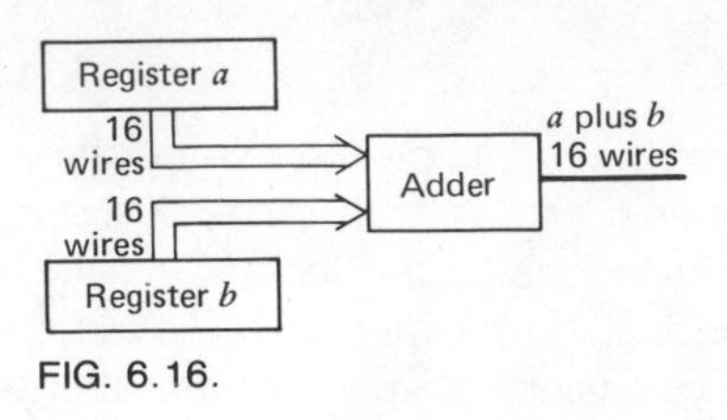

FIG. 6.16.

a	b	C_i	Sum	C_o
0	0	0	0	0
0	0	1	1	0
0	1	0	1	0
0	1	1	0	1
1	0	0	1	0
1	0	1	0	1
1	1	0	0	1
1	1	1	1	1

FIG. 6.17.

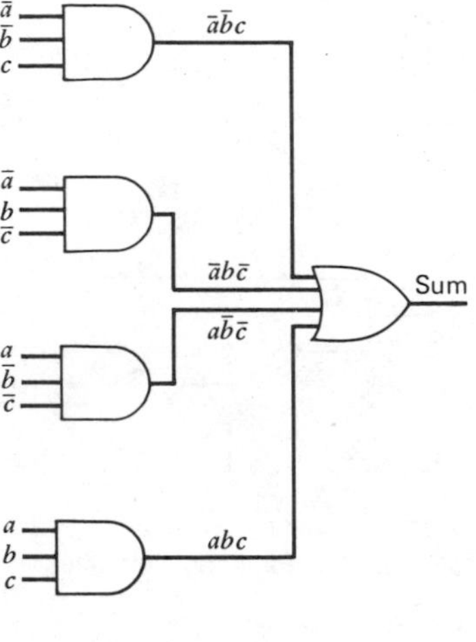

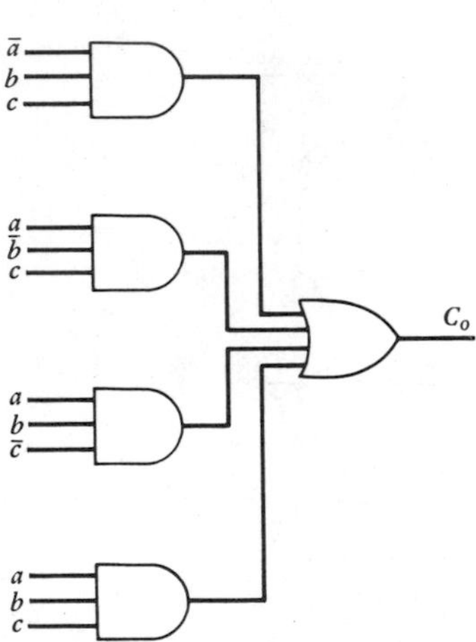

FIG. 6.18.

The addition of two 16-bit numbers could be accomplished as shown schematically in Figure 6.16. Sixteen wires would come from each register of the numbers to be added and would connect to the adder. Sixteen 1-bit adders would be required. The arithmetic values of 1 and 0 would be transferred by a high and low voltage level, respectively.

The logic circuit for one of the 1-bit adders will be considered. Such an adder could operate between the second space in register *a* and the second space in register *b*. Since the addition of two binary numbers is done bit by bit, some provision must be made for carry input and carry output across bit spaces. For example, the addition of 1 in the second digit position of input *a* to 1 in the second digit position of input *b* would result in 10. The 1 must be carried into the third digit position and then be added to the appropriate inputs *a* and *b*. Each 1-bit adder must contain a provision for carry input and carry output.

The truth table for a 1-bit adder of two numbers *a* and *b* is shown in Figure 6.17. The inputs are *a* and *b*. The carry inputs to the adder are represented by C_i. The sum of the 1-bit adder is simply sum and the carry outputs of the adder are C_o. It is noted that the carry output of one digit position would be carry input to the next higher digit position.

Two logic circuits are required for the 1-bit adder as shown in Figure 6.18. One circuit is for the sum and the other for the carry. A series of these 1-bit adders would be required for a full binary adder.

6.3.4/Logic Circuit Simplification

The Boolean expressions written from truth tables are often cumbersome and can be reduced. There are several postulates and theorems in Boolean algebra that provide the mathematical justification for the simplification of logic circuits. The application of these Boolean principles can be accomplished either directly by manipulating the algebraic expressions, or by utilizing related graphical techniques such as the Karnaugh map and the Quine-McCluskey method. The ultimate goal of the simplification effort is to remove unnecessary logic blocks from the circuit.

Some basic postulates and theorems of Boolean algebra are given without proof in Table 6.5. The proofs of the theorems are readily available in more advanced references. The objective of the exercise is to briefly expose the reader to some logic circuit simplification through the application of Boolean algebra. A more rigorous treatment usually is presented in subsequent engineering courses.

The first set of theorems relate to the AND operator and are given as theorems 1 through 6. Theorems 7 through 13 deal with OR operations. Theorems 15 through 20 have various special names; however, they can be identified in general as simplification theorems. The symbols *x*, *y*, and *z* may represent either single or groups of Boolean variables. Also note that $x \cdot y$ is equivalent to xy. The theorems are used to reduce some Boolean expressions to an equivalent, but simpler, equation. Some examples of logic circuit simplifications are given as follows:

1. Simplify $\bar{a}a + bb$

$$aa + bb = 0 + bb \qquad \text{Th. 4}$$
$$= 0 + b \qquad \text{Th. 3}$$
$$= b \qquad \text{Th. 7}$$

Consequently, the expression $\bar{a}a + bb$ is equivalent to b.

2. Simplify $abc + ab(\bar{c} + d)$

$$abc + ab(\bar{c} + d) = ab(\bar{c} + c + d) \qquad \text{Th. 14}$$
$$= ab(1 + d) \qquad \text{Th. 10}$$
$$= ab$$

3. Simplify the carry circuit used in the binary adder.

$$C_o = \bar{a}bc + a\bar{b}c + ab\bar{c} + abc$$
$$= c(ab + a\bar{b}) + \bar{a}bc + ab\bar{c} \qquad \text{Th. 14}$$
$$= ca + \bar{a}bc + ab\bar{c} \qquad \text{Th. 15}$$
$$= c(a + \bar{a}b) + ab\bar{c} \qquad \text{Th. 14}$$
$$= c(a + b) + ab\bar{c} \qquad \text{Th. 17}$$

TABLE 6.5. Theorems of Boolean Algebra

	AND Operations
1.	$x \cdot 0 = 0$
2.	$x \cdot 1 = x$
3.	$x \cdot x = x$
4.	$x \cdot \bar{x} = 0$
5.	$x \cdot y = y \cdot x$
6.	$x \cdot (y \cdot z) = x \cdot y \cdot z$
	OR Operations
7.	$x + 0 = x$
8.	$x + 1 = 1$
9.	$x + x = x$
10.	$x + \bar{x} = 1$
11.	$\bar{\bar{x}} = x$
12.	$x + y = y + x$
13.	$x + (y + z) = x + y + z$
	Simplifications
14.	$x \cdot y + x \cdot z = x \cdot (y + z)$
15.	$x \cdot y + x \cdot \bar{y} = x$
16.	$x + x \cdot y = x$
17.	$x + \bar{x} \cdot y = x + y$
18.	$x \cdot y + \bar{x} \cdot z + y \cdot z = x \cdot y + \bar{x} \cdot z$
19.	$\overline{x \cdot y} = \bar{x} + \bar{y}$
20.	$\overline{x + y} = \bar{x} \cdot \bar{y}$

$= ca + cb + ab\bar{c}$ Th. 14

$= a(c + \bar{c}b) + cb$ Th. 14

$= a(c + b) + cb$ Th. 17

$= ac + ab + cb$ Th. 14

Note that several different approaches might be used in simplifying a logic circuit. The Boolean approach to circuit minimization is often dependent on the designer's experience and luck. As mentioned previously, more systematic approaches are available.

PROBLEMS

6.18. Perform the following operations using binary addition.

(a) 101001 + 1110
(b) 11101 + 1001
(c) 1101 + 1010

6.19. Perform the following binary subtraction.

(a) 11011 − 1010
(b) 11101 − 111
(c) 1001 − 110

6.20. Perform the following binary multiplication.

(a) 1011 × 110
(b) 10101 × 1010
(c) 1101 × 101

6.21. Perform the following binary division.

(a) 101101 ÷ 110
(b) 11101 ÷ 101
(c) 111 ÷ 11

6.22. Convert the following binary numbers to decimal equivalents.

(a) 11011
(b) 11010
(c) 10111

6.23. Convert the following decimal numbers to binary equivalents.

(a) 415
(b) 122
(c) 15

6.24. Perform the following arithmetic in the decimal system, then convert to the binary system and perform the arithmetic operations. Compare the results in the decimal system.

(a) $420 + 150$
(b) $171 \div 25$
(c) 14×250
(d) $135 - 42$

6.25. Write Boolean expressions for the following truth tables.

(a)

a	b	B
0	0	0
0	1	1
1	0	1
1	1	0

(b)

a	b	c	B
0	0	1	1
0	1	0	1
1	0	0	1
1	0	1	0

6.26. Draw logic circuits for the following Boolean expressions:

(a) $B = abc + ac$
(b) $B = ac + ab + bc$
(c) $B = \bar{a}b + b$
(d) $B = (a + b)(c + d)$

6.27. A builder wants to activate a light in a hall from three different locations. Assume that all three switches are down initially and that the light is off. The light goes on when any one switch is moved or when all three switches are moved. Determine the truth table, write the Boolean expression and draw the logic circuit.

6.28. An alarm system is to be installed with four different stations. The stations can be represented as toggle switches. The alarm is to ring only when the switches are alternately up and down. Write the truth table, the Boolean expression, and sketch the logic circuit.

6.29. Simplify the following expressions.

(a) $a(b + c) + (ab + ac)d$

(b) $\bar{a}\bar{b}\bar{c} + a\bar{b}\bar{c} + \bar{a}\bar{b}c + ab\bar{c}$
(c) $\bar{a}\bar{d} + a\bar{b}\bar{d} + \bar{a}\bar{c}d + \bar{a}bcd$
(d) $ac\bar{d} + ae + \bar{b}c\bar{d} + \bar{b}e + ad\bar{e}$

6.30. Simplify the Boolean expression required for the three-way light switch described in Problem 6.27. Draw the simplified logic circuit.

6.31. The expression $Cat = (a + b)(h + g)$ can be made from AND and OR gates. Derive the Boolean expression that is equivalent yet only NOR gates are used. Sketch both logic circuits.

6.32. A Boolean expression is given as:

$$M = abc + a\bar{b}c + c\bar{a}b + b\bar{a}\bar{c}$$

Determine a simplified equation that includes only two AND operators, one NOT, and one OR operator. Sketch the logic circuit.

CHAPTER SEVEN

FLUIDS AND ENGINEERING

7.1/Introduction

All matter appears in the form of a solid, liquid, or vapor. Solids have strong molecular bonds that limit material deformation, whereas the attraction between molecules in liquids and vapors is weaker and the matter deforms readily. Liquids and vapors are commonly grouped together and are called fluids. Fluids such as water that experience only slight changes in volume under pressure are said to be incompressible. Air and other gases expand and contract appreciably and are called compressible fluids. All forms of matter deform in some predictable manner when subjected to forces, temperature changes, or other influences.

The study of fluid mechanics deals with the condition of rest or motion of fluids under the influence of forces. The laws of Newton, conservation of mass, conservation of energy, equilibrium equations, and mathematical manipulations are some of the common tools used by engineers in the field of mechanics. These tools can describe or predict the behavior of fluids and can relate such things as deformation, stress, velocity, acceleration, pressure, temperature, and energy.

7.2/Fundamental Terms

There are some basic terms commonly used in the formulation and solution of fluid problems that need to be defined. Density ρ

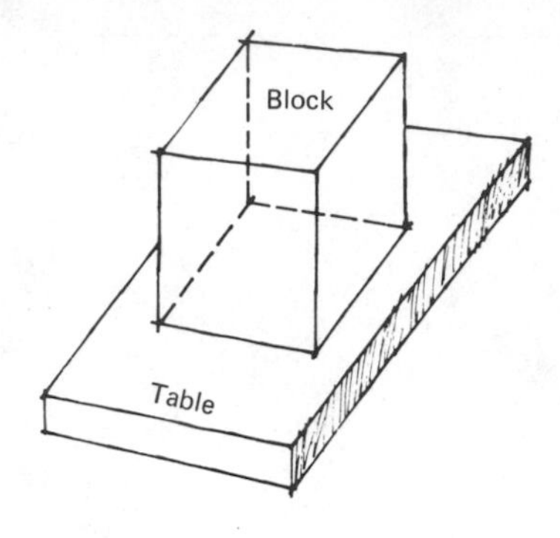

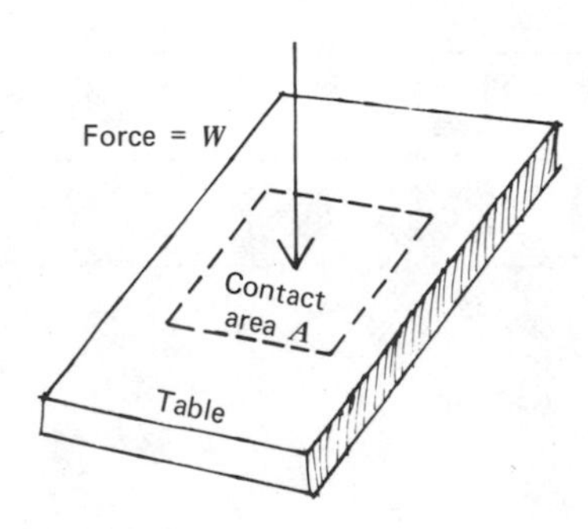

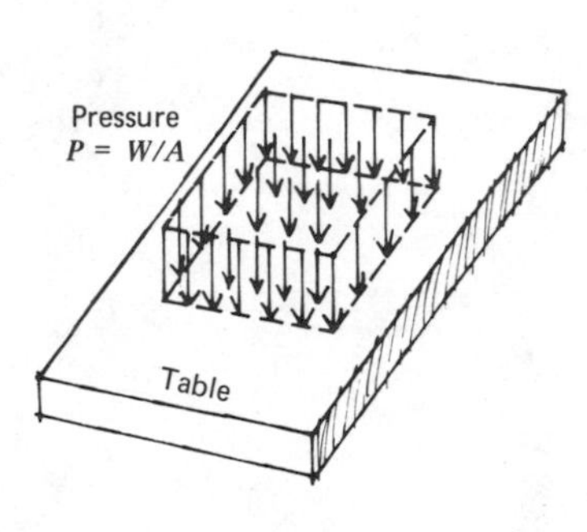

FIG. 7.1. Pressure is defined as force per unit area.

is mass per unit volume and is often measured in units of pounds mass per cubic foot (lbm/ft^3) or kilograms per cubic meter (kg/m^3). Specific weight γ is weight per unit volume and is expressed in units of force over volume; such as pounds per cubic foot (lb/ft^3) or Newtons per cubic meter (N/m^3). Specific volume is the inverse of specific weight. Specific gravity (s.g.) is the ratio of the density or specific weight of the solution to the density or specific weight of water.

$$\text{s.g.} = \frac{\rho}{\rho_{\text{water}}} = \frac{\gamma}{\gamma_{\text{water}}}$$

In vernacular speech, the term "pressure" is frequently used very loosely to describe such things as one's emotional state, coercion, or perhaps physical force. Pressure is rigorously defined as force per unit area and is commonly expressed in units of pounds per square inch (psi) or Newtons per square meter (N/m^2). In engineering, it is important to define pressure properly and separate it from force and energy. For example, if a cube of either steel or water of weight *W* were resting on a horizontal surface, as shown in Figure 7.1, the force exerted on the table by the block would be equal to the total weight *W* of the cube. The pressure on the table would be distributed over the block surface at the interface and would be equal to the total force divided by the contact area, that is, $P = W/A$.

Most everyone accepts the fact that temperature can be referenced to either a Fahrenheit or a Celsius scale. Pressure can also be measured with respect to two different references. Pressure measured relative to a reference of absolute zero pressure as found in outer space is called absolute pressure. A pressure referenced to the surrounding atmospheric pressure is called gage pressure. The numerical difference between absolute and gage pressure is due to the pressure created by the weight of the atmosphere. As shown in Figure 7.2, gage pressure can be converted to absolute pressure under standard atmospheric conditions by adding 14.7 psi to the gage pressure. Pressure measurements will be identified in this book by adding the word gage or absolute. For example, 150 psi absolute is an absolute pressure whereas 150 psi gage is a gage pressure.

The term "head" refers to the length of a vertical column of fluid usually of unit cross section whose total weight produces a

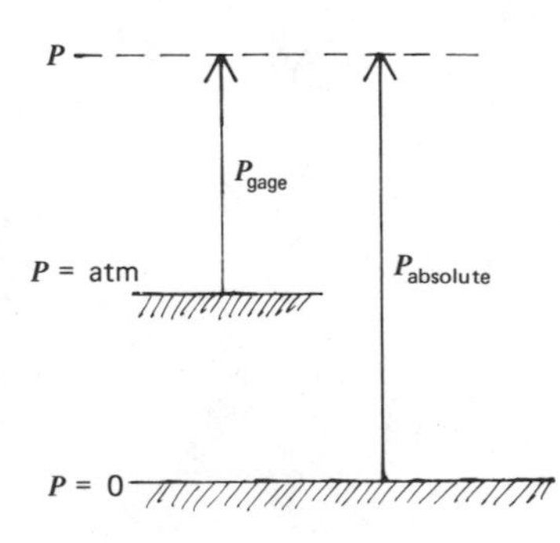

FIG. 7.2. The relationship of gage and absolute pressure.

pressure at the base of the column. Since head is easily converted to pressure, the two terms are frequently used interchangeably. For example, a column of air extending vertically from the earth at sea level through the atmosphere into outer space where the pressure is absolute zero would exert a force on the earth equivalent to the weight of the column of atmosphere. If the column of atmosphere were one square inch in cross section, then the sea level pressure would be 14.7 psi absolute as shown in Figure 7.3. The atmospheric pressure at an elevation higher than sea level would be less since the length of the overhead column of atmosphere would be less. The atmospheric pressure decreases as elevation increases. The term head could refer to columns of fluid other than air, such as water or mercury.

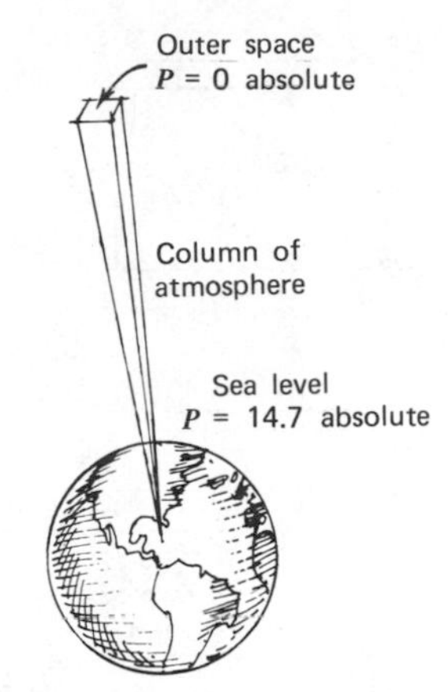

FIG. 7.3. A head of atmosphere.

One might wonder if the identification of gage and absolute pressure is worth the bother. How significant is a pressure differential of one column of atmosphere or 14.7 psi? Atmospheric pressure pushing down on an average-sized human head (A = 44 in.^{2}) at sea level results in a vertical force of roughly 650 lb. Also, if atmospheric pressure were reduced only by a factor of 2, then the fresh-water lakes would boil at a temperature of roughly 175F rather than 212F. If atmospheric pressure were reduced drastically, then some materials, such as Nylon, would experience so-called "outgassing" and the material would lose mass to the environment. This phenomenon has been experienced in spacecraft. The sea level atmospheric pressure of 14.7 psi absolute is seldom a negligible amount when compared to other pressures commonly found in a problem.

7.3/Fluid Statics

Fluids respond differently than solids when subjected to external forces as shown in Figure 7.4. As a structural mass, an incompressible fluid can provide reliable support only when loaded in compression. Although there are a few liquids that can develop a minimal amount of tension, liquids separate when tension forces tend to pull the fluid continuum apart. Solids and fluids differ significantly in their response to shearing forces that attempt to twist or slide parallel layers of material relative to each other. As shown in Figure 7.5, a shearing force F applied to an elastic solid results in a distortion. If the shearing force were removed, the elastic solid would return to its original unstretched position unless the

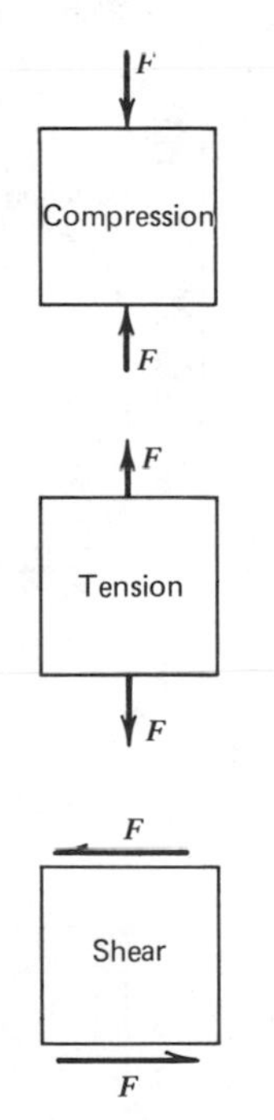

FIG. 7.4. Shear and tension forces in most fluids are considered negligible.

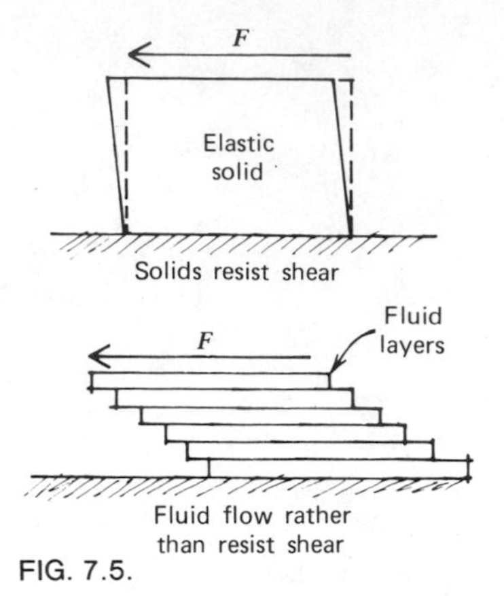

FIG. 7.5.

applied force exceeded the elastic limit of the solid and caused some permanent deformation. A fluid behaves differently than a solid in that it cannot resist the action of a shearing force and the fluid would not return to an original configuration. A fluid is essentially inelastic and deforms continually or flows when subjected to shearing forces. Once a fluid begins to flow there are some internal friction forces that develop between moving fluid layers.

The analysis of internal and external forces associated with a fluid continuum at rest is simplified by the fact that the fluid can experience compression only. At the interface of a liquid and a solid such as a container, dam or other submerged object, the inability of a liquid to resist shear eliminates any tangential forces or components of force. The resultant force at the liquid-solid interface must be perpendicular (normal) to the surface. Also, since the solid surface cannot pull on the liquid and induce tension, the normal force must be compressive. The interactive forces at a liquid-solid interface can be represented on a free-body diagram as normal compression as shown in Figure 7.6.

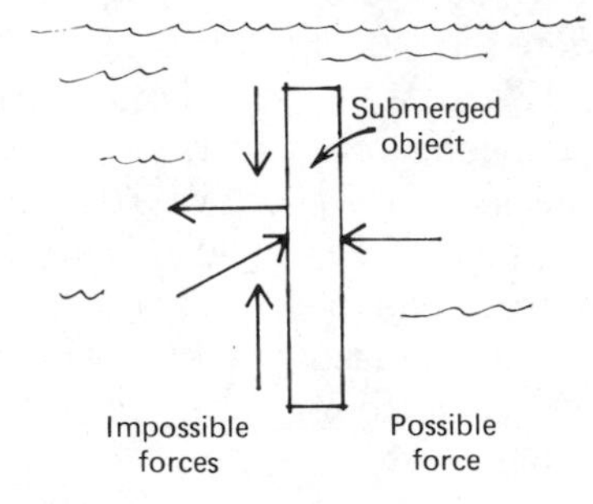

FIG. 7.6. Fluid forces act normal to submerged surfaces.

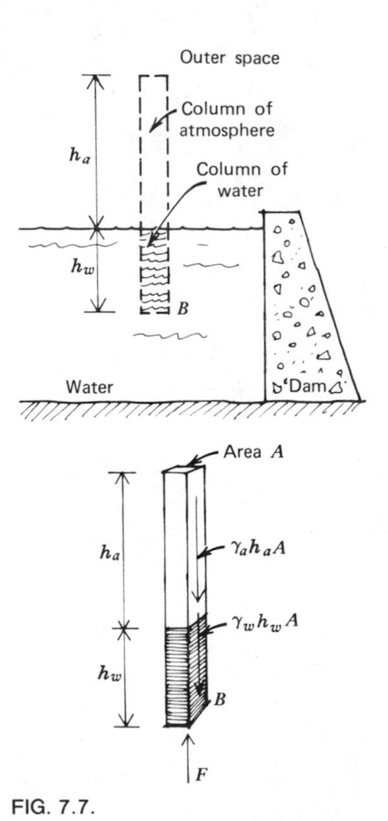

FIG. 7.7.

The magnitude of the internal pressure at any arbitrary point in a static fluid, such as point B in Figure 7.7, can be determined by applying the equations of equilibrium to the forces that act on an overhead column of fluid. A vertical column of both water and atmosphere with a cross-sectional area of A is removed from the open reservoir. Since there are no shearing forces in a static fluid, the tangential forces along the edge of the column are zero and the total weight of the water and air above the point B is supported by an upward normal force F. The weight of the water in the column is $\gamma_w h_w A$, where γ_w is the specific weight of water, h_w the length of the water column, and A the cross sectional area. Similarly, the dead weight of the atmosphere is $\gamma_a h_a A$. Since the fluid is at rest, the resultant force in the fluid column must be zero, therefore from equilibrium conditions

$$F = \gamma_w h_w A + \gamma_a h_a A$$

The absolute pressure at point B can then be found by dividing the force F by the cross-sectional area A, which results in the equation

$$P = \frac{F}{A} = \gamma_w h_w + \gamma_a h_a$$

If gage pressure were desired, then the column of atmosphere would not be included and the gage pressure at some arbitrary point B would be due to the water column only and

$$P = \gamma_w h_w$$

As evident from the above equations, internal pressure in a static fluid depends on the depth, but not the volume. The pressure at a depth of 10 ft in a large reservoir of water is the same as in a small reservoir at the same depth or even the same as the pressure at the base of a 10-ft vertical tube filled with water.

Now that the magnitude of the pressure at any arbitrary point can be found, what about the direction? The pressure at a point in a fluid at rest is equal in all directions and is said to be hydrostatic, as shown in Figure 7.8. This condition of hydrostatic pressure results from the absence of shear and can be proven by equilibrating forces on a small element of static fluid. The concept is, however, intuitively acceptable and the rigorous proof will be covered in subsequent courses.

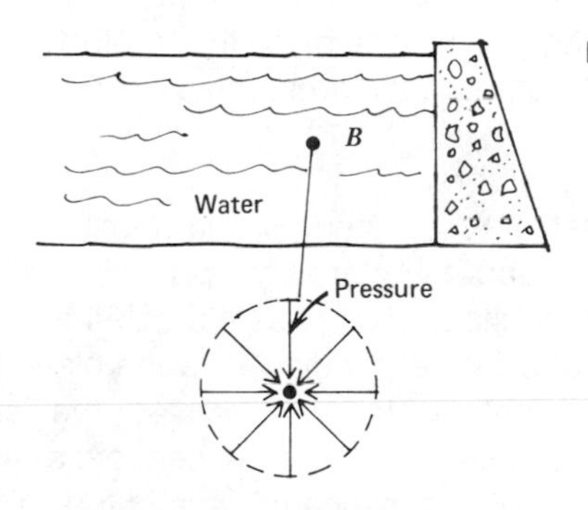

FIG. 7.8. Hydrostatic pressure.

7.3.1/Submerged Objects

Forces on objects submerged in an incompressible fluid such as water can be determined by multiplying the pressure of the water by the area on which the pressure acts. If the submerged surface is horizontal, as shown in Figure 7.9, then the depth is constant and the pressure distribution is uniform across the area. The total force F is the product of the uniform pressure times the area and acts at the centroid.

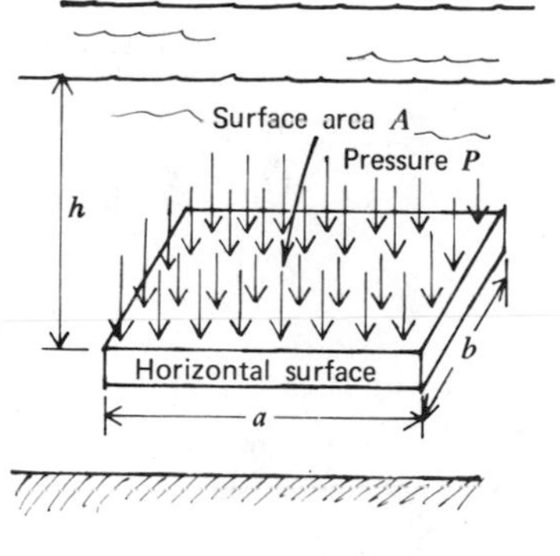

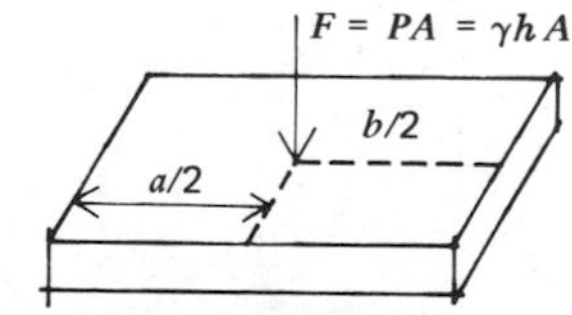

FIG. 7.9. The resultant force on a horizontal submerged surface.

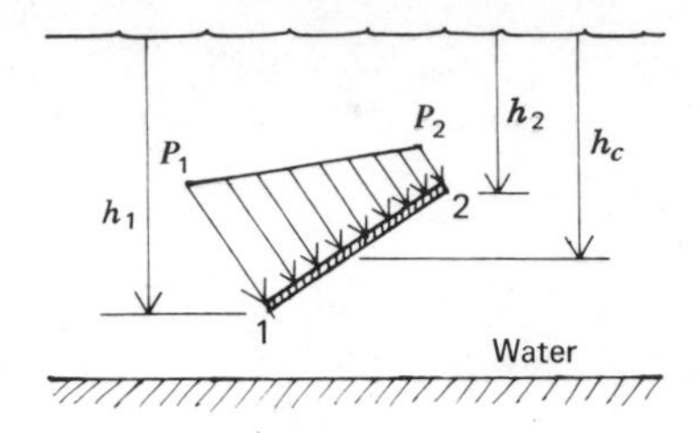

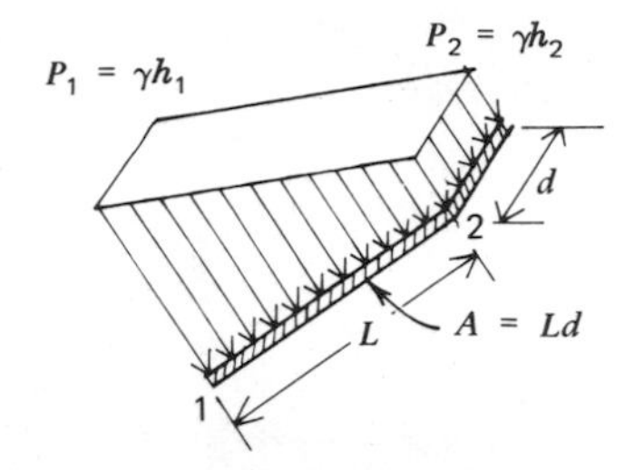

FIG. 7.10. The pressure prism on an inclined submerged surface.

If the submerged area is not horizontal, then the pressure varies linearly with the depth and can be pictorially represented as a trapezoid or so-called pressure prism. The magnitude of the resultant force is the volume of the pressure prism and the resultant acts at the centroid of the prism. To determine the resultant force acting on one side of a submerged plate, consider a plane area A of length L and uniform width d, as shown in Figure 7.10. The pressure P_1 at depth 1 is γh_1 and extends uniformly across the width of the plate. At depth 2, the pressure $P_1 = \gamma h_2$ and also extends across the plate width. The volume of the pressure prism is the average pressure, $\frac{1}{2}(P_1 + P_2)$, times the surface area A. The magnitude of the resultant force on the plate is equal to the volume of the pressure prism, therefore,

$$F = \left(\frac{P_1 + P_2}{2}\right)A = \left(\frac{h_1 + h_2}{2}\right)\gamma A$$

If one defines

$$\left(\frac{h_1 + h_2}{2}\right) = h_c$$

then the magnitude of the resultant force on a submerged plate of constant width can be written as

$$F = \gamma h_c A$$

where γ is the specific weight of the liquid, h_c the depth to the centroid of the plate, and A is the surface area on which the pressure acts.

The location of this resultant force must be selected such that the moment of the distributed force system is equivalent to the moment of the resultant force about any arbitrary point. It can be shown from statics that this resultant must pass through the centroid of the pressure prism and be perpendicular to the plate as shown in Figure 7.11. Don't confuse the two centroids that are discussed. The resultant force passes through the centroid of the pressure prism. The magnitude of this resultant is computed using the average depth of the plate, which is referred to as the centroid of the plate. The two centroids are generally not identical.

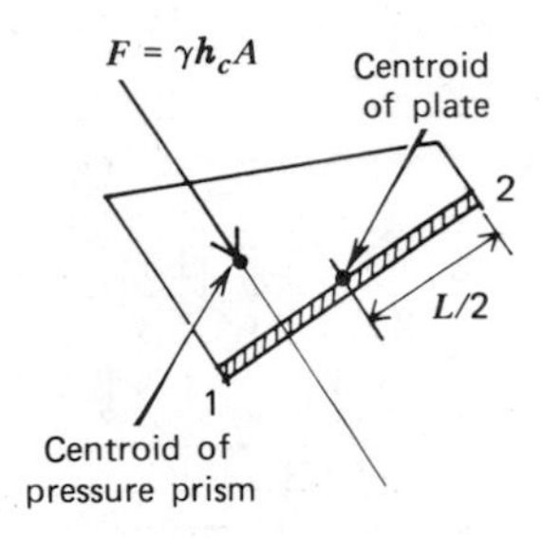

FIG. 7.11. The resultant force on an inclined submerged surface.

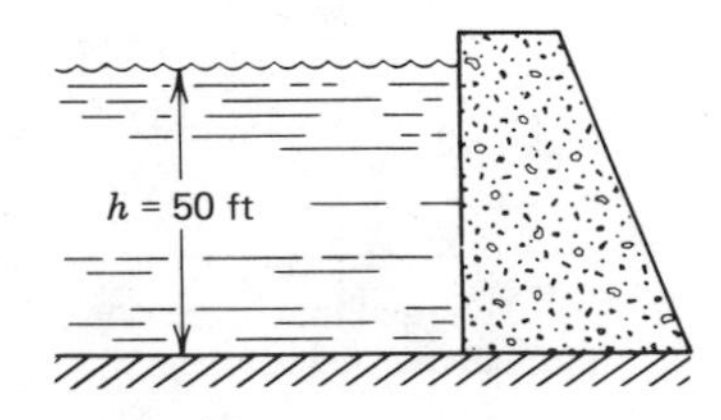

EXAMPLE PROBLEM 7.1

Determine the force of the water on a dam that is 100 ft wide and has a water depth of 50 ft.

Solution

The gage pressure P_1 at the water surface is zero measured relative to the atmosphere. The pressure P_2 can be found to be

$$P_2 = \gamma h_2 = (62.4)(50) = 3120 \text{ psf}$$

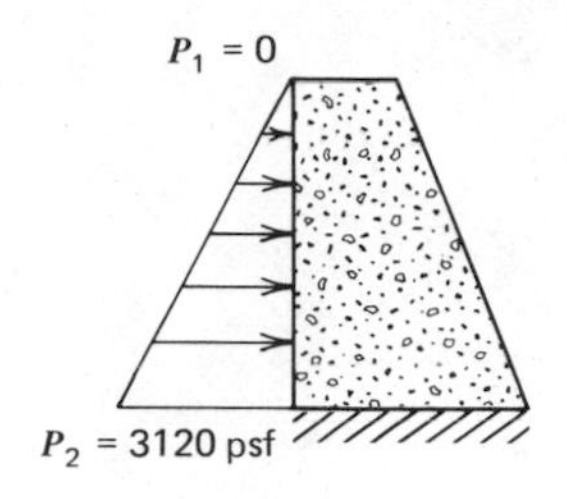

The magnitude of the resultant force can be found as the volume of the pressure prism

$$F = \left(\frac{P_1 + P_2}{2}\right)A = \left(\frac{3120}{2}\right)(50)(100) = 7.8(10)^6 \text{ lb}$$

This resultant acts at the centroid of the pressure prism, which is one-third up from the bottom and at the midpoint of the 100-ft span.

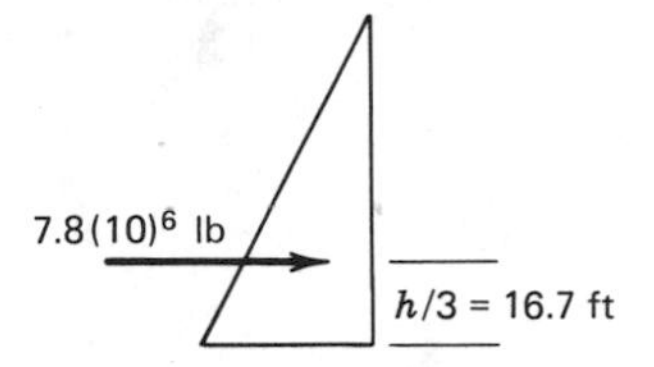

An alternative method is to find the pressure at an average depth and then multiply by the area to find the force.

$$F = \gamma h_c A = (25)(62.4)(50)(100) = 7.8(10)^6 \text{ lb}$$

EXAMPLE PROBLEM 7.2

A gate 6 ft high and 4 ft wide is placed in a dam as shown. The gate is hinged at A and rests against a sill at point B. Determine the magnitude and location of the resultant force exerted on the gate by the water when the water level is 25 ft.

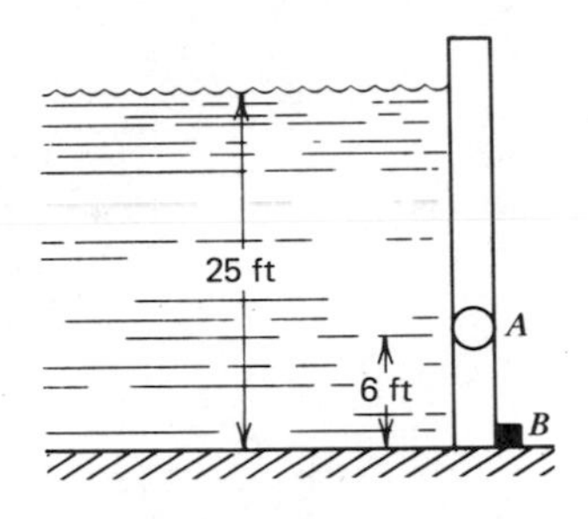

Solution

The magnitude of the resultant force is the volume of the pressure prism. The pressures that form the sides of the trapezoid are computed as

$$P_A = \gamma h_A = (62.4)(19) = 1185.6 \text{ psf}$$

$$P_B = \gamma h_B = (62.4)(25) = 1560 \text{ psf}$$

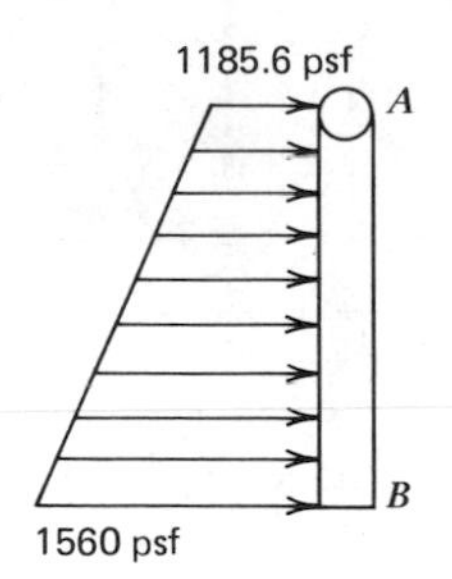

The magnitude of the resultant force follows as

$$F = \left(\frac{1185.6 + 1560}{2}\right)(24) = 3.29(10)^4 \text{ lb}$$

This same resultant force could also be computed by

$$F = \gamma h_c A = (62.4)(25.3)(24) = 3.29(10)^4 \text{ lb}$$

The point of application of the resultant force is at the centroid of the trapezoid. For convenience, represent the trapezoid as the sum of a triangle and a rectangle.

The magnitude of F_1 and F_2 are computed as

$$F_1 = (1/2)(374.4)(24) = 4492.8 \text{ lb}$$

$$F_2 = (1185.6)(24) = 28454.4 \text{ lb}$$

The location of the resultant force can be found by taking moments about the point B.

$$Fd = F_1 \frac{h}{3} + F_2 \frac{h}{2}$$

Upon substitution, the lever arm distance of the resultant force is

$$d = 2.86 \text{ ft}$$

7.3.2/Buoyancy

Some of the principles of fluid statics were applied to problems before the time of Christ. An ancient Greek scholar named Archimedes was challenged by his king to determine the authenticity of the crown. The king suspected that someone had fraudulently alloyed the precious metals. As part of Archimede's solution, he formulated the principle of buoyancy, which states that a body wholly or partially immersed in a fluid is buoyed up with a force equal to the weight of the fluid displaced by the body.

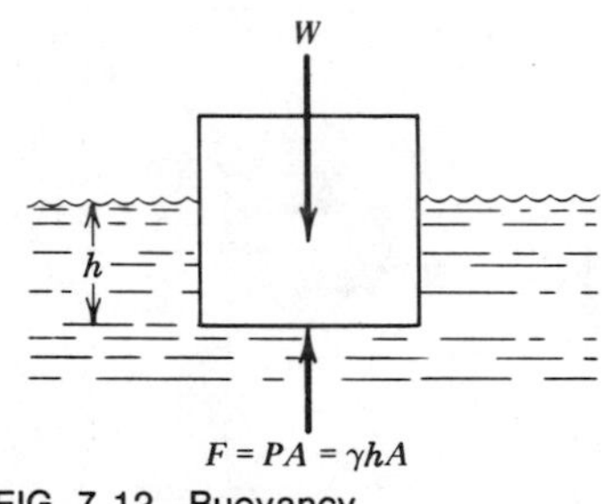

FIG. 7.12. Buoyancy.

Although Archimedes supposedly observed the principle while entering a swimming pool, his statement can be verified using concepts of fluid mechanics. Consider a rectangular block partially immersed in water and in static equilibrium, as shown in Figure 7.12. The downward force is the weight of the block W. The upward buoyancy force F on the bottom surface of the block is

$$F = PA = \gamma hA$$

where A is the horizontal surface area, γ is the specific weight of the fluid, and h is the depth. There are no shear forces along the sides. Since the block is motionless, the summation of forces in the vertical direction must be zero, therefore

$$W = F$$

and upon substitution

$$W = \gamma hA = \gamma(\text{volume immersed})$$

The volume of the block immersed is hA and is equal to the volume of the incompressible fluid that is displaced. Since the volumes of both the immersed portion of the body and the displaced fluid are equal, bodies that have a specific weight less than that of the fluid will float and bodies that have a specific weight that exceeds that of the fluid will sink. If the weight exceeds the maximum buoyancy force, then the block moves downward. Nevertheless, the upward force on either the floating or sinking body is equal to the weight of the displaced fluid. Hence, the principle of Archimedes is an expression of the basic notions of fluid statics. How could you use these concepts to determine the density of a rigid body? Furthermore, if a body begins to sink will it continue until it reaches the bottom or will it reach an equilibrium position at some other depth?

EXAMPLE PROBLEM 7.3

A spherical underwater exploration capsule has a radius of 3 ft and weighs 10 000 lb. The capsule is completely immersed in fresh water and suspended by a single cable as shown. Determine the tension in the cable.

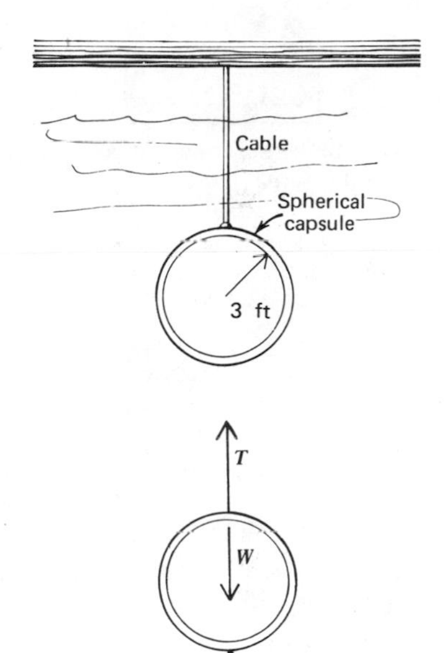

Solution

A first step is to draw a FBD and identify the tension force in the cable T, the capsule weight W, and the bouyancy force F.

$$F = (\text{volume of sphere})\gamma_W$$

$$F = \frac{4}{3}\pi r^3 \gamma_W = \frac{4}{3}\pi(3)^3 62.4 = 7057 \text{ lb}$$

Since the capsule is in equilibrium, the summation of vertical forces must be zero.

$$\Sigma F = T - W + F = 0$$

$$T = W - F = 10\,000 - 7057$$

$$T = 2943 \text{ lb}$$

7.3.3/Hydraulic Systems

An incompressible fluid confined in a hydraulic system provides a convenient medium to transfer force and power. Since the beginning of the century, power systems have played a key role in the development of agricultural machinery, construction, manufacturing, industrial processes, transportation, and recreation. Most everyone is familiar with hydraulic jacks, power steering, backhoes, tractors, presses, and other applications of hydraulics.

The working fluid in a hydraulic system could be either a liq-

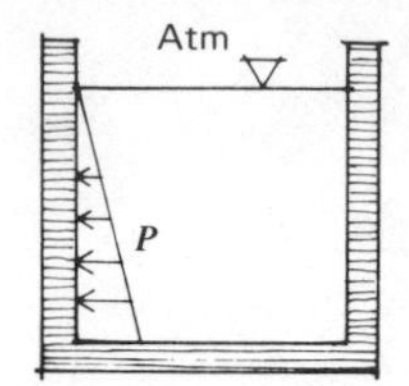

FIG. 7.13. Pressure due to the weight of the fluid.

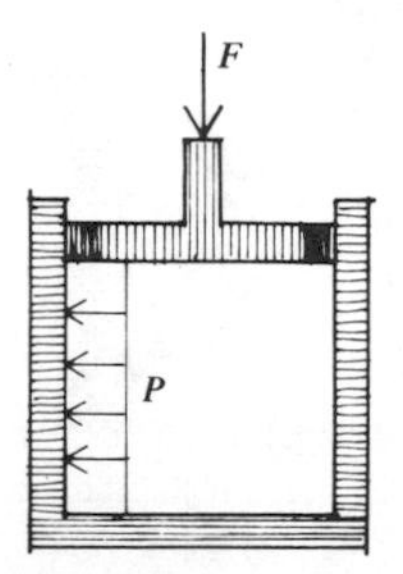

FIG. 7.14. Pressure due to an applied external force.

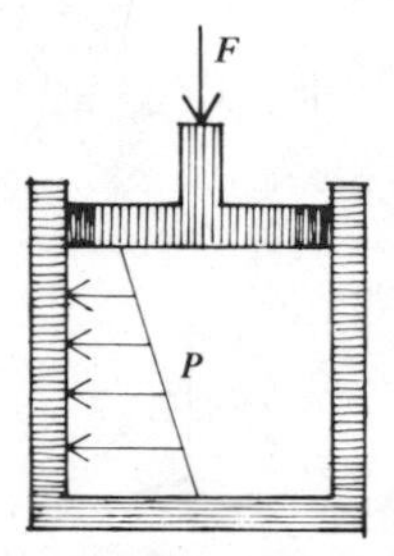

FIG. 7.15. Pressure due to both fluid weight and an applied force.

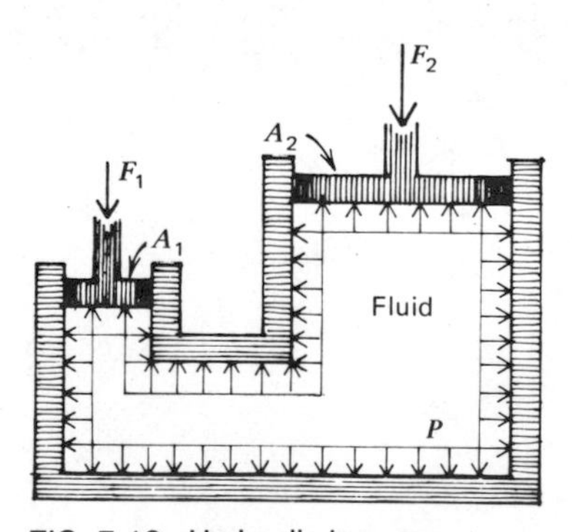

FIG. 7.16. Hydraulic leverage.

uid or a gas. Most hydraulic liquids have a petroleum base. Aqueous hydraulic liquids usually have small droplets of oil suspended throughout the medium to inhibit rust and provide lubrication. In addition to the physical properties of fluids commonly manipulated in fluid mechanics, a hydraulics design engineer must be concerned with other fluid characteristics that include foaming, demulsification, corrosion, cost, and availability. Although the hydraulic fluid might contain several additives, the principles of fluid statics apply to both aqueous and petroleum based liquids. Problems related to the design of power systems provide meaningful application of the fundamentals of fluid statics.

In an open cylinder, the gage pressure in the fluid at any depth is proportional to the length of the overhead column of fluid, shown in Figure 7.13. If the cylinder is fitted with a piston and an external force is applied that pushes on the entrapped fluid, then an additional uniform pressure, shown in Figure 7.14, is superimposed at every point throughout the medium. In the seventeenth century, Pascal stated that pressure applied to an enclosed fluid is transmitted undiminished to every portion of the fluid and the walls of the containing vessel. If the fluid is incompressible, such as water or oil, then the pressure induced by the piston is transmitted simultaneously throughout. In the case of a compressible fluid such as air, there is some time lag since the pressure change propagates through the medium at the speed of sound.

The total pressure at any point is equal to the static pressure due to the fluid weight plus the pressure induced by the piston, shown in Figure 7.15. The magnitude of pressure along only one side is shown, but one must remember that the pressure is equal in all directions at any given point. Frequently, the pressure resulting from the weight of the fluid is negligible when compared with the induced pressure.

Pascal's principle can be used to achieve a hydraulic leverage whereby an applied force is multiplied and a larger output force is available. In Figure 7.16, an input force F_1 is applied to a piston of area A_1 and the pressure F_1/A_1 is induced throughout the fluid. The resulting force F_2 on the second piston (neglecting pressure change for differences in height) is the product of the pressure F_1/A_1 times the area A_2, therefore,

$$F_2 = \left(\frac{A_2}{A_1}\right) F_1$$

where the ratio of the areas is the mechanical advantage of the hydraulic system. Hydraulic leverage is similar to a mechanical lever system in which the mechanical advantage is the ratio of the force lever arms. Hydraulic leverage explains why a cork forced into a bottle sometimes pushes out the bottom. The force on the bottom of the bottle is equal to the cork force times the ratio of the areas. Does the resultant force on a dam increase significantly when boats are on the reservoir?

Although forces can be multiplied through hydraulic leverage, the input work cannot be increased by the system. The work input would be equal to the applied force times the distance the piston traveled. The output work would be the product of the output force and the distance the piston moved. Since energy can neither be created nor destroyed, the most efficient system theoretically possible could only convert all the input energy to the output energy according to the first law of thermodynamics.

EXAMPLE PROBLEM 7.4

A man weighs 180 lb and stands on the 2-in. diameter piston as shown. Determine the diameter of the large piston required if the man is to support the weight of his 4000 lb car.

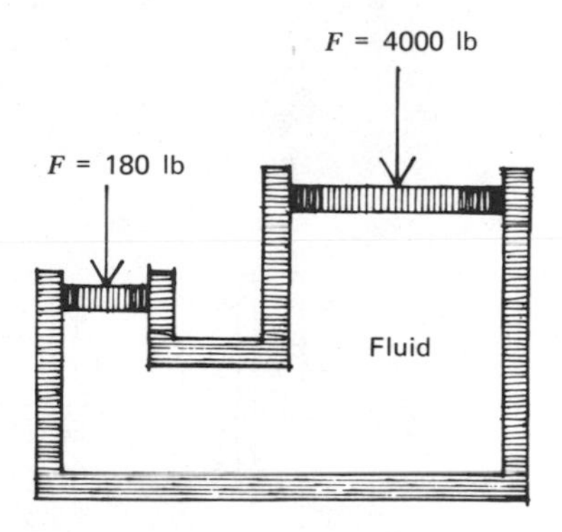

Solution

Assume that the pressure is uniform throughout the fluid, then

$$\frac{F_2}{A_2} = \frac{F_1}{A_1}$$

and

$$A_2 = \frac{F_2 A_1}{F_1} = \frac{(4000)}{180}\,\pi(1)^2$$

$$A_2 = 69.81 \text{ in.}^2$$

$$d_2 = 9.42 \text{ in.}$$

EXAMPLE PROBLEM 7.5

A hydraulic cylinder with a 4-in. diameter bore is used in the hoisting mechanism of a dump truck. The cylinder assembly is pin connected at both ends. If the dump bed and its load combined weigh 25 000 lb, determine the cylinder pressure required to support the load in the position shown.

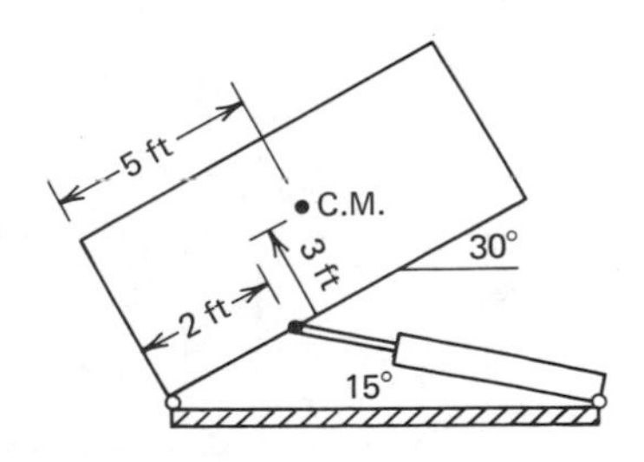

Solution

A FBD of the dump is drawn that identifies the forces. Since moments will be taken about point A, the pin reactions are not shown. The moment about point A can be written as

$$M_A = (25\,000)(5 \cos 30) - F(2 \sin 45°)$$

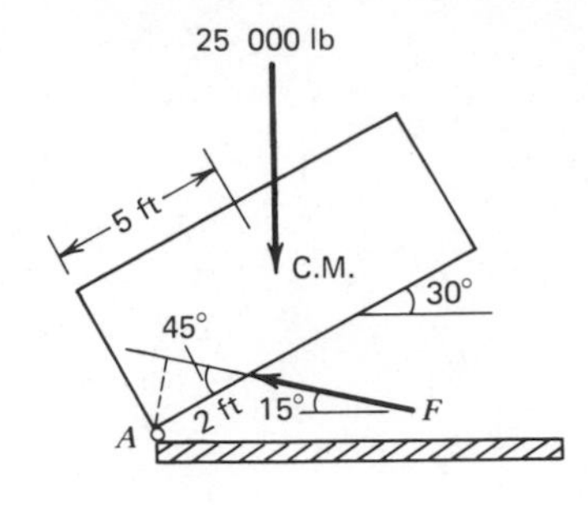

Since $M_A = 0$ for equilibrium, the force F can be determined as

$$F = 76\ 546.5 \text{ lb}$$

The cylinder pressure can be determined by dividing the applied force by the area of the cylinder.

$$P = \frac{F}{A} = \frac{76\ 546.5 \text{ lb}}{\pi 4} = 6091.4 \text{ psi}$$

7.3.4/Manometers

Most everyone is familiar with the measurement of temperature using a thermometer that contains a column of fluid such as mercury. The fluid expands or contracts according to the changes in ambient temperature. The linear displacement of the fluid column is calibrated to correspond to temperature in degrees Fahrenheit or Celsius.

A column of incompressible fluid, such as water or mercury, can also be used to make precise measurements of pressure. Such a device is called a manometer. The displacement of the fluid in a manometer is governed by the fundamental principles of mechanics. There are no moving mechanical parts that might introduce error; however, temperature, capillarity, surface tension and changes in fluid properties are possible sources of error and might require some consideration for extreme accuracy. To introduce the concept, consider an open reservoir of water as shown in Figure 7.17. A small tube is attached at a depth h and the reservoir water is free to flow into the tube and eventually come to equilibrium. When at rest, the top of the water in the manometer will be at the same elevation as the water in the reservoir. The pressure at the base of the thin column of water is identical to that at the same elevation anywhere in the large reservoir. Can you accept the fact that the water pressure depends on the vertical depth and not the total volume of the surrounding water? If the manometer were filled with a fluid that had twice the specific weight of water, then the manometer fluid level would only be half as high as the level of the water in the reservoir, shown in Figure 7.18.

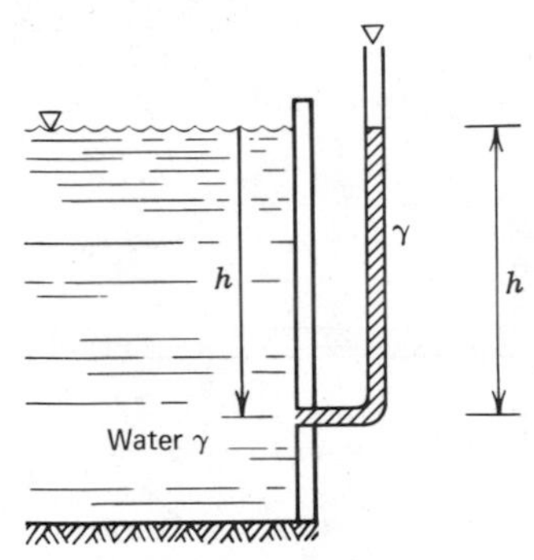

FIG. 7.17. A manometer measures pressure.

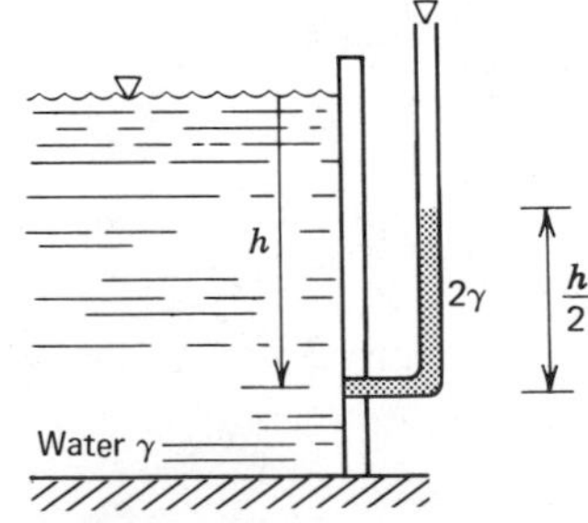

FIG. 7.18.

The concepts of fluid statics can be applied to a manometer and an equation that relates pressure, elevation, and specific weight can be determined. Consider a typical U-tube manometer attached to a pipe, pressure vessel, or some other closed container. One end of the manometer is open to the atmosphere. The pressure at the point of attachment is some unknown P_x. The container fluid fills part of the U-tube manometer and has a specific weight γ_f, shown in Figure 7.19. The second fluid in the manometer has a specific weight of γ_m and extends above the fluid interface a distance h_2. The pressure at point A is equal to the pressure due to the weight of the column of fluid above plus the pressure P_x.

$$P_A = P_x + \gamma_f h_1$$

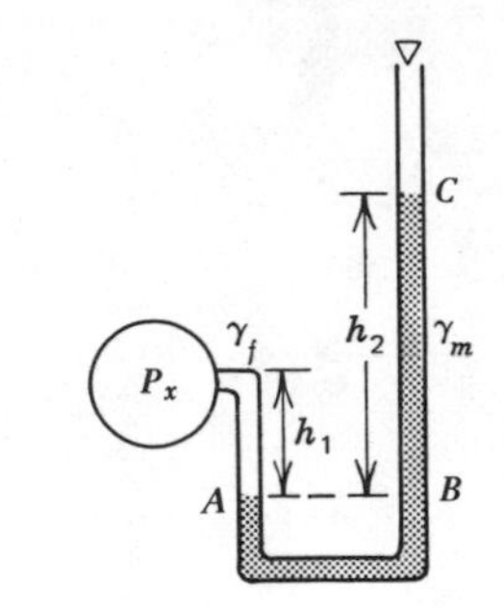

FIG. 7.19.

The absolute pressure at point B is equal to the pressure produced by the column of manometer fluid plus the column of atmosphere. If the atmospheric pressure is selected as the zero reference, then the gage pressure at point B is

$$P_B = h_2\gamma_m$$

Since the points A and B are at the same elevation, the pressures must be equal, therefore

$$P_A = P_B$$

and

$$P_x + h_1\gamma_f = h_2\gamma_m$$

The unknown gage pressure P_x can now be found as

$$P_x = h_1\gamma_m - h_1\gamma_f$$

For a closed container and an open manometer, the gage pressure P_x could be converted to absolute by adding the atmospheric pressure. If both the container and the manometer were open, then atmospheric pressure would cancel when equating P_A and P_B and the P_x given in the equation would still be gage pressure. Perhaps the safest way to solve monometer problems is to apply the general equation of statics rather than a special manometer formula.

EXAMPLE PROBLEM 7.6

The enclosed water tank contains air and water. The manometer is filled with mercury (s.g. = 13.57). Determine the reading of the pressure gage P_x.

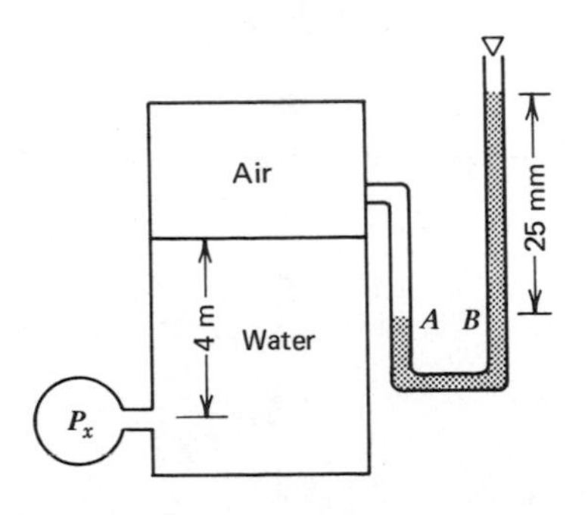

Solution

The specific weight of mercury, γ_m, is found as the product of the specific gravity and the specific weight of water, γ.

$$\gamma = 9.8(10)^3 \text{ N/m}^3$$

$$\gamma_m = (13.57)(9.8(10)^3) = 1.33(10)^5 \text{ N/m}^3$$

The gage pressure at point B due to the column of mercury is

$$P_B = \gamma_m h = 1.33(10)^5\,(0.025) =$$

$$3.32(10)^3\ \mathrm{N/m^2} = 3.32\ \mathrm{kPa} \qquad (1\ \mathrm{Pa} = 1\ \mathrm{N/m^2})$$

The pressure at point A is equal to the pressure at point B since they are both at the same elevation. The weight of the column of air over point A is negligible since air weighs approximately 1/1000 that of water. The gage pressure P_x is then the sum of the air pressure on the water surface plus the pressure from the overhead column of water.

$$P_x = P_B + \gamma h$$

$$P_x = 3.32 + (9.8)(4) = 42.5\ \mathrm{kPa}$$

PROBLEMS

7.1. (a) How many inches of water are equivalent to a standard atmospheric pressure of 14.7 psi absolute?

(b) How many millimeters of mercury (s.g. = 13.56)?

(c) How many millimeters of water?

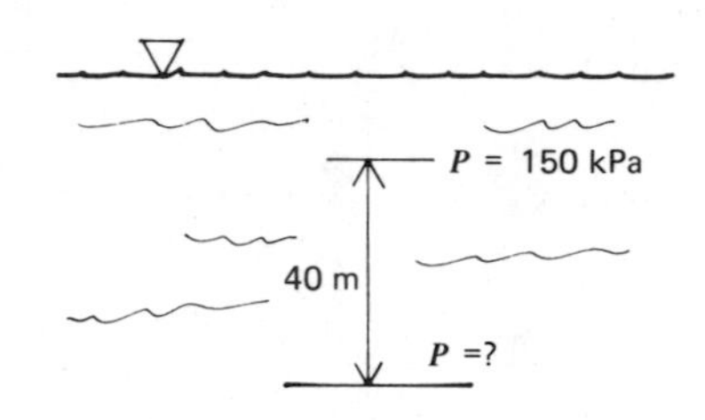

7.2. The pressure at a certain depth in the ocean is 150 kPa and the specific weight of the salt water is 10 kN/m^3. Determine the pressure at a depth 40 m below.

7.3. An open container at sea level contains oil, fresh water, sea water, and mercury as shown. Determine the absolute pressure at the bottom of the tank. The specific gravities are given.

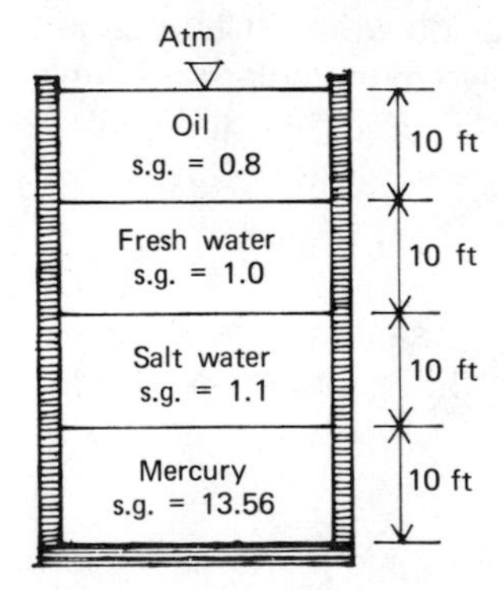

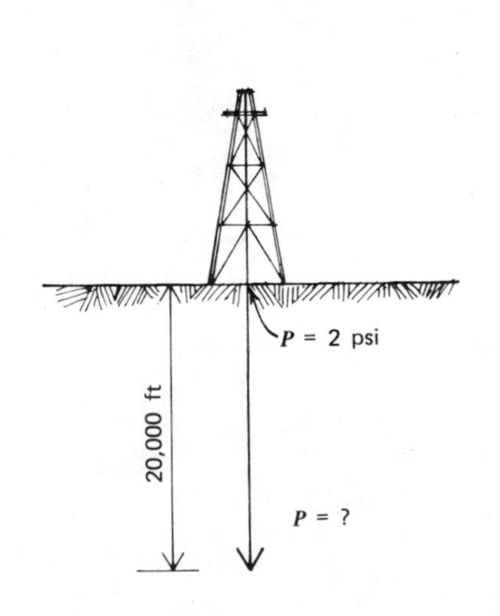

7.4. A representative of an oil company estimated that the costs of ultradeep drilling operations were roughly proportional to the square of the depth. Bit temperatures below 20 000 ft exceed 500 F. A drilling fluid is poured into the drill pipe to cool and lubricate the bit. Assume that the drilling fluid is a water-mud solution with a specific weight of 72 lb/ft^3. Estimate the gage pressure at a depth of 20 000 ft. Assume that the fluid has a pressure of 2 psi gage at ground surface.

7.5. The water in a reservoir is 55 ft deep. Determine the resultant force of the water on a 1-ft vertical strip of the dam.

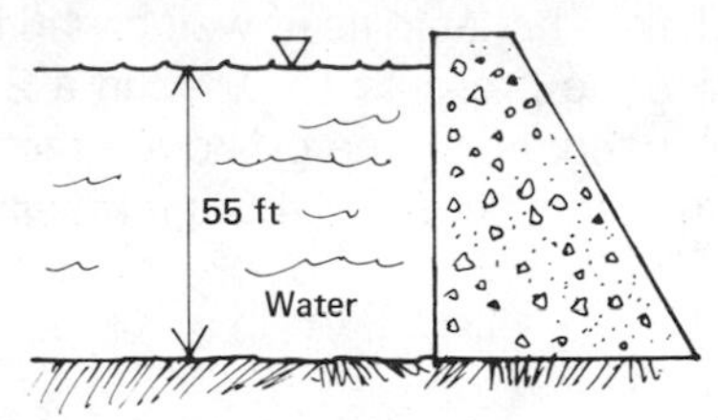

7.6. A small rectangular plate 1 m high and 0.5 m wide is used to dam an irrigation stream. A layer of silt 0.25 m deep with a specific weight twice that of water has accumulated in the bottom. Determine the magnitude of the resultant force exerted on the plate by both the silt and the water when the water is level with the top of the plate.

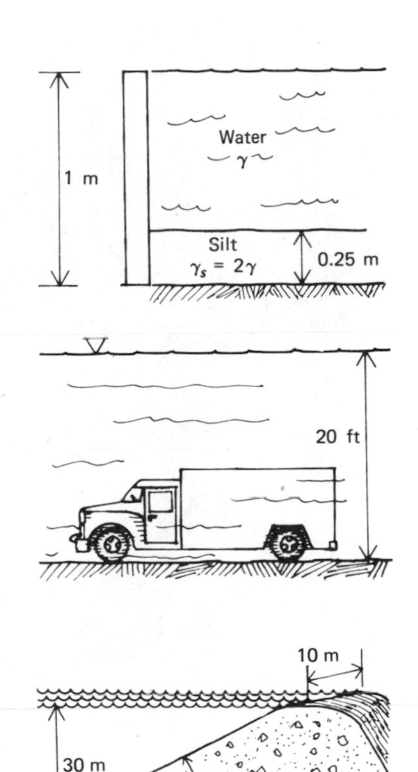

7.7. During an accident, a heavy truck left a bridge and quickly came to rest on the floor of a lake 20 ft deep. In the court case that followed, an attorney argued that the door must have jammed mechanically or the driver could have escaped. Estimate the resultant force of the water on the door. Assume that the rectangular door is 2.5 ft wide and 5 ft high. The door is vertical and the bottom edge of the door is 2 ft above the lake bottom.

7.8. A swimming pool has dimensions of 30 m x 10 m x 2 m and is filled with water. Determine the resultant force of the water on (a) the bottom, (b) a side, and (c) an end.

7.9. An earth dam 10 m wide is designed such that the weight of the water contributes a vertical force on the dam and increases the friction force at the base of the fill. Determine the total vertical force of the water on the dam. Assume water weighs 9.8 kN/m^3.

7.10. The army must be prepared to provide floating landing pads for helicopters in combat situations. To accomplish this, inflatable rubber floats are covered with AM2 landing mat. The rubber floats have a submergible thickness of 2 ft and are negligible in weight. The AM2 mats are aluminum panels that weigh 6.6 lb/ft^2. Determine the minimum dimensions of a square float that would support a 5000-lb helicopter on a swamp where muddy water has a specific weight of 70 lb/ft^3.

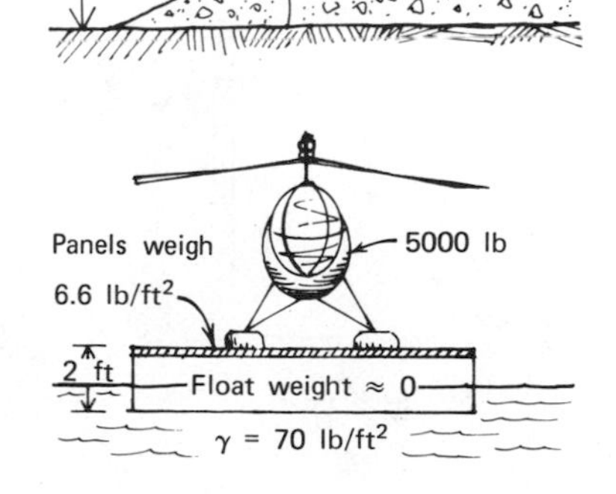

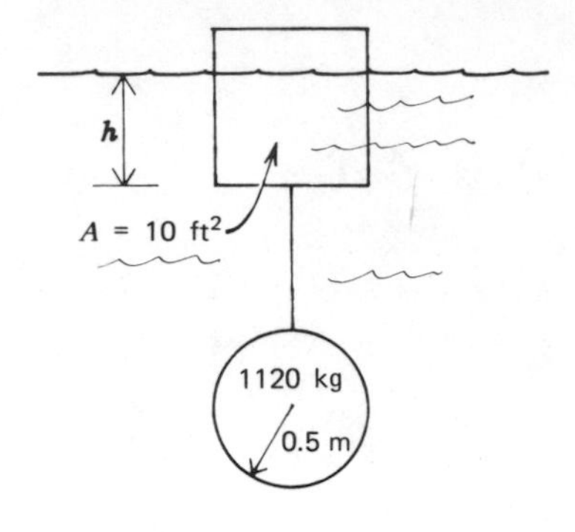

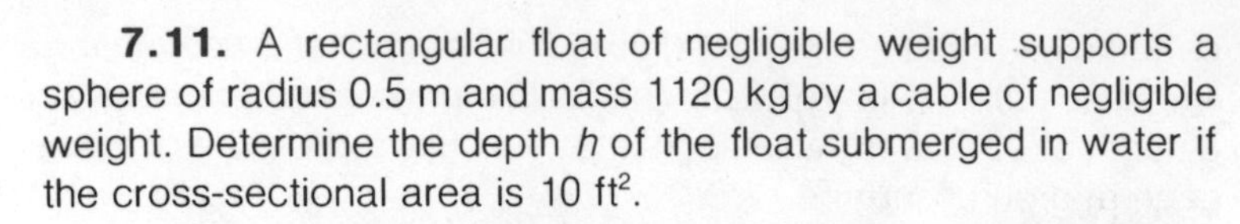

7.11. A rectangular float of negligible weight supports a sphere of radius 0.5 m and mass 1120 kg by a cable of negligible weight. Determine the depth h of the float submerged in water if the cross-sectional area is 10 ft^2.

7.12. A hydraulic cylinder is used to raise a horizontal table covered uniformily with 18 bales of hay that weigh 65 lb each on an automatic hay hauling machine. The table itself weighs 4.6 lb/ft^2 and is 9 ft square. Determine the pressure required in a 2-in. diameter cylinder to raise the loaded table. The cylinder attaches to the table at a point 4 ft out from the hinge and 4.5 ft from either side.

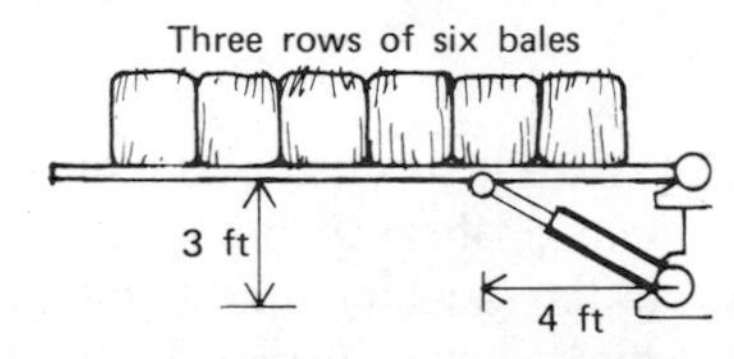

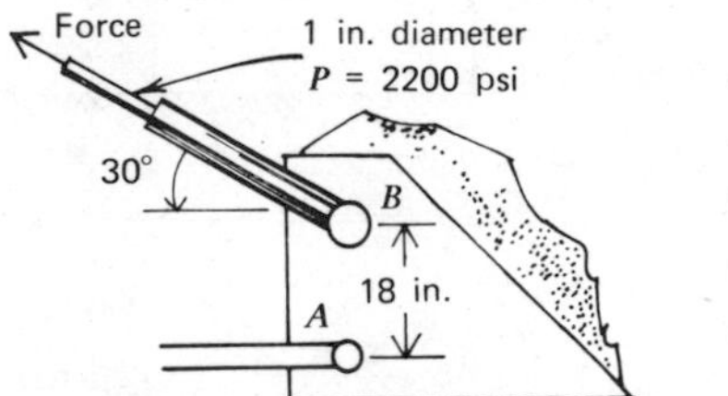

7.13. A front-end loader is pin connected at point A to the frame of a tractor. A 1-in. diameter hydraulic cylinder is attached to the bucket at point B, which is directly above point A. If the hydraulic system develops a pressure of 2200 psi, determine the maximum moment available from the cylinder to rotate the bucket about point A when in the position shown.

7.14. Compute the minimum pressure in a vertical forklift cylinder with a 3-in. bore that is required to lift a load of 2500 lb.

7.15. A 100-mm radius cylinder filled with water is fitted with a piston of negligible weight. A force F is applied to the piston. The manometer attached is filled with mercury (s.g. = 13.57) and is displaced as shown. Determine the force F.

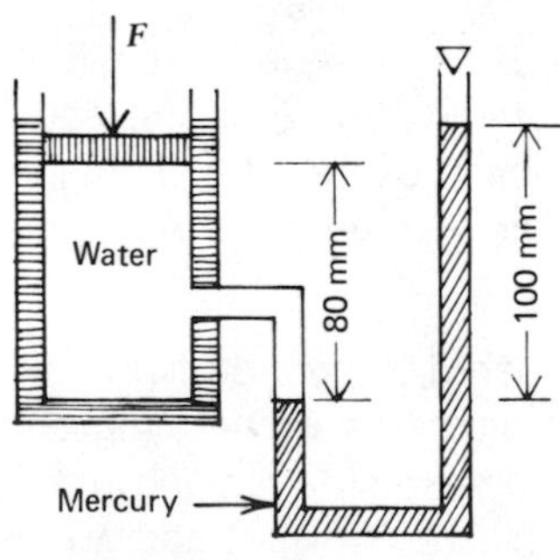

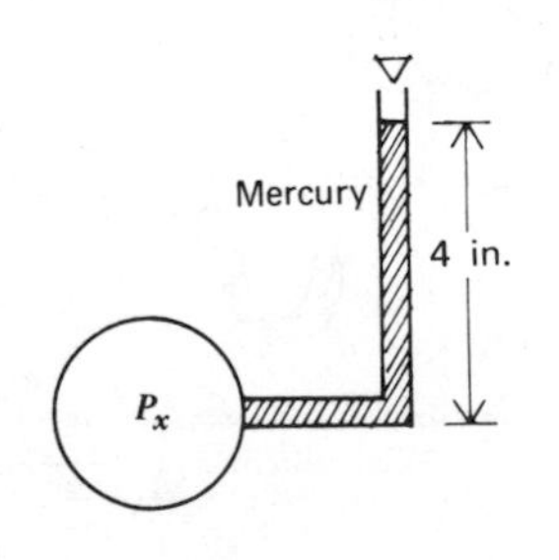

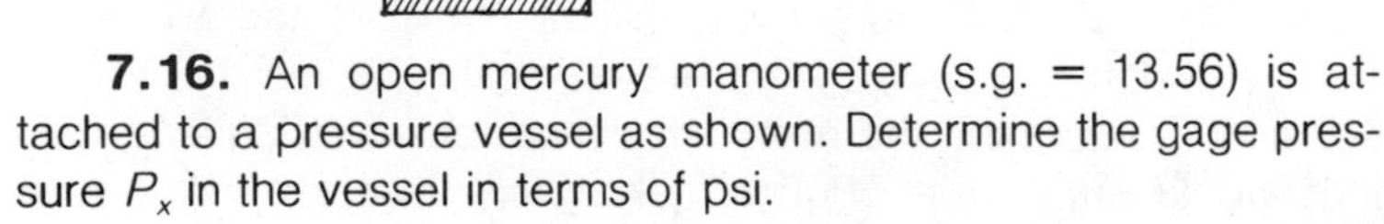

7.16. An open mercury manometer (s.g. = 13.56) is attached to a pressure vessel as shown. Determine the gage pressure P_x in the vessel in terms of psi.

7.17. A pipe has an internal pressure of 150 psi gage. Determine the elevation h of an inclined manometer filled with water attached to the pipe.

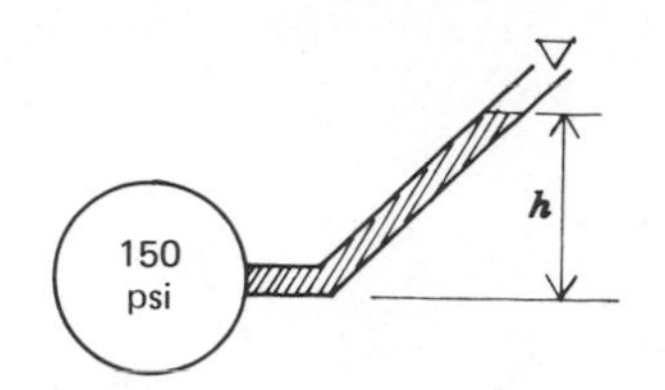

7.18. Tanks A and B contain water and are open to the atmosphere. The manometer has mercury (s.g. = 13.56) between the water columns. Determine the depth of the water in tank A.

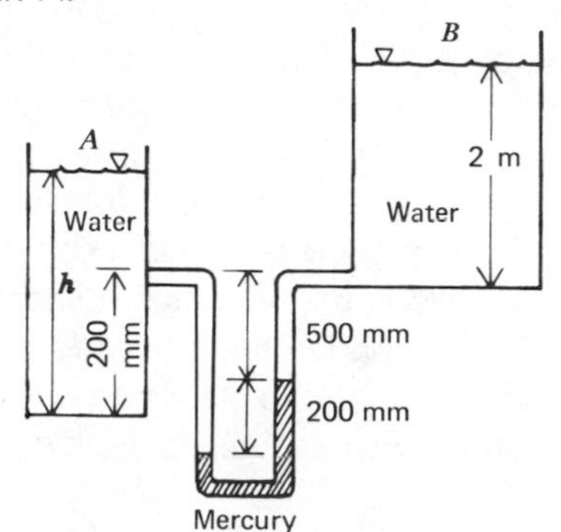

7.19. The tank contains air and water. Determine the gage pressure reading P at A. The manometer fluid is mercury.

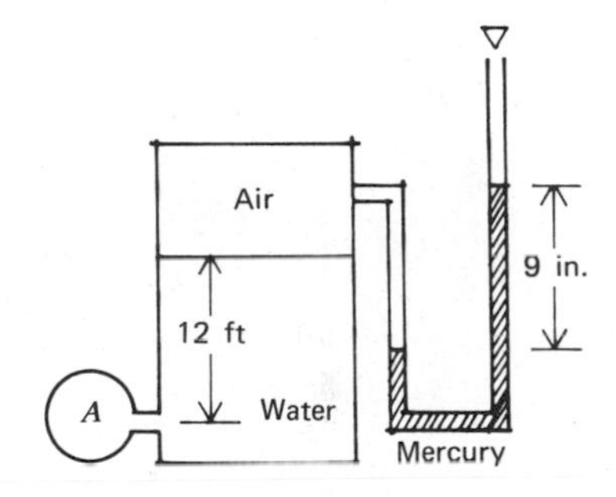

7.4/Fluid Flow

If the resultant force that acts on a volume of fluid is nonzero, then the fluid is not in static equilibrium and the fluid must flow in the direction of the resultant force. For example, as shown in Figure 7.20, the force of gravity pulls water downhill unless countered by a rock, dam, or some other structure that provides an equilibrating force. Petroleum flows through a horizontal pipe in the direction of the unbalanced force produced by the pump. The pipe wall provides a force equal and opposite to the force resulting from the fluid pressure and there is no flow in the radial direction. The winds blow in response to the unbalanced atmospheric forces that occur as a consequence of pressure differentials. The human heart must provide sufficient force to push the blood through the circulatory system of our body.

The study of fluid flow includes a variety of concepts that overlap multiple disciplines and extend in sophistication from undergraduate introductory courses to graduate programs on the

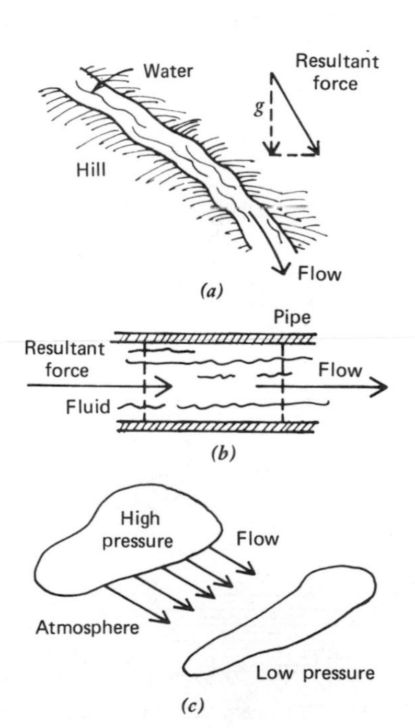

FIG. 7.20. Fluids flow in response to forces caused by *(a)* gravity, *(b)* pressure differentials in pipes, and *(c)* atmospheric pressure differentials.

frontiers of research. Liquids and other incompressible fluids are frequently introduced at the entry level since they maintain a constant density during the flow and their analysis is a logical extension of the application of the fundamentals of mechanics as applied to solids. Gases and other liquids with weak molecular cohesion experience density changes and require equations that include the interrelationship of thermodynamic properties such as pressure, temperature, volume, and density in addition to the laws of mechanics. In the remaining sections of this chapter, characteristics of incompressible fluid flow and some fundamental equations will be introduced. Each student is expected to learn the skills necessary to solve some meaningful problems that relate to the flow of liquids in pipes. Subsequent courses in fluid mechanics that usually require prerequisite skills in mathematics, dynamics and thermodynamics will eventually broaden the student's problem-solving capability to include such areas as open-channel flow, turbines, jet propulsion, shock waves, multiple pipe systems, and fluid measurements.

The flow of fluids, especially water, has challenged humans for several centuries. There are several archealogical evidences that suggest that early agrarian people diverted water for the irrigation of crops. Excavated ruins of the Minoan civilization indicate that flowing water was one of the many luxuries included in the palace of Cnossus in 1500 B.C. Roman engineers constructed aquaducts in the neighborhood of 200 B.C. Early Aztecs also built culinary water systems in South America. It is suspected that their early achievements were based on the art rather than science. The present science that describes the motion of fluids is a conglomerate of empirical and theoretical information that has been accumulated primarily since the fifteenth century. During this development, the results from actual experiments that have been used to define empirical equations and the predictions from theoretical calculations have steadily become closer together. At the present time, engineers are experiencing good agreement between theory and reality.

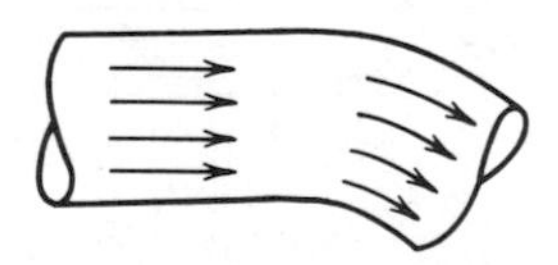

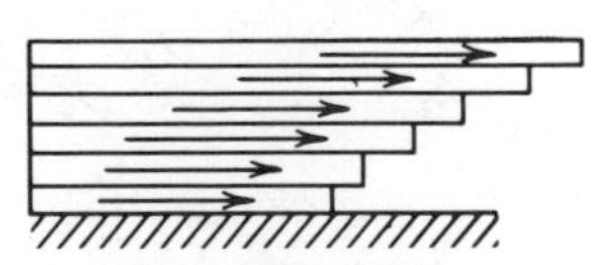

FIG. 7.21. Laminar flow.

Fluid motion is categorized into two general flow regimes. The laminar regime includes flows where the fluid particles translate along parallel paths, as shown in Figure 7.21. By analogy, visualize the flow of traffic along a multilaned freeway. Each car could be visualized as a fluid particle. A laminar traffic flow would allow the cars to travel at different speeds, around curves, and some could even stop; however, none could change lanes. Passing, the changing of lanes, tailspins, and U-turns would be analogous to the mixing of fluid particles in turbulent flow. The turbulent regime refers to flows characterized by the random motion of fluid particles, shown in Figure 7.22.

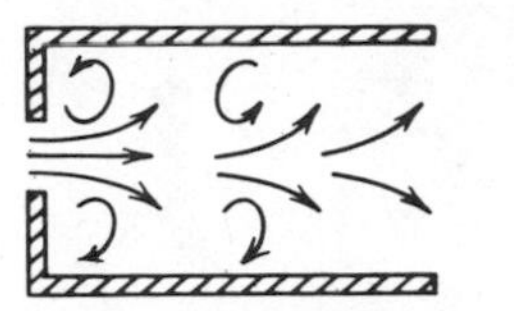

FIG. 7.22. Turbulent flow.

The frictional resistance to flow that exists internally in fluids or at boundary surfaces is called viscosity. The viscosity of a fluid can be determined by timing a sample as it flows under a constant head through a standard orifice at a constant temperature, shown in Figure 7.23. Thick fluids have large internal shear stresses that restrict the flow, consequently the time required to complete the sample flow is long and the fluid is said to have a high viscosity. The fluids that flow more readily have lower viscosity.

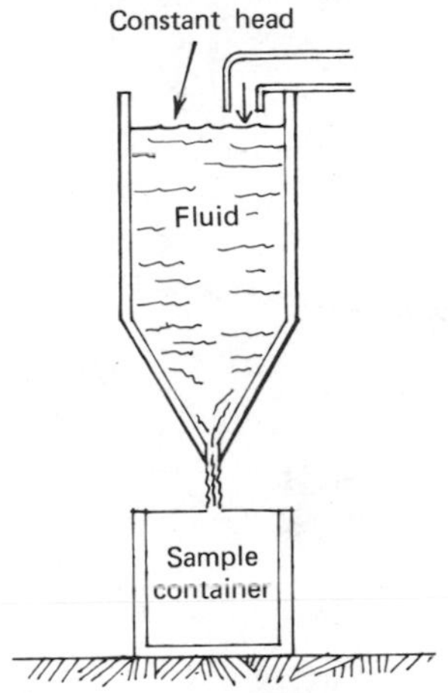

FIG. 7.23. A device to measure the viscosity of fluids.

The flow of a real fluid through a pipe could be laminar or turbulent and would also depend upon the viscosity of the fluid. The velocity of fluid particles would vary across the cross section of the pipe. Fluid particles in contact with the static pipe wall would have zero velocity. The velocity of the fluid would then increase as one moved from the pipe wall to the centerline of the flow. Typical velocity profiles for real fluids flowing in pipes are shown in Figure 7.24.

The mathematical complexity of the analysis of fluid flow can be simplified if the real fluid is replaced with an imaginary or ideal fluid. An ideal fluid is assumed to have no internal friction between layers of fluid and is therefore said to be inviscid. Also, there is no friction between the fluid and boundary walls to decrease the fluid velocity. Consequently, there is no special boundary layer. There would be no turbulence nor dissipation of energy resulting from friction. A typical velocity profile of an ideal fluid flowing in a pipe is shown in Figure 7.25. The velocity is uniform across the

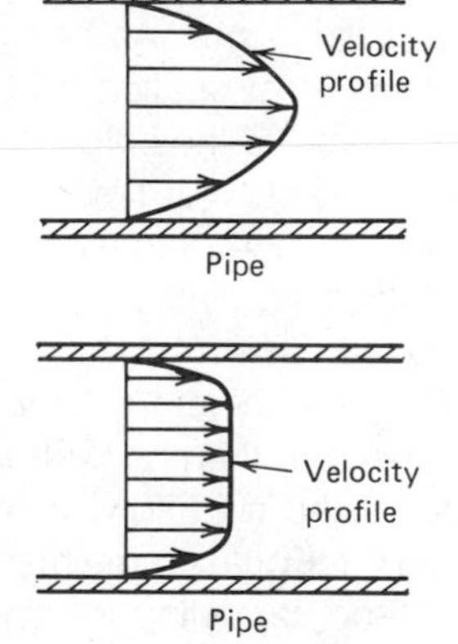

FIG. 7.24. Velocity profiles for a real fluid flowing in a pipe.

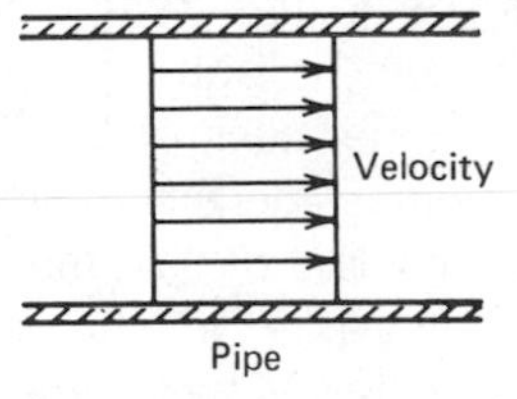

FIG. 7.25. Velocity profile for an ideal fluid.

cross section. The pressure variation is hydrostatic and the volume of fluid sliding through the frictionless pipe is similar to a rigid block moving on a smooth plane. The simulation of the real fluid with an ideal fluid is typical of other simplifying models used in engineering. The validity of such a simplification is sufficient to handle many of the fluid problems that are routinely solved by engineers. However, the use of an ideal fluid is a judgment decision that must be made by the analyst.

7.4.1/Continuity Equation

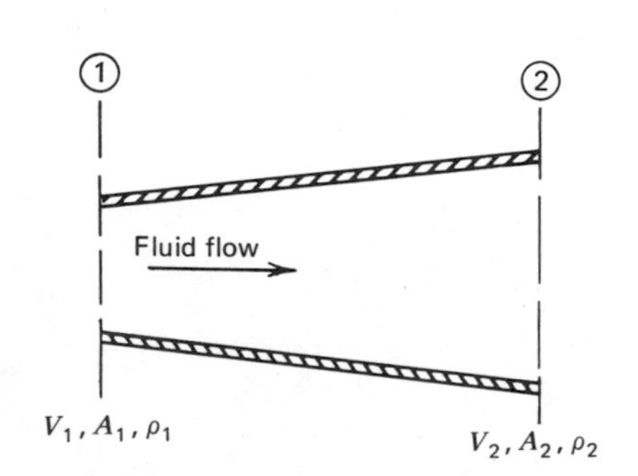

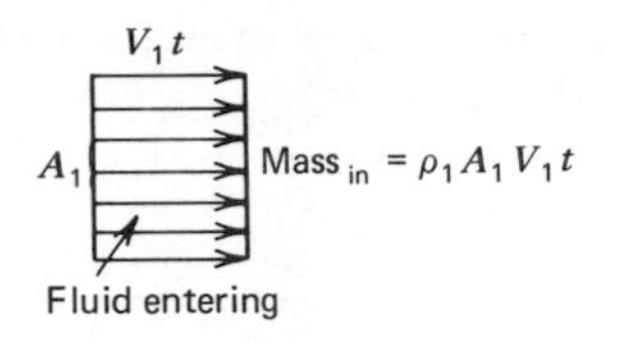

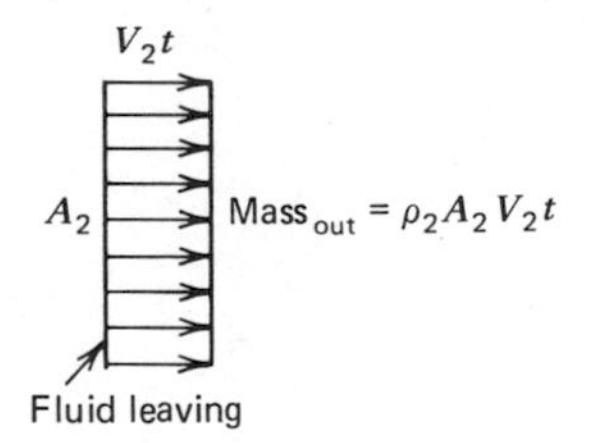

FIG. 7.26. Steady flow of an ideal incompressible fluid.

If a valve is suddenly opened and fluid begins to flow into an empty pipe, there is considerable turbulence and the flow is said to be unsteady. After some time, the churning random type motion of fluid particles disappears and the fluid settles into a steady flow. The condition of steady flow requires that all fluid variables such as pressure, density, and velocity remain constant with respect to time at any arbitrary location. One must be careful to separate steady flow from uniform flow. In contrast to uniform flow, steady flow does not require that the flow variables are identical at every location in the continuum. For example, in steady flow the velocity of water in a pipe may vary along the length of a pipe as the pipe diameter changes, whereas uniform flow would require the velocity to be constant along the length.

In Figure 7.26, assume steady flow of an ideal incompressible fluid through the conductor. Since the flow is steady, the mass of the fluid contained in the pipe section is constant. The pipe wall doesn't expand or contract to allow any accumulation or reduction in fluid. Since the law of conservation of mass states that matter can be neither created nor destroyed, the mass of the fluid flowing in must be equal to the mass flowing out. As shown in Figure 7.26, the mass of the fluid flowing in at section 1 during time t is the mass density ρ_1 times the volume of the entering fluid $A_1 V_1 t$. The volume of fluid that enters the section in time t is the volume of a cylinder with cross sectional area A_1 and length $V_1 t$. To help visualize $V_1 t$ as the length of the fluid cylinder, assume that the entering fluid velocity is 10 ft/s. If t were equal to 2 s, then all the fluid within a distance of 20 ft from the entrance would flow into the section. The mass flowing in can then be written as $\rho_1 A_1 V_1 t$. In a similar manner, the mass of the fluid flowing out in the same period of time t would be $\rho_2 A_2 V_2 t$. Equating the mass of the flow in to the mass of the flow out and canceling the time t, the so-called continuity equation follows as

$$\rho_1 A_1 V_1 = \rho_2 A_2 V_2$$

Since the flow is incompressible, $\rho_1 = \rho_2$ and the density can be canceled from both sides of the equation. Also, since the entrance and exit sections were chosen arbitrarily, the continuity equation would be the same for any other sections that might be selected.

In general, the continuity equation for incompressible flow can be written as

$$Q = AV = \text{constant}$$

where Q is the so-called volume flow rate and has dimension such as cubic meters per second (m^3/s) or cubic feet per second (ft^3/s). The continuity equation is an expression of the conservation of mass and is fundamental in the solution of fluid problems. It provides a simple relationship between fluid velocity and the cross sectional area of the conductor. In a steady flow, the velocity decreases when the area increases. Also, a decrease in the cross-sectional area causes an increase in the fluid velocity in order for Q to remain constant.

EXAMPLE PROBLEM 7.7

The velocity of water flowing steadily through a 100 mm diameter pipe is 8 m/s. Determine the velocity of the water when the pipe diameter is 200 mm.

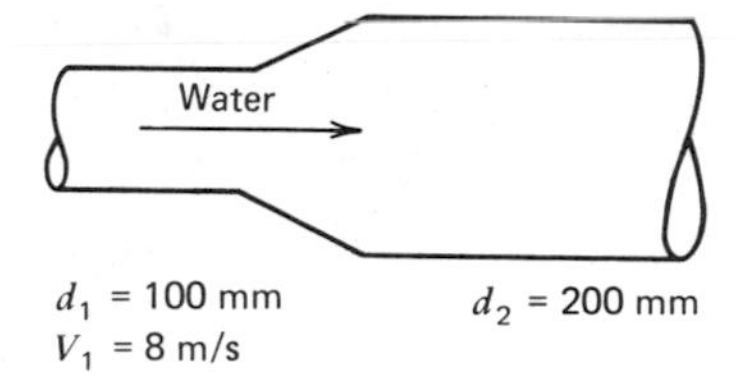

Solution

The volume flow rate can be determined as

$$Q = V_1 A_1 = (8)\frac{\pi(0.1)^2}{4} = 0.063 \text{ m}^3/\text{s}$$

Since the flow rate is constant and

$$V_1 A_1 = V_2 A_2$$

$$V_2 = \frac{Q}{A_2} = 2 \text{ m/s}$$

EXAMPLE PROBLEM 7.8

Water is flowing through a 6-in. diameter irrigation pipe at the rate of 600 gpm. Determine the diameter of the reducer necessary to increase the velocity by a factor of 2.0. Assume a steady flow and a full pipe.

Solution

The flow rate can be determined as

$$Q = [(600 \text{ gal/min})(\text{ft}^3/7.48 \text{ gal})(\text{min}/60 \text{ s})] = 1.337 \text{ ft}^3/\text{s}$$

Since the area $A_1 = \pi r^2$, and $r = \frac{1}{4}$ ft, then

$$V_1 = \frac{Q}{A_1} = 1.337 \text{ ft}^3/\text{s} \left(\frac{16}{\pi}\right) = 6.809 \text{ ft/s}$$

The flow rate is constant and $V_2 = 2V_1$, therefore

$$A_2 = \frac{Q}{V_2} = \frac{(1.337)}{(2)(6.809)} = 0.098 \text{ ft}^3$$

and the diameter is

$$d_1 = 4.24 \text{ in.}$$

7.4.2/Bernoulli's Equation

An expression that relates pressure and velocity in fluids is called the Bernoulli equation. It can be developed by using Newton's laws of motion or by applying the pinciple of conservation of energy. In this introductory treatment, the concept will be presented on the basis of conservation of energy. The energy to be considered will appear in the form of kinetic, potential and so-called pressure energy.

The steady flow of an ideal fluid through a typical pipe section is shown in Figure 7.27. In some small time t, the mass of the fluid that will flow into the pipe at section one is given by

$$M_{in} = A_1 L_1 \rho_1$$

where A_1, L_1, and ρ_1 are the cross-sectional area, length and density of the entering fluid cylinder, respectively. The kinetic energy of the mass cylinder that will enter in time t can then be written as

$$\text{KE} = \frac{1}{2} M_{in} V_1^2$$

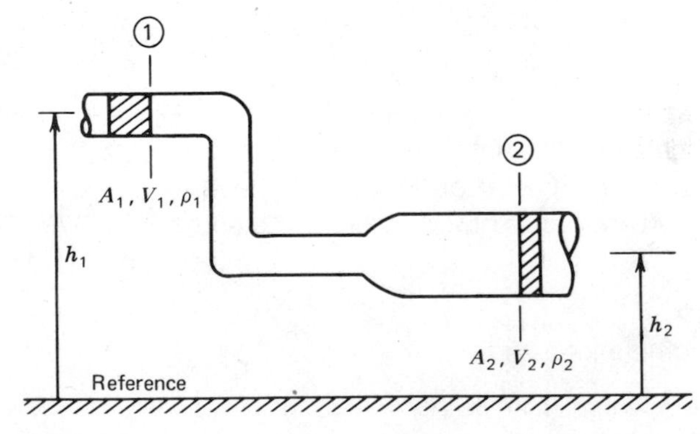

FIG. 7.27.

where V_1 is the entering velocity. In our case, the velocity profile at any cross section along the pipe is constant due to the ideal fluid assumption and the velocity of each entering fluid particle is the same. In a real fluid, the velocity of the entering mass would be an average velocity.

The mass cylinder entering also has potential energy with respect to some arbitrarily selected reference line. This type of energy is called potential since it could be recovered if the fluid mass were allowed to fall through the distance h_1 to the reference line. The potential energy of the entering mass can be written as

$$\text{PE} = M_{in} g h_1$$

where $M_{in}g$ is the weight of the mass element and h_1 is the elevation to the mass center.

There is another form of energy in a flowing fluid that is associated with the internal pressure pushing fluid particles downstream that is commonly referred to as flow work or pressure energy. A pressure acting on a surface area results in a force. The product of force times the distance through which the force moves has been defined previously as work and has the same units as energy. In fact, the first law of thermodynamics equates work and energy when the change in internal energy of the system is zero.

In order to understand pressure energy, consider again the mass of fluid that will enter the pipe section in Figure 7.27. The total force that pushes on the leading or upstream surface of the fluid element is the product of the average pressure P_1 and the cross-sectional area of the pipe A_1 as shown in Figure 7.28. If one assumes that the force P_1A_1 is constant as it pushes the mass cylinder through a short distance L_1 into the pipe section, then the work done is the product of force times distance and can be written

$$W = P_1A_1L_1$$

Although the pressure is hydrostatic or uniform in all directions at any point in the fluid stream, there is no work done in any direction other than the direction of flow since the fluid displacement is assumed to be one dimensional. The equation expressing the work done by the entering fluid can be manipulated. Since $M_{in} = \rho_1 A_1 L_1$, the term $A_1L_1 = M_{in}/\rho_1$ can be substituted and the flow work or pressure energy follows as

$$W = \text{Pressure energy} = \frac{P_1 M_{in}}{\rho_1}$$

The above description of pressure energy is intended to

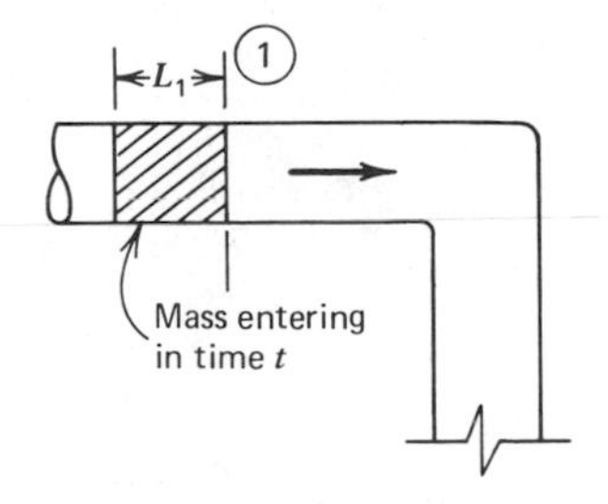

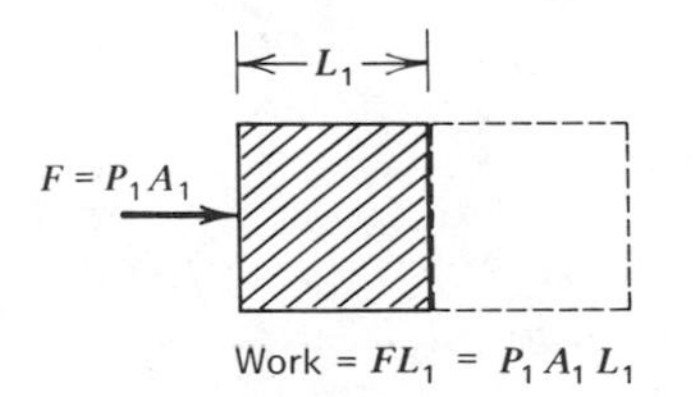

FIG. 7.28. Flow work.

assist in the visualization of the term. More rigorous and complete developments can be followed to derive the same term, but they require other prerequisite skills that have intentionally been avoided in this text. One must be careful to understand the difference between pressure energy and kinetic energy. In static fluids, both the pressure energy and kinetic energy are zero since there is no motion. In open jets the fluid stream is not constrained and there is no fluid pressure. The pressure energy is therefore zero, however, the kinetic energy associated with the moving mass is certainly nonzero.

The total energy of the fluid entering the pipe section in Figure 7.27 is the sum of the kinetic, potential and pressure energies and can be written

$$\text{Total energy in} = \frac{1}{2} M_{in} V_1^2 + M_{in} g h_1 + \frac{P_1 M_{in}}{\rho_1}$$

In a similar manner, the total energy at the output section could be derived as

$$\text{Total energy output} = \frac{1}{2} M_{out} V_2^2 + M_{out} g h_2 + \frac{P_2 M_{out}}{\rho_2}$$

Since an ideal fluid flows through the pipe, there is no friction between fluid layers or between the fluid and the pipe walls. There is no energy dissipated as the ideal fluid moves through the pipe, therefore the energy at section one must equal the energy at section two. Since the flow is steady, the mass flowing in during some time t must equal the mass flowing out, therefore $M_{in} = M_{out}$. Upon equating the two energies and canceling the mass, the equation follows as

$$\frac{1}{2} V_1^2 + g h_1 + \frac{P_1}{\rho_1} = \frac{1}{2} V_2^2 + g h_2 + \frac{P_2}{\rho_2}$$

If one divides both sides of the equation by g, and recognizes that $\gamma = g\rho$, then a popular form of Bernoulli's equation follows as

$$\frac{1}{2} \frac{V_1^2}{g} + h_1 + \frac{P_1}{\gamma_1} = \frac{1}{2} \frac{V_2^2}{g} + h_2 + \frac{P_2}{\gamma_2} = \text{constant}$$

Note that the units of each term are those used to measure head, such as feet or meters.

There are some interesting phenomena that can be partially explained by Bernoulli's equation. For convenience, if one assumes that the elevation change in a fluid flow is negligible, then

$$\frac{1}{2} \frac{V^2}{g} + \frac{P}{\gamma} = \text{constant}$$

Consequently, as the velocity of a fluid such as water or air increases, the pressure must decrease and vice versa. This explains why airplanes fly, shower curtains suck in when the water flows, canvas convertible tops puff up when the car drives down the road, and particles are drawn out the open windows of a moving vehicle. An increase in velocity reduces the pressure and a pressure differential results.

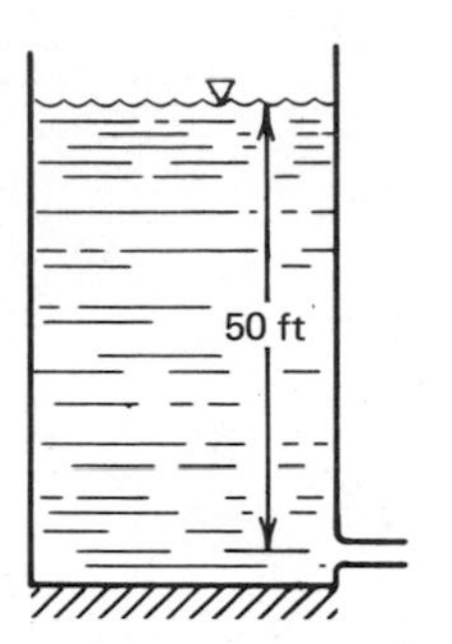

EXAMPLE PROBLEM 7.9

Compute the velocity of the water as it flows out through a cylindrical hole located 50 ft below the surface of the large open reservoir.

Solution

Assume that water is an ideal fluid and that energy is conserved. Bernoulli's equation is written to equate the energy at the reservoir surface to the energy in the pipe outlet at the base.

$$\frac{1}{2}\frac{V_1^2}{g} + h_1 + \frac{P_1}{\gamma_1} = \frac{1}{2}\frac{V_2^2}{g} + h_2 + \frac{P_2}{\gamma_2}$$

The gage pressure at both the top of the reservoir and the outlet is zero, therefore,

$$P_1 = P_2 = 0$$

Since the velocity of the reservoir is zero at the open surface, $V_1 = 0$. Also, $h_2 = 0$ if the reference line is selected to pass through the outlet. The nonzero terms remaining can be written

$$h_1 = \frac{1}{2}\frac{V_2^2}{g}$$

or

$$V_2 = \sqrt{2gh_1}$$

This is known as Torricelli's law. Upon substitution of $h_1 = 50$ ft, it follows that

$$V_2 = 56.7 \text{ fps}$$

EXAMPLE PROBLEM 7.10

The flow rate of water through the reducer is steady at 15 cfs. The pressure in the 8-in. diameter pipe at section 1 is 50 psi. Determine the pressure in the 4-in. diameter pipe at point 2.

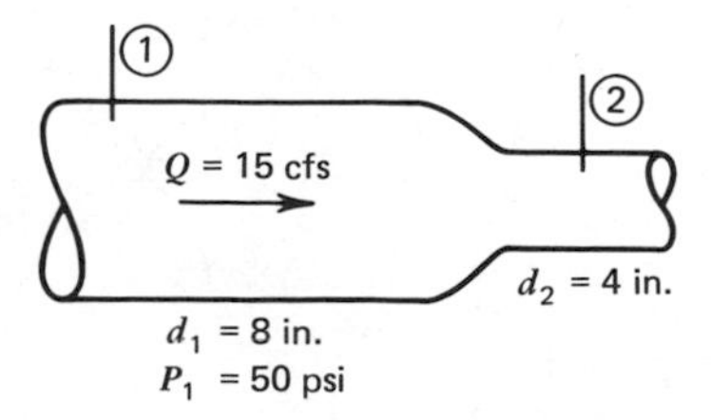

Solution

The velocity of the water at section 1 can be determined from the continuity equation.

$$Q = V_1 A_1$$

$$V_1 = \frac{Q}{A_1} = \frac{15}{(\pi/4)[8/12]^2} = 42.97 \text{ fps}$$

The velocity at section 2 can also be determined from the continuity equation.

$$V_2 = \frac{Q}{A_2} = \frac{15}{(\pi/4)[6/12]^2} = 76.39 \text{ fps}$$

The Bernoulli equation can be written with $h_1 = h_2 = 0$ and $\gamma_1 = \gamma_2 = \gamma$ as

$$\frac{V_1^2}{2g} + \frac{P_1}{\gamma} = \frac{V_2^2}{2g} + \frac{P_2}{\gamma}$$

Upon substitution of $\gamma = 62.4$ lb/ft^3, $V_1 = 42.97$ fps, $P_1 = 7200$ lb/ft^2, and $V_2 = 76.39$ fps, it follows that

$$P_2 = 3334.8 \text{ psf or } 23.2 \text{ psi}$$

PROBLEMS

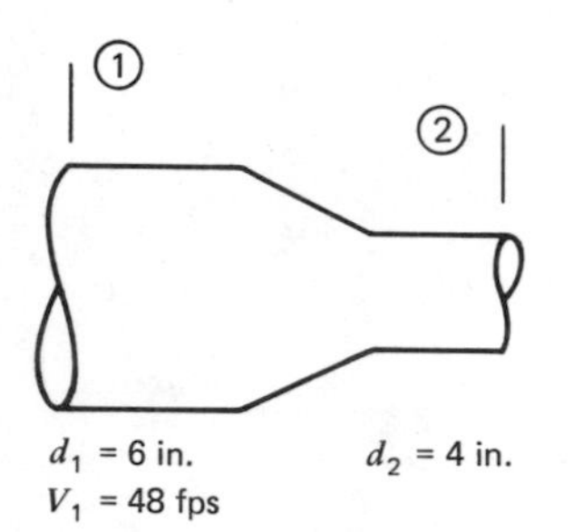

7.20. Water flows through a reducer and the flow is steady. If the water velocity is 48 fps in the 6-in. diameter pipe at section 1, determine the water velocity at section 2 in the 4-in. diameter pipe.

7.21. A bullet punctures a round 0.5 in. diameter hole in an open water storage tank at a depth of 20 ft. Estimate the maximum flow rate of water through the hole.

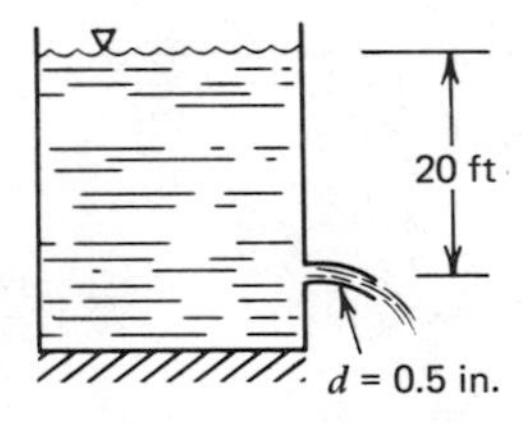

7.22. An 800 ft x 400 ft rectangular sewage lagoon is built for a small community and is initially empty. If the effluent were to flow steadily through the 2 ft diameter pipe at 8 fps, estimate the minimum time required to fill the lagoon to a depth of 6 ft.

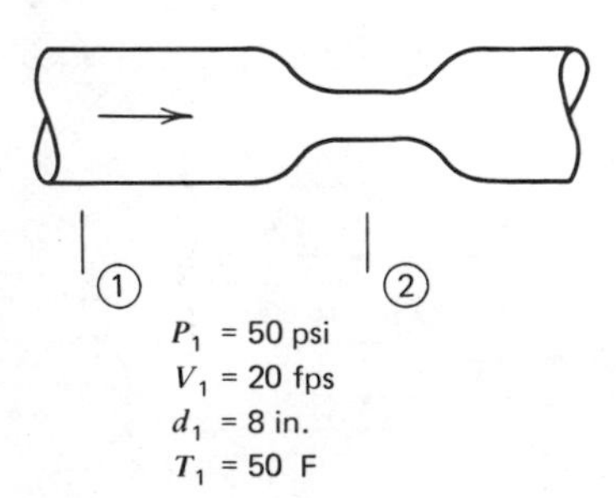

7.23. If the pressure of flowing water drops below the vapor pressure, vapor pockets form and collapse and damage the pipe wall and produce undesirable vibrations. This process is called cavitation. At 50 F the vapor pressure of water is 0.18 psi absolute. Assume steady flow and determine the minimum diameter of the flow meter at section 2 that will avoid cavitation.

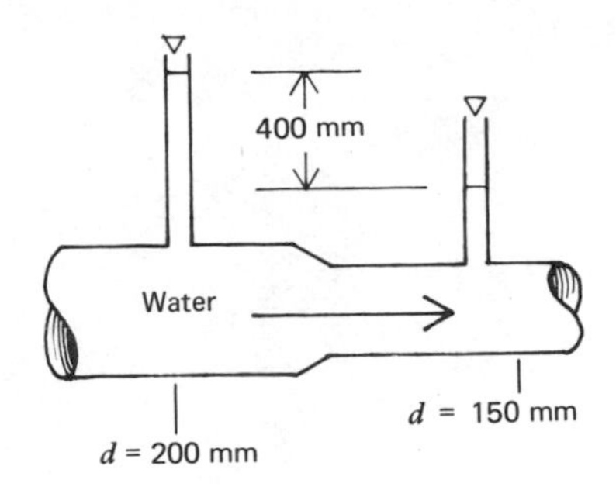

7.24. The manometers are both filled with water. Determine the flow rate of the water through the line. Assume steady flow.

7.25. Determine the minimum depth of a water reservoir that will provide a flow rate of 1200 gpm in a 4-in. diameter horizontal irrigation pipe that opens to the atmosphere.

7.26. A 4-in. diameter culinary water line carries a steady flow of 30 gpm to a housing development on a hill. If a minimum water pressure of 16 psi is needed on top of the hill, estimate the water pressure in the line at the bottom of the hill.

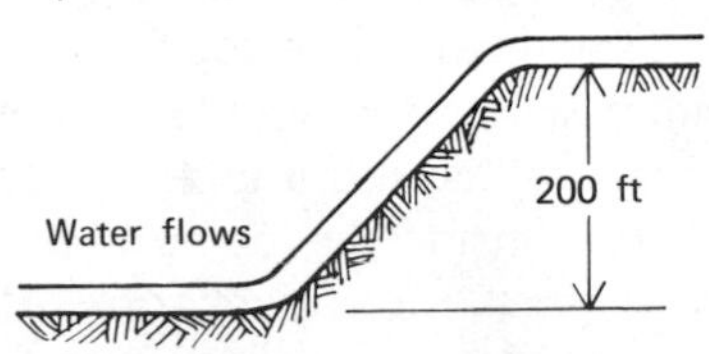

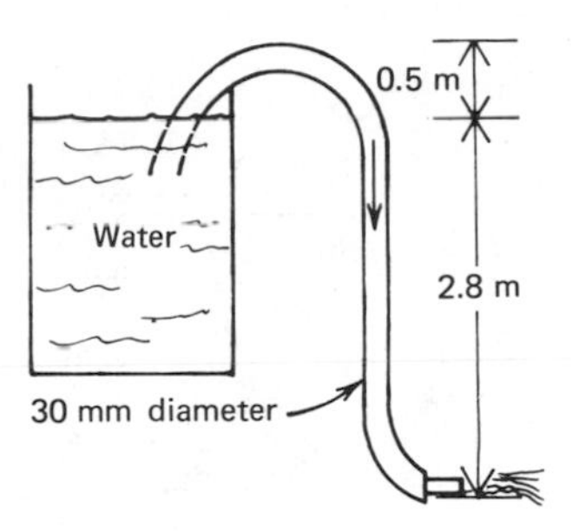

7.27. A siphon is used to drain water from a tank. The hose is 30 mm in diameter, the outlet end is 2.8 m below the water surface and the hose bend is 0.5 m above the water surface. Calculate the flow rate of water through the hose at that instant.

7.28. Water flows at 10 000 gpm through the pipe. The pressure in the 8-in section is 150 psi. Determine the pressure and velocity in the 10-in. pipe.

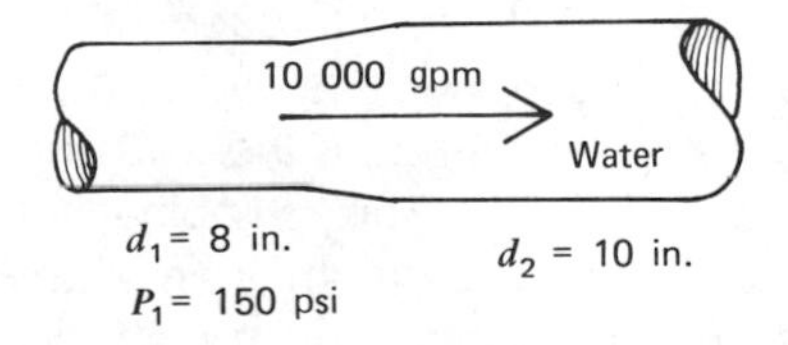

7.4.3/Hydraulic Pumps

The total hydraulic energy at any cross section along a pipe through which an ideal fluid flows is constant and equivalent to the sum of the potential, kinetic, and pressure energy. In gravity flows, the total hydraulic energy never exceeds the initial potential energy that the fluid particles had before they started to fall continuously through the pipe. In many systems, the fluid velocity or pressure required greatly exceeds the values obtainable from gravity and an auxiliary energy supply is required. A pump is a device that transfers mechanical, electrical, or perhaps heat

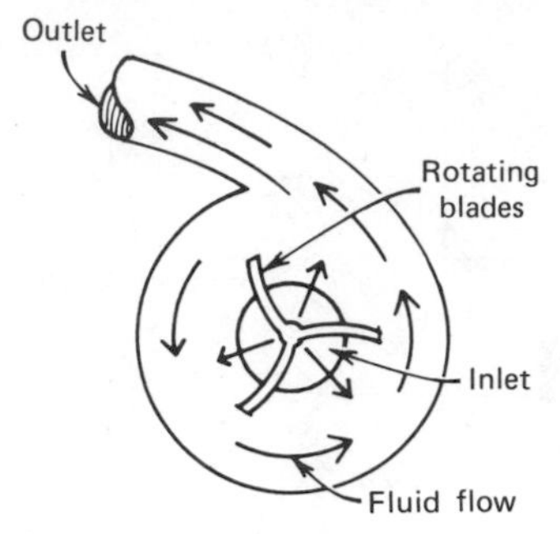

FIG. 7.29. Nonpositive displacement pump.

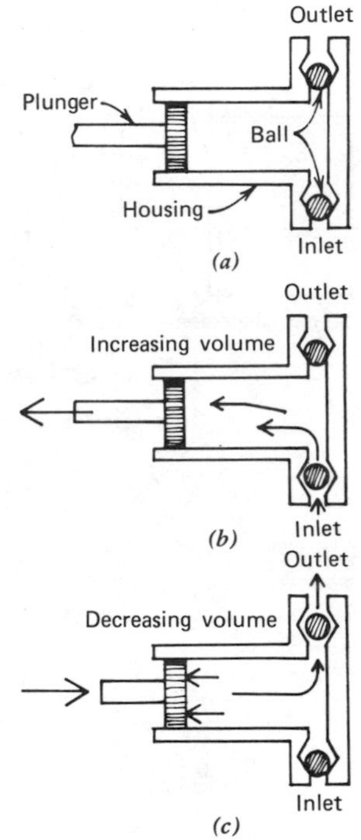

FIG. 7.30. *(a)* A positive displacement pump. *(b)* Fluid enters. *(c)* Fluid is pushed out.

energy from an outside source to the total hydraulic energy of the fluid.

Although there are several different types of pumps, the internal mechanisms responsible for the transfer of fluid are categorized either as a nonpositive displacement or a positive displacement pump. Nonpositive displacement pumps are usually used for large volume flows at relatively low pressures. The fluid enters the pump chamber and is accelerated by rotating impeller blades as shown in Figure 7.29. Rather than trap discrete volumes of fluid and then transfer them through the pump chamber, a nonpositive displacement pump pushes on parts of the fluid continuum and thereby increases the hydraulic energy of the complete volume. An eggbeater, stirring spoon, paddle wheel, and propeller are examples of nonpositive displacement devices.

A positive displacement pump divides the continuous flow at the inlet into discrete volumes of fluid that are trapped and then moved through the pump chamber. Inside the pump chamber, energy is transferred mechanically to the fluid packages. At the chamber outlet, the fluid flow continues with an increased hydraulic energy that remains constant in the absence of friction or other energy inputs.

A cylinder fitted with a piston as shown in Figure 7.30 is a typical positive displacement pump. When the plunger is pulled back, the pressure inside the cylinder is less than that of the fluid and the inlet ball valve opens to allow fluid to enter. After the chamber has filled, an external force applied to the plunger increases the fluid pressure and moves the fluid through the outlet ball valve with increased energy. The work done by the plunger force moving through a distance is transferred to the fluid continuum and increases the hydraulic energy. This pumping action results in a reciprocating motion. The human heart is an example of such a pump and the pulse rate is an indicator of the reciprocating flow.

An alternate positive pump that would provide steady flow would be either a vane pump or the gear type pump shown in Figure 7.31. In both cases, a discrete volume of fluid is captured at the inlet, energized as it passes through the pump chamber and then forced through the outlet. Positive displacement pumps are frequently used in power systems where pressure rather than volume flow is of primary importance.

The mechanical energy transferred by a hydraulic pump to a fluid stream is contributed continuously as the fluid particles pass through the pump chamber. Rather than a single surge of energy, a pump must transfer energy at some particular rate in order to provide a steady flow of fluid. The rate at which energy is added is defined as power and is measured in terms of energy units per unit of time, such as ft·lb/s, Btu/day or joules/s. The power capability of a pump is often expressed in terms of horsepower, where

$$1 \text{ hp} = 550 \text{ ft}\cdot\text{lb/s}$$

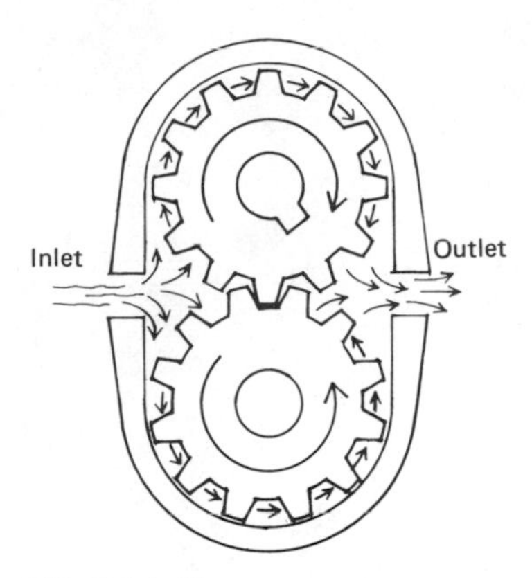

FIG. 7.31. Vane or gear-type pump.

In practice the total energy transferred to the fluid by the pump is less than the energy input to the pump as a consequence of leakage and friction. The pump efficiency is defined as the ratio of the pump output power to the pump input power. Frequently, the power input to a pump is expressed in terms of horsepower whereas the output power is expressed in terms of pressure and volume flow rate at the discharge. This fraction is always less than one and is often multiplied by a hundred in order to express efficiency as a percent. The efficiency of a pump can be expressed as

$$\text{Pump efficiency} = \frac{\text{output power}}{\text{input power}} \times 100$$

The pumps in this text will be assumed to be 100 percent efficient unless specified otherwise.

A pump has been described as a source of energy for a fluid system. A turbine is a mechanical device that extracts energy from a fluid flow and converts the energy to some other form such as electricity. Again, since a turbine extracts energy continuously at some particular rate, the term power best describes the turbine output. A turbine can be analyzed in a manner similar to that of a pump, realizing that power is extracted rather than added to the system.

The Bernoulli equation can be modified to include the energy input from a pump or the extraction of energy by a turbine. In Figure 7.32 a pump is inserted into the flow of an ideal fluid through pipe. Without the pump, the total energy at section one would equal the total energy at section 2. Since energy is conserved, the combined kinetic, potential, and pressure energy at section 1 plus the energy added by the pump must equal the total kinetic, potential and pressure energy at section 2. In equation form, the energy balance per pound of fluid can be written as

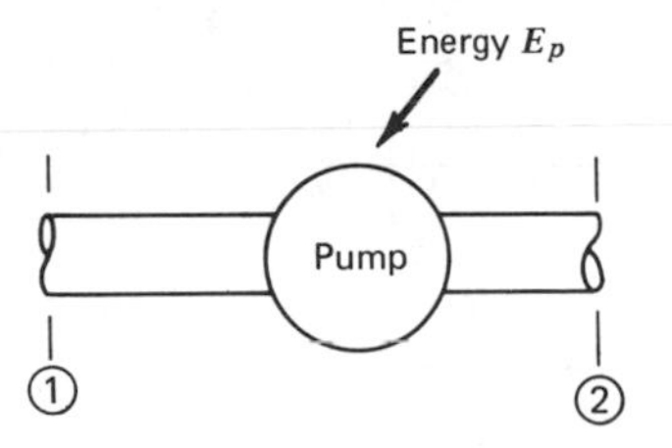

FIG. 7.32. A pump adds energy to the flow.

$$\left(\frac{1}{2}\right)\frac{V_1^2}{g} + h_1 + \frac{P_1}{\gamma} + E_p = \left(\frac{1}{2}\right)\frac{V_2^2}{g} + h_2 + \frac{P_2}{\gamma}$$

The mechanical energy added by the pump is transferred into some combination of kinetic, potential and pressure energy at section 2. In some cases, such as large irrigation systems, the pump energy is converted primarily into kinetic energy and the fluid flow rate is increased. In hydraulic systems that are typically used on construction equipment, the pump energy is transferred to pressure energy.

There might be some question about the units of the pump energy E_p. In Bernoulli's equation, the units of each term are usually feet, inches, or meters and represent a so-called head. How could pump energy have units of feet or meters? If, for example, one multiplies feet by pounds and divides by pounds, then E_p becomes energy per pound of fluid as follows

$$E_p = \text{ft} = \frac{\text{ft}\cdot\text{lb}}{\text{lb}} = \frac{\text{energy}}{\text{lb of fluid}}$$

The pump power can be determined by multiplying E_p by the weight of the fluid that flows through the pump per unit time. This weight flow rate can be written as $Q\gamma$, where Q is the volume flowrate and γ is the weight per unit volume. The power equation follows as:

$$\text{Pump power} = E_p Q\gamma = (\text{ft}\cdot\text{lb}/\text{lb})(\text{ft}^3/\text{s})(\text{lb}/\text{ft}^3) = \text{ft}\cdot\text{lb}/\text{s}$$

Since the power of both pumps and turbines is frequently expressed in terms of horsepower, it follows that:

$$\text{Pump power} = \frac{E_p Q\gamma}{550}\ \text{hp}$$

The power equation for a pump applies equally well to a turbine.

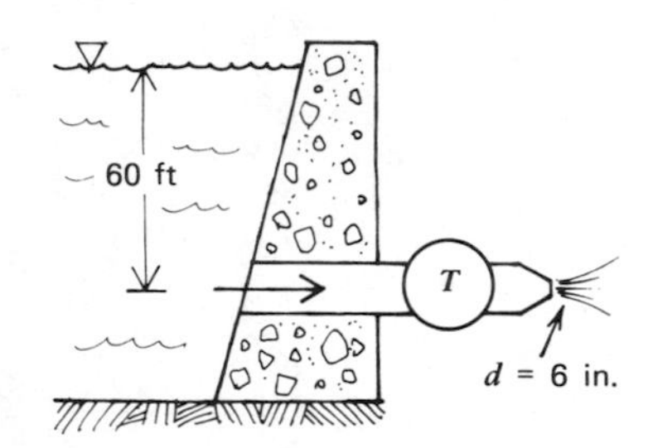

EXAMPLE PROBLEM 7.11

Water flows steadily at 5 cfs through the turbine and out the nozzle into the atmosphere. Determine the energy extracted from the flow by the turbine.

Solution

The surface of the reservoir is identified as section 1. At the surface, the velocity is zero, the gage pressure is zero, and the elevation is 60 ft. The nozzle exit is chosen as section 2. Since the elevation reference is taken as the nozzle centerline, $h_2 = 0$. The gage pressure at the nozzle exit is zero. The area and exit velocity can be determined as shown. Bernoulli's equation follows as

$$h_1 = E_T + \frac{1}{2}\frac{V_2^2}{g}$$

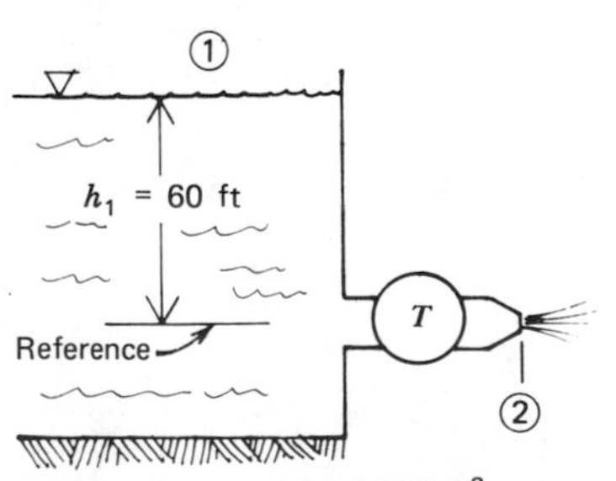

where the zero terms have been dropped. Upon substitution, the turbine energy per pound of fluid can be found as

$E_T = 49.9$ ft·lb/lb

The horsepower of the turbine is

$$\text{hp} = \frac{Q\gamma E_T}{550} = \frac{(5)(62.4)(49.9)}{550} = 28.3$$

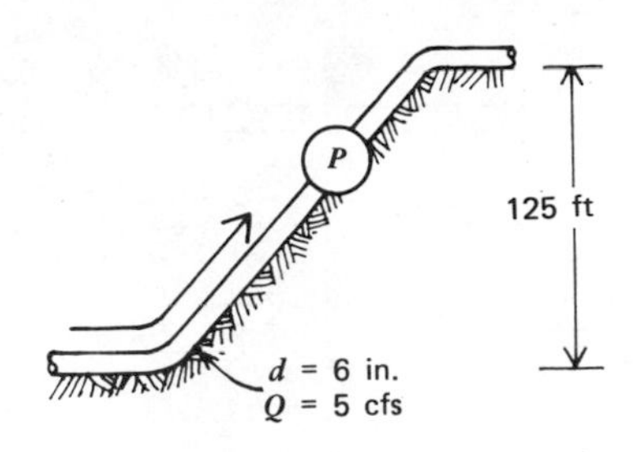

EXAMPLE PROBLEM 7.12

A 6-in. diameter culinary water line provides water at a flow rate of 5 cfs to a subdivision on a hill. The elevation difference is 125 ft. Determine the pump horsepower required to provide a pressure on top of the hill equal to the water pressure at the base of the hill.

Solution

Assume water to be an ideal fluid and write Bernoulli's equation to relate the energy at the base of the hill to the energy on top the hill.

$$\frac{1}{2}\frac{V_1^2}{g} + h_1 + \frac{P_1}{\gamma} + E_p = \frac{1}{2}\frac{V_2^2}{2} + h_2 + \frac{P_2}{\gamma}$$

Since the pipe is uniform and the flow rate is constant, then

$$Q = V_1 A = V_2 A$$

and $V_1 = V_2$. Also, the problem states that $P_1 = P_2$ and $\gamma =$ constant for an incompressible flow. Bernoulli's equation reduces to

$$E_p = h_2 - h_1 = 125 \text{ ft}$$

The pump horsepower can now be computed as

$$\text{Power} = \frac{Q\gamma E_p}{550} = \frac{(5)(62.4)(125)}{550} = 70.9 \text{ hp}$$

PROBLEMS

7.29. A submerged pump is located 120 ft below ground surface in a 4-in. diameter well casing. Determine the pump horsepower required to provide continuous flow of 20 gpm at an elevation of 30 ft above the ground through a 2-in. diameter pipe with a pressure of 5 psi gage.

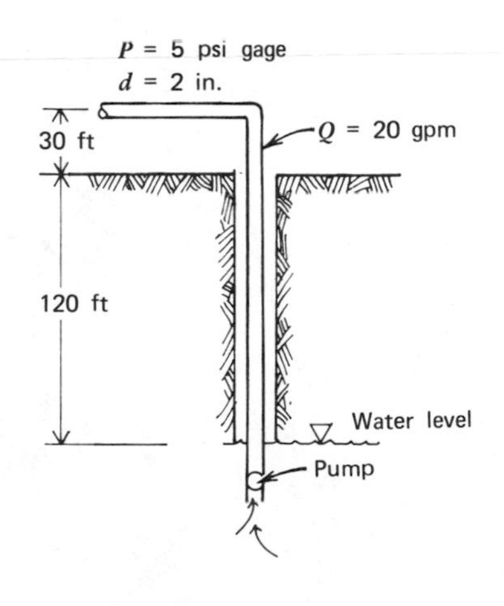

7.30. Raw sewage dumped into a small lake has caused an overgrowth of one-celled plant life that gradually depleted the supply of free oxygen. The fish have died and the lake is in an advanced state of eutrophication. In order to revitalize the lake, it is proposed that the cold oxygen-poor water near the bottom

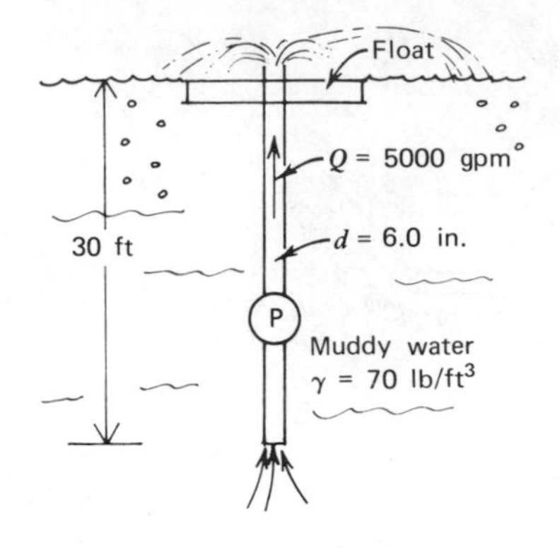

be pumped to the surface and aerated. The cold oxygen enriched water bubbled across the water surface would then sink back down beneath the thermocline and help restore the oxygen. Estimate the minimum horsepower required to pump 5000 gpm from a depth of 30 ft through a 6-in. diameter pipe to the lake surface. Assume that the muddy water weighs 70 lb/ft^3 and neglect friction losses.

7.31. A 4-in. diameter pipe carries water from one reservoir to another as shown. Determine the pump horsepower required to maintain a steady flow of 2.0 cfs. The reservoir levels remain constant.

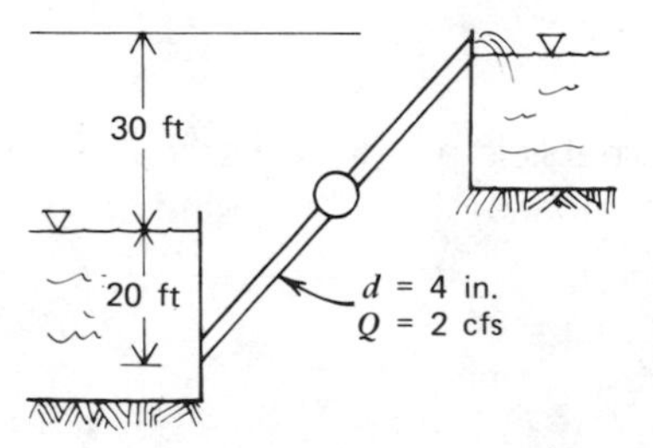

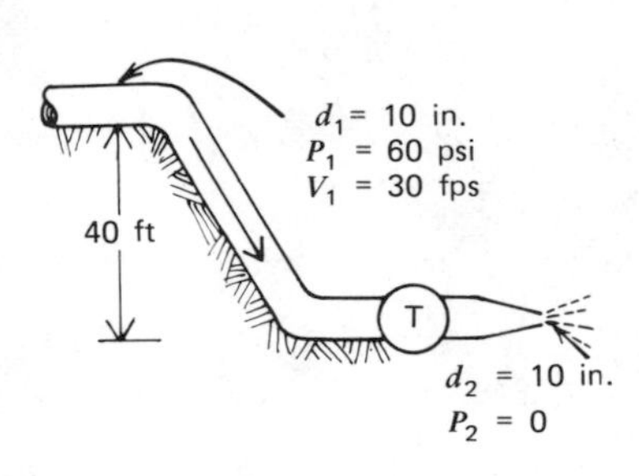

7.32. Water flows steadily through the system. Compute the horsepower output of the turbine. Assume that the turbine is 87 percent efficient.

7.33. A windmill is used to pump water through a 2-in. diameter pipe a vertical distance of 40 ft into a large open tank. If a cow drinks approximately 10 gal of water per 24-h day, estimate the windmill horsepower required to water 800 head of cattle.

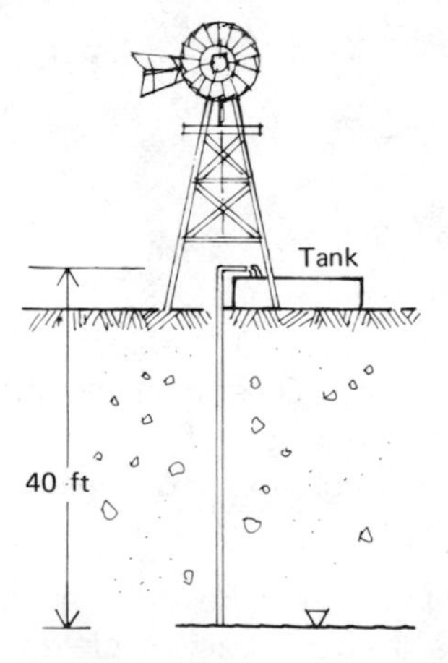

7.34. A steady flow of 200 l/s of water must be pumped through a 300-mm pipeline from a reservoir over a hill through a vertical distance of 50 m. Determine the pump horsepower required to maintain a pressure of 150 kPa at the top of the hill.

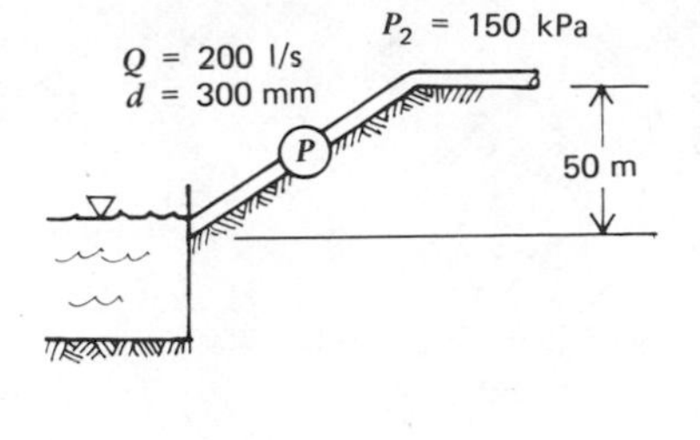

7.35. A 10-kW pump is placed in the water line as shown. Determine the velocity of the water through the 50-mm diameter nozzle.

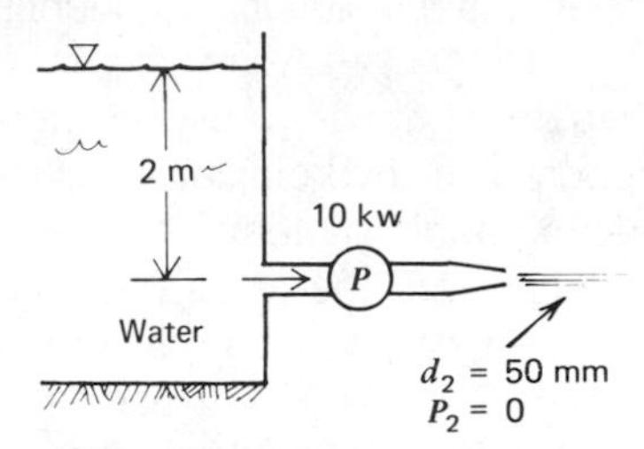

7.36. Determine the turbine output in kilowatts. Water flows steadily at 0.5 m^3/s.

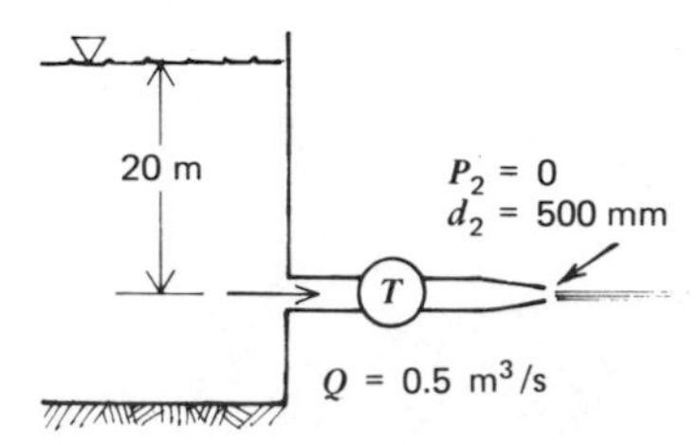

7.37. Water flows steadily at 2 cfs through the pipe. The manometer fluid is also water. Determine the pump horsepower.

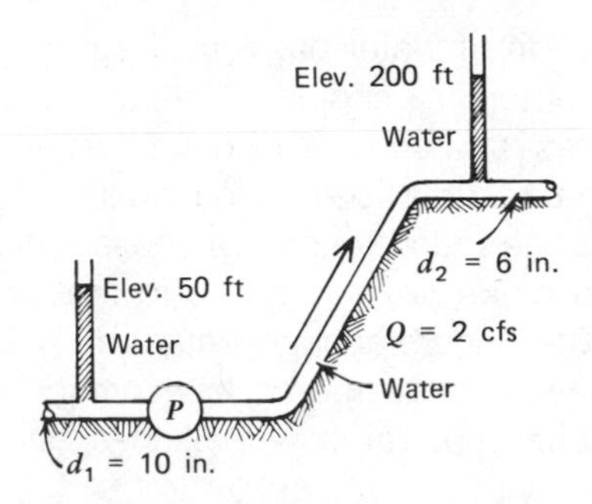

7.5/Water Engineering and the Ecosystem

Water is essential to life. It is a basic ingredient to virtually every commercial and industrial enterprise. Life's pleasures owe much to water. It serves in countless ways to make our surroundings more attractive and livable. A safe dependable supply for domestic purposes and for carrying away domestic wastes makes water a vital element in the maintenance of personal and public health.

The accomplishment of practically every societal goal is directly or indirectly dependent on water. Thus, the initiation of activities for the enhancement of our economic, social, and environmental well-being translates into a myriad of water uses—each having its peculiar requirement in terms of quantity, quality, location, and timing of need. As society's numbers, aspirations, and priorities change over time, water needs change accordingly. Consequently, there is an ever-present engineering challenge in planning, designing, and building the array of needed works, and in devising the management systems that extend the utility of any given supply and assure cost-effective solutions to identified water problems.

Engineering plans and management options must be fitted to the peculiar physiographic, topographic, hydrologic, and climatologic situations and to the demographic patterns that prevail. No two river basins are alike. Neither are the cultural and economic patterns that are superposed on river basins and that determine water demand patterns. Therefore, the engineering approach for convoluting water availability from a natural system into a match up with social demands is different in every case. There are no "off-the-shelf" solutions to water problems. Innovation, good judgment, and skillful application of principles of engineering analysis and project evaluation are required for each situation.

The approach to the solution of water problems begins with an understanding of the potential supply. Natural supply sources—streams, rivers, lakes, and underground aquifers—are subject to wide fluctuations that relate to short and long term precipitation patterns. Ways must be found to iron out these natural excesses and deficiencies to make flows conform to the more uniform user requirements. The understanding of natural hydrologic systems, the forces that drive them, and how prescribed physical perturbations will alter natural flow balances is the domain of engineering hydrology.

Hydrologic design provides the takeoff point for hydraulic design. Hydrologic analysis tells us what the water "loadings" might be. Hydraulic engineering draws from the theory and applications of fluid mechanics in the design of physical works to achieve prescribed levels of control and regulation. This may involve reservoirs to store water when nature provides abundantly for release when nature gives grudgingly or not at all. Normally, pipelines, canals, tunnels, flumes, pumps, etc. are utilized in conveying needed quantities from location of occurrence to place of use. Facilities and devices to eliminate or neutralize the potentially harmful effects of chemical, physical, or biological constituents in water are often required also. Calculating the forces, velocities, depths, and pressures to which these structures will be subjected, and then designing structures of such materials and configuration that they will function efficiently, dependably, economically, and safely is the responsibility of the hydraulic engineer.

Hydrologic and hydraulic principles underpin all water engineering specialties. Important among these is that of irrigation engineering. Water used in the production of food and fiber constitutes the largest single water use. Areas requiring irrigation are extensive and are found in every continent. The scope of irrigation engineering encompasses a basic understanding of soil-plant-water systems and extends to the planning, design, and construction of farm-scale and project-scale physical works to regulate, convey, and distribute water. The application of sound engineering principles and practices to on-farm water management is essential to a permanent and profitable agriculture under irrigation. The irrigation engineer must carefully appraise the soils, crops, climate, water supply, water quality, topography, labor and energy supply, and other economic factors in selecting a method of water application. Then the engineer must apply the hydraulic principles appropriate to the method selected (surface, subsurface, overhead sprinkler) to arrive at the most cost-effective design and layout.

An extremely important dimension of water engineering relates to the protection/enhancement of water quality. Recognition that certain diseases can be transmitted through water; that aquatic environments and recreational opportunities deteriorate with deteriorating water quality; and that degraded water quality results in a restriction of the number of uses that can be satisfied from a given supply, places a heavy responsibility on sanitary/environmental engineers to find solution to water quality problems. Comprehensive water quality management programs must consider all water uses including public water supply, propagation of fish and other aquatic life, recreation, industrial, agricultural, and any other legitimate use. As with other water engineering problems, the determination of the best management measures for use in protecting/enhancing water quality is a site specific consideration. In addition to management structures and facilities previously mentioned, other measures for treatment of water and wastewater must be utilized by the water quality engineer. These might include sedimentation and filtration, aeration, water softening, and chlorination on the "clean water" side; grit chambers, settling tanks, sludge digestors, chemical-treatment works, trickling filters, sludge-drying beds, oxidation ponds, and land application on the "dirty water" side.

Water engineering, regardless of speciality, provides a good springboard to resources planning in general. The water engineer is a vital team member in the comprehensive planning and management of resources. Engineering considerations must be blended with a substantial body of other economic, legal, institutional, political, and other inputs in the formulation of plans, the evaluation of alternatives, and the balancing of interests.

APPENDIX A

THE SELECTION OF A CALCULATOR

Selecting an appropriate calculator must be done early in an engineering career. Often, a student must select one with the capability to perform functions that he or she doesn't yet completely understand. Differences in logic, memory, cost, and special functions make the selection rather frustrating. This appendix is intended to assist in the purchase of a calculator. Once a decision is made, the operator's manual supplied with the calculator provides instructions on its use.

A serious-minded engineer should purchase a quality calculator at the beginning of his or her career. The many hours spent solving problems as a student and the continued professional use as an engineer certainly justify the initial cost.

Engineers should purchase a so-called "scientific" calculator. In addition to the arithmetic keys, the scientific model has a keyboard that includes exponents such as [$\sqrt{x}$], [x^2], and [y^x]. Trigonometric functions such as [sin], [cos], and [tan] are essential. Other special keys that put numbers in and out of storage, compute logarithms, invert numbers, convert numbers to scientific notation, change the algebraic sign of a number, and other special function keys are frequently used by engineers. The need of a more expensive "programmable" calculator depends on the student. Many students do well without a programmable calculator, yet instructors of upper-division classes are promoting their use on particular problems. Innovative students can spend several delightful hours programming problems of their own interest.

The logic of an electronic calculator relates to the sequence of keypunching that must be followed to perform calculations. Algebraic logic and reversed Polish notation (RPN) are the two main types. Some characteristics of each are discussed in the following paragraphs.

A.1/Algebraic Logic Systems

An algebraic system allows the individual to punch keys in the same order as the numbers and operators appear in written form in an equation as one reads from left to right. The problem $5 \times 2 = 10$ would be completed by punching the sequence [5] [×] [2] [=] and then 10 would appear on the display.

In chained operations, the answer depends on the order or hierarchy of arithmetic operations. For example, $6 + 2 \times 3 = 24$ if one adds first and then multiplies. However, $6 + 2 \times 3 = 36$ if one multiplies first and then adds. This hierarchy is usually specified by adding parentheses, such as $(6 + 2) \times 3 = 24$ and $6 + (2 \times 3) = 36$. One computes the operation(s) inside the parentheses first. Without parentheses, one must understand the hierarchy of the algebraic system.

Calculators based on a *simple algebraic system* have no hierarchy and calculate in the order the numbers are entered. For example, $6 \times 2 + 3 \times 5 = 75$ would be keyed as

[6] [×] [2] [+] [3] [×] [5] [=]

and 75 would appear on the display. Most calculators have a hierarchy and compute multiplication and division before addition and subtraction. An *algebraic system with hierarchy* would compute $6 \times 2 + 3 \times 5 = 27$. Since multiplication is completed prior to addition, 6×2 would be added to 3×5 inside the calculator.

Calculators that use algebraic logic usually have both right and left parenthesis keys along with an equal key. An equation involving parentheses should be keyed in just as it is written. The computer completes the operation inside the parentheses first. If there are multiple sets of parentheses, the computer completes the innermost operations first and then works outward. For example, $6 \times (2 + 3) \times 5 = 150$ and $(6 \times 2) + (3 \times 5) = 27$.

The use of the equal key [=] in algebraic logic systems tells the calculator to complete all pending operations. For example, in an algebraic system with hierarchy, one would compute $2 + 4 \times 3 = 14$ by punching [2] [+] [4] [×] [3] [=]. The calculator would compute 4×3 and then add 2. However, using the equal sign twice, one could punch in [2] [+] [4] [=] [×] [3] [=] and 18 would appear on the display. The equal sign completes the addition of 2 and 4 and then multiplies the sum by 3.

A.2/RPN Logic Systems

The reverse Polish notation (RPN) system follows a logic system developed by the Polish mathematician Lan Lukasiewicz in 1959. The keystroke sequence places the operator after the operand(s) in a so-called "postfix" notation. There are no parenthesis nor equal keys as found on calculators that use algebraic logic. In RPN calculators, the key sequence for adding $8 + 2 = 10$ would be [8] [enter] [2] [+] and 10 would appear on the display. The [enter] key is used to separate the first two numbers. The pressing of the operator, in this case [+], completes the pending operation and stores the results in a memory stack.

A typical four-stack memory of an RPN system is shown in Figure A.1, where X is the display or first memory and T is the top. In the case of $8 + 2 = 10$, if one pressed [8], then the number 8 would appear on the display, which is register X. If one then pressed [enter] the number 8 would be advanced to memory Y. The pressing of [2] would place the number 2 in memory X and the pressing of [+] would place the sum 10 in register X. This sequence is illustrated in Figure A.2.

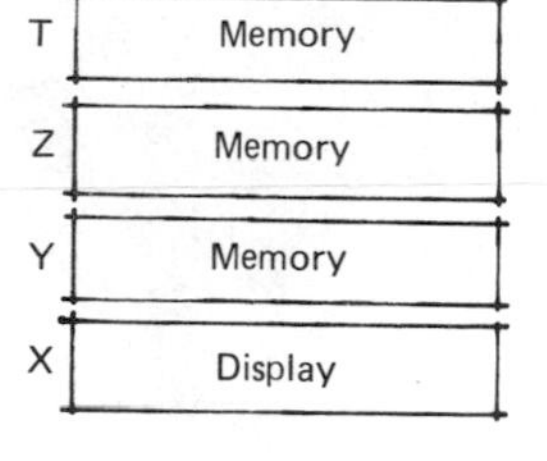

Fig. A.1.

T				
Z				
Y		8	8	
Display	8	8	2	10
key	[8]	[enter]	[2]	[+]

FIG. A.2.

T						
Z						
Y		8	8		10	
Display	8	8	2	10	4	40
key	[8]	[enter]	[2]	[+]	[4]	[x]

FIG. A.3.

The memory stack routine for $(8 + 2) \times 4 = 40$ or [8] [enter] [2] [+] [4] [×] is shown in Figure A.3.

Several RPN logic systems have keys such as [R↓] that move the numbers in the stack downward or [x ⇆ y] that transfer numbers between registers. These special keys provide flexibility, allow intermediate results to be checked and facilitate the correction of errors in chained problems.

In the selection of a calculator, one must choose between the algebraic and RPN logic systems. Any advantages of one system over another are primarily a matter of taste. Either system will complete engineering problems.

EXAMPLE PROBLEM A.1

$8 + 2 \times 3 = \ ?$

Algebraic logic

[8][+][2][×][3][=] Display 30

RPN logic

[8][enter][2][+][3][×] Display 30

EXAMPLE PROBLEM A.2

$$\frac{(8 + 3)(2 + 5)}{(4 \times 2)} = \ ?$$

Algebraic logic

[(][8][+][3][)][×][(][2][+][5][)][÷]

[(][4][×][2][)][=] Display 9.625

RPN logic

[8][enter][3][+][2][enter][5][+][×] [4][enter][2][×][÷]

Display 9.625

EXAMPLE PROBLEM A.3

$$\frac{[(2 \times 5) + 3] \times 2}{15} = ?$$

Algebraic logic

[( ] [( ] [2] [×] [5] [ )] [+] [3] [ )] [×]

[2] [÷] [15] [=]   Display 1.7333

RPN logic

[2] [enter] [5] [×] [3] [+] [2] [×] [15] [÷]   Display 1.7333

APPENDIX B

TRIGONOMETRY

Trigonometry is a branch of mathematics that deals with the relationship between sides and angles of triangles. A triangle that has one angle equal to 90 degrees is a *right triangle,* as shown in Figure B.1. The side opposite the right angle is called the hypotenuse. If one arbitrarily identifies an angle θ as shown, then the ratio of the side opposite the angle θ over the hypotenuse is defined as the sine of angle θ and is written as *sin* θ. The ratio of the adjacent side over the hypotenuse is defined as *cos* θ. *Tan* θ is the ratio of the opposite side over the adjacent side and *cot* θ is the inverse of tan θ. These definitions are given in Table B.1 and should be memorized.

The sine, cosine, tangent, or cotangent of some specified angle, say θ, is a dimensionless number that represents the ratio of the lengths of the sides of the right triangle. Some selected values are given in Table B.2. For example, the sine of 30 degrees is 0.5 (sin 30° = 0.5) and implies that the ratio of the length of the side opposite the angle of 30° divided by the hypotenuse length is always 0.5 in any right triangle, regardless of the size. The ratio of 0.5 may represent opposite and hypotenuse

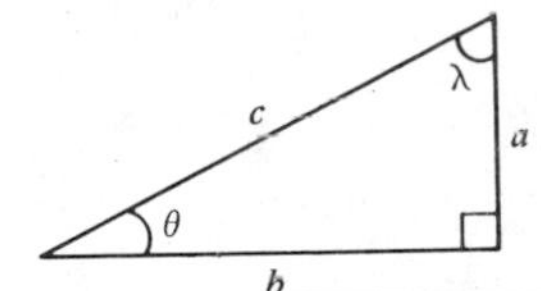

FIG. B.1.

TABLE B.1. Trigonometric Functions for a Right Triangle

$$\text{Sin } \theta = \frac{\text{opposite side}}{\text{hypotenuse}} = \frac{a}{c}$$

$$\text{Cos } \theta = \frac{\text{adjacent side}}{\text{hypotenuse}} = \frac{b}{c}$$

$$\text{Tan } \theta = \frac{\text{opposite side}}{\text{adjacent side}} = \frac{a}{b}$$

$$\text{Cot } \theta = \frac{\text{adjacent side}}{\text{opposite side}} = \frac{b}{a}$$

TABLE B.2. Common Trigonometry Functions and Angles

Angle	*Sin*	*Cos*	*Tan*	*Cot*
0	0	1	0	∞
30	1/2	$\sqrt{3}/2$	$\sqrt{3}/3$	$\sqrt{3}$
45	$\sqrt{2}/2$	$\sqrt{2}/2$	1	1
60	$\sqrt{3}/2$	1/2	$\sqrt{3}$	$\sqrt{3}/3$
90	1	0	∞	0

side lengths of 3 and 6 ft, 1000 and 2000 ft or 400 and 800 m, however, the angle remains at 30°.

If the appropriate side lengths of a right triangle are known, then the angles can be found using *inverse trigonometric functions*. If sides a and b are known in Figure B.1, then the angle θ could be determined as the angle whose tangent is the ratio a over b. In equation form, one would write $\theta = \tan^{-1}(a/b)$. In a similar way, $\theta = \sin^{-1}(a/c)$ identifies the angle θ whose sine is length a divided by length c.

The Pythagorean theorem states that the sum of the squares of the legs of a right triangle is equal to the square of the hypotenuse. In reference to Figure B.1, the theorem is written as

$$a^2 + b^2 = c^2$$

The sum of the three angles in any triangle must equal 180°. In a right triangle, one angle is always 90°, therefore, the other two must sum to 90°. Referring to Figure B.1,

$$\theta + \lambda = 90°$$

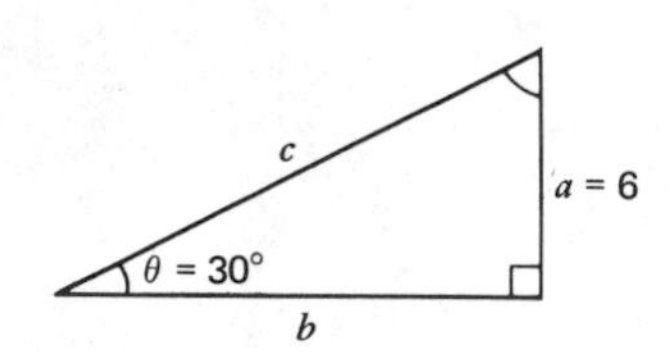

EXAMPLE PROBLEM B.1

For the triangle given, find the hypotenuse c and the side b.

Solution

Find c: $\sin\theta = \dfrac{a}{c}$

$$c = \frac{a}{\sin\theta} = \frac{6}{\sin 30°} = \frac{6}{0.5} = 12$$

Find b: $\cos\theta = \dfrac{b}{c}$

$$b = c(\cos\theta) = 12(\cos 30) = 12(0.866) = 10.39$$

Check using Pythagorean theorem:

$$a^2 + b^2 = c^2$$

$$c = \sqrt{a^2 + b^2} = \sqrt{6^2 + 10.39^2} = 12$$

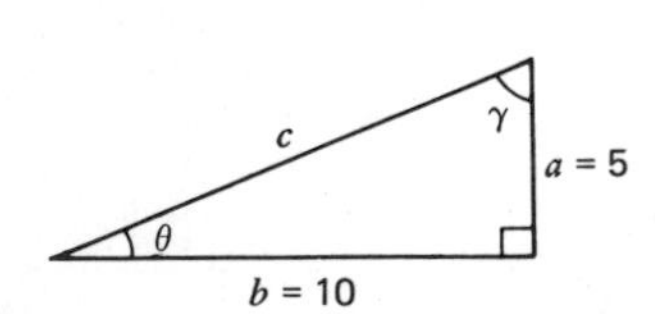

EXAMPLE PROBLEM B.2

For the triangle given, find the angles θ and λ. Also find the side c.

Solution

Find θ: $\tan\theta = \dfrac{a}{b}$

$$\theta = \tan^{-1}\frac{a}{b} = \tan^{-1}\frac{5}{10} = 26.5°$$

Find λ: $\lambda + \theta = 90°$

$$\lambda = 90 - \theta = 90 - 26.5° = 63.5°$$

Find c: $\sin\theta = \frac{a}{c}$

$$c = \frac{a}{\sin\theta} = \frac{5}{\sin 26.5} = 11.2$$

EXAMPLE PROBLEM B.3

Determine the value of side c in the right triangle of Example Problem B.1. Use a calculator.

Solution

The value of side c can be determined using the sine function:

$$\sin\theta = \frac{a}{c} \quad \text{or} \quad c = \frac{a}{\sin\theta} = \frac{6}{\sin 30°}$$

Using algebraic logic to determine c, one presses the sequence

[6] [÷] [3] [0] [sin] [=]

and the answer appears on the display

$c = 12.00$

Using RPN logic to determine c, one presses

[6] [enter] [3] [0] [sin] [÷]

and on the display

$c = 12.00$

EXAMPLE PROBLEM B.4

Determine θ for the triangle given in Example Problem B.2. Use a calculator.

Solution

The angle θ can be found using the inverse tangent function

$$\tan^{-1}\left(\frac{5}{10}\right) = \theta$$

Using algebraic logic, one punches

[5] [÷] [1] [0] [INV] [tan]

and $\theta = 26.5°$.

Using RPN logic, the sequence is

[5] [enter] [1] [0] [÷] [$\tan^{-1}$]

and $\theta = 26.5°$.

APPENDIX C

ENGINEERING ECONOMY

Many engineers must consider cost in their judgment decisions. Since many engineering designs involve large sums of money, that must be recovered over several years, some understanding of compound interest is essential. A knowledge of interest formulas is also helpful in the management of one's personal affairs, whether related to engineering, the purchase of a house, automobile or some type of financial investment.

The variables that are used in engineering economy are defined as follows:

n = *number* of payment *periods*

P = a *present* sum of money

F = a *future* sum of money at the end of n periods

A = a *uniform payment* or receipt made at the end of each period for a total of n periods

i = *compound interest* rate per period

Some interest formulas that relate n, P, F, A, and i are as follows:

$$F = P(1 + i)^n$$

$$F = A\left[\frac{(1 + i)^n - 1}{i}\right]$$

$$P = A\left[\frac{(1 + i)^n - 1}{i(1 + i)^n}\right]$$

$$A = P\left[\frac{i(1+i)^n}{(1+i)^n - 1}\right]$$

EXAMPLE PROBLEM C.1

A present sum of \$100 is invested with interest at 9 percent compounded annually. Determine the amount accumulated after 25 years.

Solution

Given: $P = 100$, $i = 0.09$, $n = 25$.
Find: F.

$$F = P(1+i)^n$$

$$F = 100(1 + 0.09)^{25} = \$862.30$$

EXAMPLE PROBLEM C.2

Determine the annual year-end payment for 5 years that is necessary to repay a present loan of \$5000 if the interest is at 11 percent.

Solution

Given: $P = \$5000$, $i = 0.11$, $n = 5$.
Find: A.

$$A = P\left[\frac{i(1+i)^n}{(1+i)^n - 1}\right]$$

$$A = 5000\left[\frac{0.11(1+0.11)^5}{(1+i)^5 - 1}\right]$$

$$A = \$1352.85$$

EXAMPLE PROBLEM C.3

Determine the future sum of Example Problem C.2.

Solution

Given: $A = \$1352.85$, $i = 0.11$, $n = 5$.
Find: F.

$$F = A\left[\frac{(1+i)^n - 1}{i}\right]$$

$$F = 1352.85\left[\frac{(1+0.11)^5 - 1}{0.11}\right]$$

$$F = \$8425.28$$

PROBLEMS

C.1. A present sum of $1000 is invested at 12 percent interest compounded annually. Determine the sum of money accumulated after 15 years.

C.2. A student must finance a present sum of $5000. Determine the annual year-end payment for 10 years if interest is 15 percent.

C.3. A student finances the present sum of $5000 for 10 years at 7 percent. Determine the future sum of money after the final payment.

C.4. A rancher has a Hereford cow presently valued at $550. The cow will yield a profit of $125 each year for 10 years. Determine the annual interest on the original investment of $550.

C.5. The addition to a football stadium will cost $5 million at the present time. Determine the annual year-end payment at 18 percent interest if the final payment must be made in 30 years.

C.6. A water system is anticipated to cost $500,000 at the present time. If the annual inflation is 9 percent, estimate the cost of the same water system if built 10 years in the future.

C.7. An annual payment of $2000 is put into a retirement fund at 8 percent compounded interest for a total of 25 years. Starting at year 26, an annual amount of $10,000 is withdrawn from the fund, which continues to earn interest at 6 percent. Determine the number of years that $10,000 may be withdrawn.

C.8. An optional item on a machine costs $500. The machine is financed at 13 percent interest for 6 years. Determine the future value of the option at the time the final payment is made.

APPENDIX D

MATERIAL PROPERTIES

TABLE D.1. Properties of Selected Metals

Material	σ Specific Weight lb/in.^{3}	σ_y Yield Strength psi	σ_t Tension Strength psi	E Modulus of Elasticity psi	α Thermal Expansion @ 68 F in./in./F	K Thermal Conductivity @ 68 F Btu/h·ft·F	ρ Electrical Resistivity @ 68 F Ω·in.
Aluminum (99.9+)	0.098	5,000	13,000	$10(10)^6$	$12.5(10)^{-6}$	128	$1.14(10)^{-6}$
Aluminum (6061-T6)	0.098	40,000	45,000	$10(10)^6$	$12.0(10)^{-6}$	97	$3.5(10)^{-6}$
Copper (99.9+)	0.323	—	37,000	$16(10)^6$	$9.0(10)^{-6}$	230	$0.67(10)^{-6}$
Brass (cold rolled)	0.316	60,000	75,000	$15(10)^6$	$9.8(10)^{-6}$	73	$2.44(10)^{-6}$
Bronze (cold rolled)	0.320	75,000	100,000	$15(10)^6$	$9.4(10)^{-6}$	48	$3.78(10)^{-6}$
Iron	0.284	—	40,000	$30(10)^6$	$6.5(10)^{-6}$	42	$3.82(10)^{-6}$
Steel (structural)	0.284	30,000	48,000	$30(10)^6$	$6.6(10)^{-6}$	29	$6.65(10)^{-6}$
Steel (0.4% C hot rolled)	0.284	53,000	84,000	$30(10)^6$	$6.6(10)^{-6}$	28	$6.73(10)^{-6}$
Steel (0.8% C hot rolled)	0.284	76,000	122,000	$30(10)^6$	$6.6(10)^{-6}$	27	$7.09(10)^{-6}$
Stainless steel (18-8 annealed)	0.286	36,000	85,000	$30(10)^6$	$9.6(10)^{-6}$	8	$27.56(10)^{-6}$
Stainless steel (18-8 cold rolled)	0.286	165,000	190,000	$30(10)^6$	$9.6(10)^{-6}$	8	$27.56(10)^{-6}$
Cast iron—gray	0.260	Brittle	25,000	$15(10)^6$	$6.6(10)^{-6}$	29	
Magnesium (99.9+)	0.063	14,000	27,000	$6.5(10)^6$	$14(10)^{-6}$	92	$1.77(10)^{-6}$
Silver (99.9+)	0.380	—	41,000	$11(10)^6$	$10(10)^{-6}$	242	$0.71(10)^{-6}$

TABLE D.2. Properties of Selected Ceramics

Material	Specific Gravity	Tension Strength 10^3 psi	Compression Strength 10^3 psi	Modulus of Elasticity 10^6 psi	Thermal Expansion @ 68 F 10^{-6} in./in.·F	Thermal Conductivity @ 68 F Btu/h·ft·F	Electrical Resistivity Ω·in.
Alumina (Al_2O_3)	4.000	35.8	412.5	52.4	5.0	16.900	—
Silicon carbide (SiC)	3.170	—	—	—	2.5	7.000	0.98 (2000 F)
Plate glass	2.500	6–6.5	90.0–180.0	10.0	4.9	0.440	$4(10)^{13}$
Borosilicate glass	2.400	—	—	10.0	1.5	0.600	—
Glass wool	0.024	—	—	—	—	0.022	—
Porcelain	2.530	2.5	49.1	—	2.6	0.900	—
Quartz	2.650	7	160.0	10.0	7.0	7.200	—
Common brick	1.750	—	1.0–20.0	—	5.0	0.420	—
Firebrick	2.100	—	—	—	2.5	0.800	—
Concrete	1.400–2.200	—	6.0	2.0	7.0	0.25–1.000	—
Ice	0.910	—	—	1.3	28.0	1.200	—

Note: Ceramic properties vary significantly. The values given are average values to be used for pedagogical purposes.

TABLE D.3. Properties of Polymers

Material	Specific Gravity	Tensile Strength 10^3 psi	Modulus of Elasticity 10^6 psi	Thermal Expansion @ 68 F 10^{-6} in./in.·F	Thermal Conductivity @ 68 F Btu/h·ft·F	Electrical Resistivity @ 68 F Ω·in.
Cellulose acetate (hard)	1.3	4.6–8.5	0.190–0.400	44–88	—	10^{10}–10^{13}
Nylon 6/6	1.15	9.0–12.0	0.40	55	0.145	10^{13}
Polyethylene	0.9	1.2–3.5	0.035	100	0.193	10^{13}
Polystyrene	1.05	5.0–10.0	0.4–0.6	35	0.048	10^{18}
PVC (rigid)	1.4	5.0–17.0	0.2–0.6	55	0.072	10^{13}
Phenol-Formaldehyde	1.3	6.0–9.0	0.5	40	0.096	10^{12}
Rubber (synthetic)	1.5	1.0–4.0	0.005–0.010	—	0.08	—
Rubber (vulcanized)	1.2	1.0–4.0	0.5	45	0.08	10^{14}
Wood (white pine)	0.5	7–12	1.4–1.6	—	0.07	—

APPENDIX E

UNIT CONVERSION FACTORS

To Convert From	*To Obtain*	*Multiply By*
Acres	ft^2	4.356×10^4
Acres	m^2	4.0469×10^3
Angstroms	cm	1×10^{-8}
Angstroms	in.	3.9370×10^{-9}
Atmospheres	kg/cm^2	1.0332
Atmospheres	$lb/in.^2$	1.4696×10^1
Atmospheres	Pa	1.0133×10^5
Barrels (petroleum, U.S.)	gal (U.S. liquid)	4.2×10^1
Bars	Pa	1×10^5
Btu	ft·lb	7.7765×10^2
Btu	hp·h	3.9275×10^{-4}
Btu	J	1.0551×10^3
Btu	kg·m	1.0751×10^2
Btu	kWh	2.9288×10^{-4}
Candelas	lm/sr	1
Centimeters	Å	1×10^8
Centimeters	ft	3.2808×10^{-2}
Centimeters	in.	3.9370×10^{-1}
Centipoises	g/(cm·s)	1×10^{-2}
Circular mils	cm^2	5.0671×10^{-6}
Circular mils	$in.^2$	7.8540×10^{-7}
Dynes	N	1×10^{-5}
Ergs	dyne·cm	1
Fathoms	ft	6
Feet	cm	3.048×10^1
Feet	in.	1.2×10^1

To Convert From	*To Obtain*	*Multiply By*
Feet	m	3.048×10^{-1}
Feet	mi	1.8939×10^{-4}
Feet	rods	6.0606×10^{-2}
Footcandles	lux	1.0764×10^{1}
Foot pounds	Btu	1.2859×10^{-3}
Foot pounds	dyne·cm	1.3558×10^{7}
Foot pounds	hp·h	5.0505×10^{-7}
Foot pounds	J	1.3558
Foot pounds	kg·m	1.3825×10^{-1}
Foot pounds	kW·h	3.7662×10^{-7}
Foot pounds	N·m	1.3558
Furlongs	ft	6.6×10^{2}
Furlongs	m	2.0117×10^{2}
Gallons (U.S. liquid)	ft^3	1.3368×10^{-1}
Gallons (U.S. liquid)	L	3.7854
Gallons (U.S. liquid)	m^3	3.7854×10^{-3}
Gallons (U.S. liquid)	pt (U.S. liquid)	8
Gallons (U.S. liquid	qt (U.S. liquid)	4
Grams	lb	2.2046×10^{-3}
Gram centimeters	kW·h	2.7241×10^{-11}
Hectares	acres	2.4711
Horsepower	Btu/h	2.5461×10^{3}
Horsepower	ft·lb/s	5.5×10^{2}
Horsepower	kW	7.4570×10^{-1}
Horsepower	W	7.4570×10^{2}
Horsepower hours	Btu	2.5461×10^{3}
Horsepower hours	ft·lb	1.98×10^{6}
Horsepower hours	J	2.6845×10^{6}
Horsepower hours	kg·m	2.7375×10^{5}
Horsepower hours	kW·h	7.4570×10^{-1}
Inches	Å	2.54×10^{8}
Inches	cm	2.54
Inches	ft	8.333×10^{-2}
Joules	Btu	9.4845×10^{-4}
Joules	ft·lb	7.3756×10^{-1}
Joules	hp·h	3.7251×10^{-7}
Joules	kg·m	1.0197×10^{-1}
Joules	kW·h	2.7778×10^{-7}
Joules	W·s	1
Kilograms	lb	2.2046
Kilograms	N	9.8067
Kilograms	slugs	6.8522×10^{-2}
Kilometers	ft	3.2808×10^{3}
Kilometers	mi	6.2137×10^{-1}
Kilometers	nmi (nautical mile)	5.3996×10^{-1}
Kilowatts	Btu/h	3.4144×10^{3}
Kilowatts	ergs/s	1×10^{10}
Kilowatts	ft·lb/s	7.3756×10^{2}
Kilowatts	hp	1.3410
Kilowatts	J/s	1×10^{3}
Kilowatt hours	Btu	3.4144×10^{3}

To Convert From	*To Obtain*	*Multiply By*
Kilowatt hours	ft·lb	2.6552×10^{6}
Kilowatt hours	hp·h	1.3410
Kilowatt hours	J	3.6×10^{6}
Knots	ft/s	1.6878
Knots	mi/h	1.1508
Liters	bushels (U.S.)	2.8378×10^{-2}
Liters	ft^3	3.5315×10^{-2}
Liters	gal (U.S. liquid)	2.6417×10^{-1}
Liters	$in.^3$	6.1024×10^{1}
Lux	lm/m^2	1
Meters	Å	1×10^{10}
Meters	ft	3.2808
Meters	in.	3.9370×10^{1}
Meters	mi	6.2137×10^{-4}
Microns	Å	1×10^{4}
Microns	ft	3.2808×10^{-6}
Microns	m	1×10^{-6}
Miles	ft	5.28×10^{3}
Miles	furlongs	8
Miles	km	1.6093
Miles	nmi (nautical mile)	8.6898×10^{-1}
Newtons	dynes	1×10^{5}
Newtons	kg	1.0197×10^{-1}
Newtons	lb	2.2481×10^{-1}
Ounces	g	2.8350×10^{1}
Ounces	lb	6.25×10^{-2}
Ounces	oz (troy)	9.1156×10^{-1}
Ounces (U.S. fluid)	cm^3	2.9574×10^{1}
Ounces (U.S. fluid)	gal (U.S. liquid)	7.8125×10^{-3}
Ounces (U.S. fluid)	$in.^3$	1.8047
Ounces (U.S. fluid)	L	2.9574×10^{-2}
Pascals	atm	9.8692×10^{-6}
Pascals	lb/ft^2	2.0885×10^{-2}
Pascals	$lb/in.^2$	1.4504×10^{-4}
Poises	g/(cm·s)	1
Pounds	g	4.5359×10^{2}
Pounds	kg	4.5359×10^{-1}
Pounds	lb (troy)	1.2153
Pounds	N	4.4482
Pounds	oz	1.6×10^{1}
Pounds	slugs	3.1081×10^{-2}
Pounds	tons (short)	5×10^{-4}
Pounds	t	4.5359×10^{-4}
Pounds per cubic foot	kg/m^3	1.6018×10^{1}
Pounds per square foot	atm	4.7254×10^{-3}
Pounds per square foot	Pa	4.7880×10^{1}
Pounds per square inch	atm	6.8046×10^{-2}
Pounds per square inch	bars	6.8948×10^{-2}
Pounds per square inch	kg/cm^2	7.0307×10^{-2}
Pounds per square inch	mm Hg	5.1715×10^{1}
Pounds per square inch	Pa	6.8948×10^{3}

To Convert From	*To Obtain*	*Multiply By*
Radians	°	5.7296×10^{1}
Radians	r (revolutions)	1.5915×10^{-1}
Slugs	kg	1.4594×10^{1}
Slugs	lb	3.2174×10^{1}
Watts	Btu/h	3.4144
Watts	ergs/s	1×10^{7}
Watts	ft·lb/min	4.4254×10^{1}
Watts	hp	1.3410×10^{-3}
Watts	J/s	1
Watt hours	Btu	3.4144
Watt hours	ft·lb	2.6552×10^{3}
Watt hours	hp·h	1.3410×10^{-3}

ANSWERS TO SELECTED PROBLEMS[1]

CHAPTER 2

2.7 735.45 N

2.8 150 lb_f

2.9 5000 N

2.10 155.42 lb_f

2.12 Yes, No

2.13 Celsius and Kelvin

2.14 68 F

2.15

C	K	R	F
10	283	542	50
−263	10	18	−442
−267	6	10	−450
− 12	261	470	10

2.16 $3.375(10)^6$ apples/day

2.17 62.43 lbm/ft^3

2.18 264.2 gal/m^3

2.19 2.471 acre/hectare, 0.4047 hectare/acre

2.20 (a) 88.5 km/h (b) 80.7 ft/s (c) 24.6 m/s

2.21 (a) 37.3 slugs (b) 554.3 kg (c) 1200 lbm

2.22 93.44 lb, 42.38 kg

2.23 22.05 $/kg

CHAPTER 3

3.2 (a) $1.76(10)^5$ TW (b) $1.76(10)^4$

3.6 $1(10)^6$ N·M

3.7 $1(10)^5$ ft·lb

3.8 0.6 hp

3.10 $6.7(10)^4$ hp

3.11 26.8 hp, 268 hp

[1]In real engineering computations, the number of significant digits retained should be such that accuracy is neither exaggerated nor sacrificed. In these solutions at the introductory level, however, rounding and significant digits are sometimes overlooked to allow the student to compare answers more readily. The emphasis is on procedure rather than number accuracy.

3.12 2400 N
3.13 420 kJ
3.14 420 kW
3.15 14 286 kg
3.16 66 667 kg
3.17 $1.19(10)^7$ kJ/s
3.19 (a) $1.803(10)^6$ hp (b) $1.42(10)^{10}$ hp·h
3.21 (a) 0.3 kW·h/day (b) 6400 times (c) $32
3.22 2.03 cycles = 1 towel (energy balance)
3.23 468.8 kJ
3.24 $4.5(10)^6$ kJ
3.25 $1.156(10)^{11}$ ft·lb_f
3.26 26.2 ft/s or 17.85 mph
3.27 $3.53(10)^{35}$ kJ
3.28 $1.91(10)^6$ N·M
3.29 14.7 kJ/kg of water vapor
3.30 $1.1(10)^{13}$ kJ
3.31 $1.53(10)^9$ kW
3.32 366.25 h/gal
3.33 Wood $<$ Natural Gas
3.34 67.8 W
3.35 (a) 40.71 kg/day
3.36 54.97 L/h
3.37 2142.8 L, $7.47(10)^7$ kJ/h
3.38 (a) 183.2 L/h (b) $7.2(10)^6$ kJ/h
3.39 20 percent
3.40 $5.33(10)^3$ kW
3.41 2.19 kg/s
3.43 (a) 1372 N·M (b) $3.457(10)^5$ kW
3.44 Approx. $4.6(10)^4$ gal
3.47 (a) 69.4 percent (b) 84.5 percent
3.48 $265.82 per year
3.52 (a) 529 bbl (b) 1571 bbl
3.53 (a) $5(10)^6$ lb (b) $1(10)^6$ lb
3.55 1840 lb
3.58 20.3 ft^3/s
3.59 $5.22(10)^6$ ton/year
3.60 $1.44(10)^8$ ton saved
3.62 $1.4(10)^8$ lb/h
3.63 $1.78(10)^6$ yd^3
3.65 9.4 percent

3.66 $3.57(10)^7$ steers

3.70 70.45^0

3.77 $3.6(10)^6$ lb of salt

3.80 $1.44(10)^6$ kg/s

3.81 14.65 percent

3.84 9.5 m^2

3.85 $32,220

3.86 $1.14(10)^8$ m^2

3.87 6.92 hp

3.88 17.14 m

3.89 2477.9 W

3.90 28.2 gal/min

3.91 26 percent

3.92 2434 m^2

3.93 18.67 J/m^2 @ sea level, 4436.7 J/m^3 in jet stream

3.94 0.405 m^2

3.95 493 kg/m^2

CHAPTER 4

4.1 $b = 86.6$, $c = 173.2$, $\theta = 30^0$

4.2 $b = 632.5$

4.3 $F_x = 234.5$ lb, $F_y = 187.2$ lb

4.4 $F_x = 1440$ lb, $F_y = 420$ lb

4.5 $F_g = 800$ lb

4.6 $R = 670.8$ N

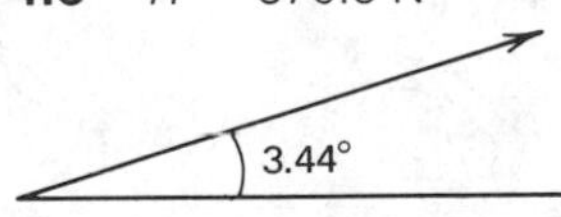

4.7 $R = 4903.6$ lb

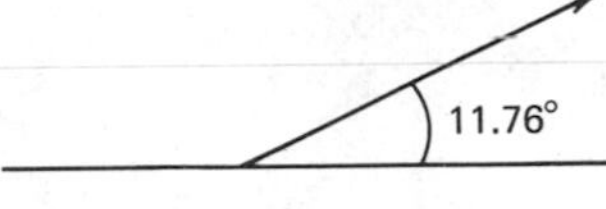

4.8 $R = 337.5$ N

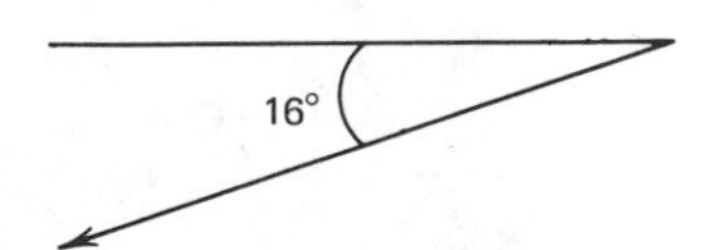

4.9 1231 lb

4.10 1612.8 in.·lb cw

4.11 183 600 ft·lb ccw

4.12 86 240 ft·lb ccw

4.13 487 378 ft·lb cw

4.14 2462.5 ft·lb ccw

4.15 M_A = 709.5 ft·lb cw, M_B = 1812.6 ft·lb cw

4.16 M_A = 544 kN·m cw, M_B = 704 kN·m ccw

4.17 R = 5600 lb, d = 6.32 ft from A

4.18 7200 ft·lb ccw, d = 4 ft up from bottom

4.19 R = 22.5 kN upward, d = 2.26 m from A

4.20 R = 56 000 lb downward, d = 18.0 ft from A

4.21 2357.6 lb

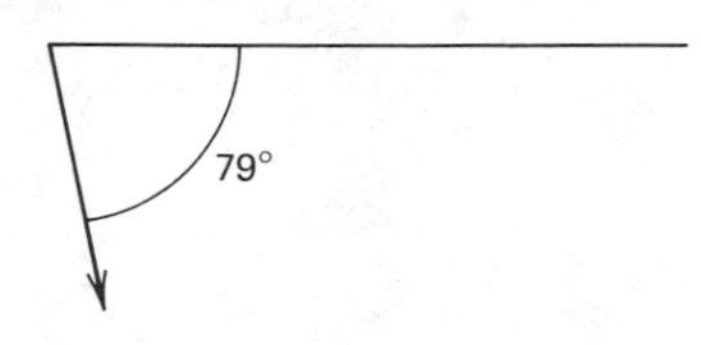

4.22 $R_x = 0$, R_y = 500 lb, M_A = 6000 ft·lb

4.23 R_{Bx} = 141.4 lb, R_{By} = 181.8 lb, R_{Ay} = 459.6 lb

4.24 $R_{Bx} = 0$, R_{By} = 300 lb, R_{Ay} = 300 lb

4.25 $R_{Bx} = 0$, R_{By} = 9000 lb, R_{Ay} = 9000 lb

4.26 T_1 = 600 lb, T_2 = 800 lb

4.27 L = 10 392 lb, D = 6000 lb

4.28 R_{Bx} = 13 333 lb, R_{By} = 8000 lb, R_{Ax} = −13 333 lb

4.29 R = −452.4 lb, M = 1131 ft·lb

4.30 2.275 m

4.31 12.66°

4.32 R_{Ax} = −2 kN, R_{Ay} = 1.33 kN, R_{By} = 2.67 kN

4.33 F_{AB} = 895.8 lb C, F_{AC} = 537.5 lb T, F_{BC} = 270.8 lb T

4.34 $F_{AB} = 0$, $F_{AC} = 0$, $F_{DC} = 0$

4.35 $F_{AB} = F_{DE}$ = 1768 lb C, $F_{AF} = F_{FE}$ = 1250 lb T, $F_{BC} = F_{CD}$ = 1061 lb C, $F_{BF} = F_{FD}$ = 707 lb C, F_{CF} = 1000 lb T

4.36 $F_{AF} = F_{DE} = 0$, F_{AB} = 1125 lb C, F_{BC} = 375 lb C, F_{CD} = 500 lb C, F_{BE} = 1000 lb C, F_{EF} = 375 lb T

4.37 F_{AB} = 333.4 N T, $F_{BC} = 0$

4.38 F_{AB} = 6.7 kN T, F_{BC} = 6.7 kN T, F_{CD} = 2.23 kN T

CHAPTER 5

5.13 $E = 22.3(10)^5$ psi

5.14 F = 99.29 kN

5.15 $E = 5.09(10)^{-4}$ in./in.

5.16 $\Delta L = 2.83(10)^{-3}$ in.

5.17 $L = 1.8(10)^4$ ft

5.18 $A = 68.4\ mm^2$

5.19 $d = 1.19$ in.
5.20 $P = 42\,411.5$ lb
5.21 $d = 6.97$ mm
5.22 $t = 0.00056$ in. (Neglect buckling)
5.23 $d = 0.460$ in.
5.25 $\sigma = 802$ psi, $\Delta L = 0.193$ in.
5.26 $\Delta L = 7.5(10)^{-5}$ in.
5.27 $d = 1.22$ in.
5.28 $\Delta L_p \approx 12\,\Delta L_B$
5.29 $d_{min} = 9.0$ mm
5.30 195 N or 43.8 lb
5.31 $A = 0.0127$ in.^{2}
5.32 $A = 0.249$ in.^{2}
5.33 $T = 3770$ lb
5.34 $\sigma = 0.347$ MPa in polyethylene
5.35 0.355 in. of concrete
5.37 $(Q/A)_{alum} = 3.34(Q/A)_{steel}$
5.38 0.010 in. of stainless steel
5.39 Yes, 10,900 psi
5.41 $1.77(10)^6$ Btu/h·ft
5.42 0.264 mm
5.43 0.0115 in.
5.44 $\sigma = -11\,380$ psi
5.45 3.7 h
5.46 97.3 percent
5.48 0.0116 ft
5.49 $\delta_H = 0.0295$ in., $\delta_v = 0.0787$ in.
5.50 $\Delta T = 1510.7$ F increase, $\Delta T = 76.9$ F decrease
5.51 8.4 h
5.52 27.9 in.
5.53 60 in.
5.58 56.3 percent
5.59 69 percent
5.60 $(Q/A)_{spw} = 19.1(Q/A)_{wall}$
5.62 $(Q/A)_c = 26(Q/A)_h$
5.63 $114.58 per year
5.64 $113.46 per year
5.65 $T = 65$ F
5.67 4569 lb
5.68 $\sigma = 0.543$ MPa, $P = 2.45$ kN
5.69 $r_{AB} = 0.629$ in., $r_{CD} = 0.593$ in.

5.70 d_{min} = 1.01 in.
5.71 $\sigma = Pr/2t$
5.77 3047 lb/lineal ft
5.79 4.37 in.
5.80 23 040 Btu/h
5.81 9 percent
5.82 5400 psi
5.83 5200 psi
5.84 8.05 in.^{2}

CHAPTER 6

6.1 0.005 mA
6.2 24 Ω, I = 0.75 A
6.3 4.16 A
6.4 $3.46(10)^{-4}$ Ω
6.5 24.44 Ω
6.6 V_{10} = 35.5 V, V_5 = 13.3 V, V_{15} = 13.3 V, V_{20} = 71.2 V
6.7 I_{20} = 1.25 A, I_{15} = 0.833 A, I_{30} = 0.416 A
6.8 $.096
6.9 $1073.8
6.10 94 h
6.11 180 Ω
6.12 234 Ω
6.13 51 Ω
6.14 0.012 mA
6.15 17 bulbs
6.16 2125 Btu/h
6.17 108 ft
6.18 108 ft
6.18 (a) 110111 (b) 100110 (c) 10111
6.19 (a) 10001 (b) 10110 (c) 11
6.20 (a) 1000010 (b) 11010010 (c) 1000001
6.21 (a) 111 Remainder 11 (b) 101 R 100 (c) 10 R 1
6.22 (a) 27 (b) 26 (c) 23
6.23 (a) 110011111 (b) 1111010 (c) 1111
6.25 (a) $B = \bar{a}b + a\bar{b}$ (b) $B = \bar{a}\bar{b}c + \bar{a}b\bar{c} + ab\bar{c}$

6.26

(a)

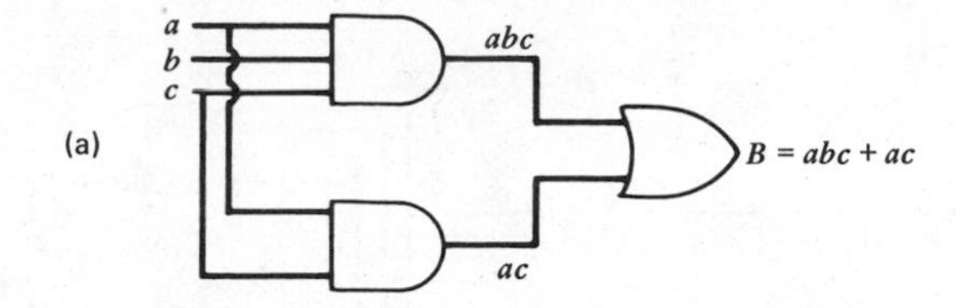

(b)

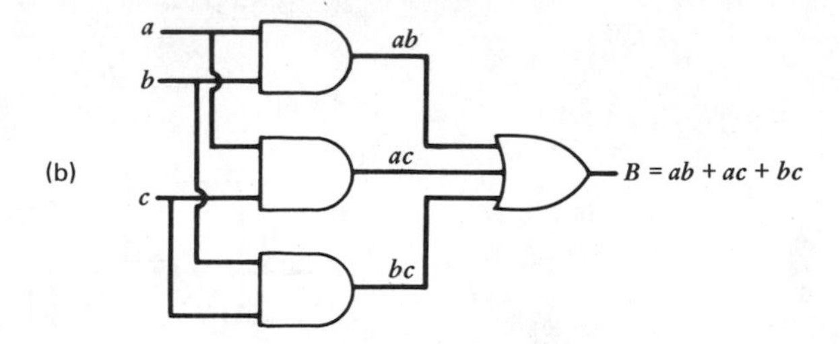

(c)

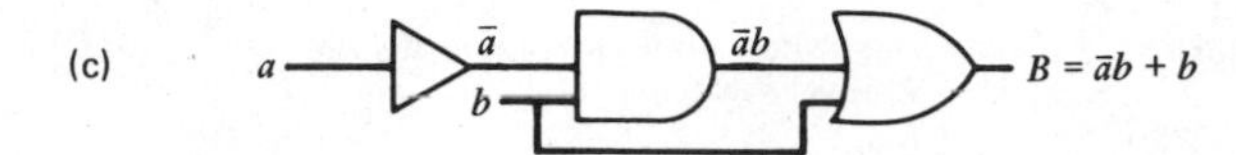

(d)

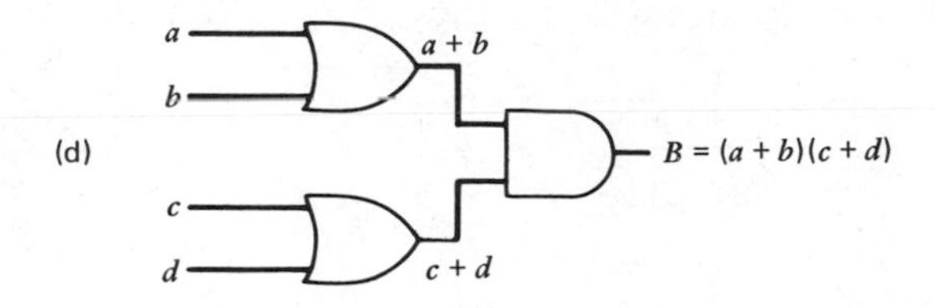

6.27

$B = \bar{a}\bar{b}c + \bar{a}b\bar{c} + a\bar{b}\bar{c} + abc$

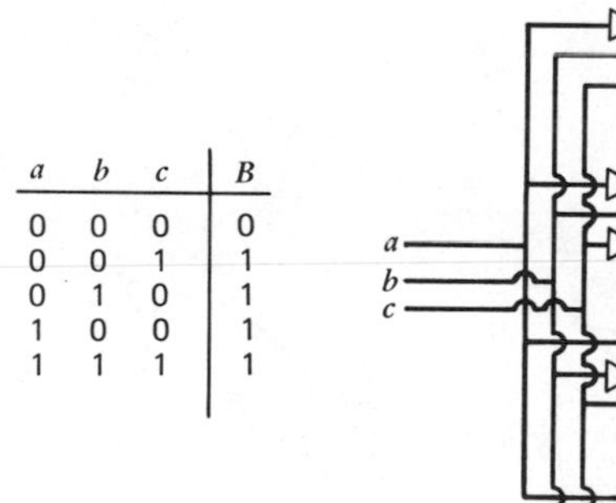

a	b	c	B
0	0	0	0
0	0	1	1
0	1	0	1
1	0	0	1
1	1	1	1

a, b, c; $\bar{a}\bar{b}c$; $\bar{a}b\bar{c}$; $a\bar{b}\bar{c}$; abc; B

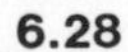

6.28

$B = a\bar{b}c\bar{d} + \bar{a}b\bar{c}d$

a	b	c	d	B
1	0	1	0	1
0	1	0	1	1

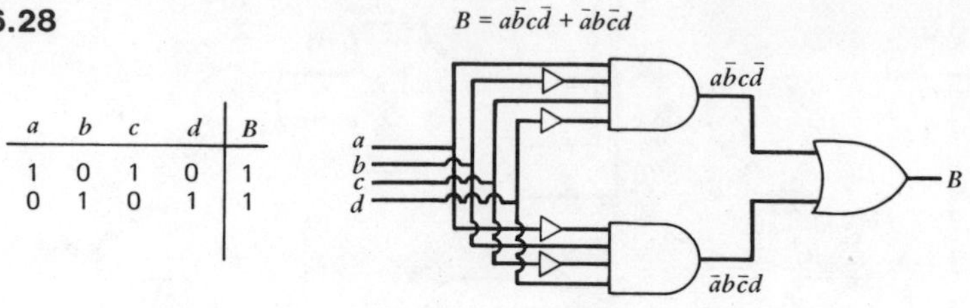

6.30

From Problem 6.27

$$B = \bar{a}\bar{b}c + \bar{a}b\bar{c} + a\bar{b}\bar{c} + abc$$
$$= c(\bar{a}\bar{b} + ab) + \bar{c}(\bar{a}b + a\bar{b})$$
$$= c + \bar{c}(\bar{a}b + a\bar{b})$$
$$= c + \bar{a}b + a\bar{b}$$

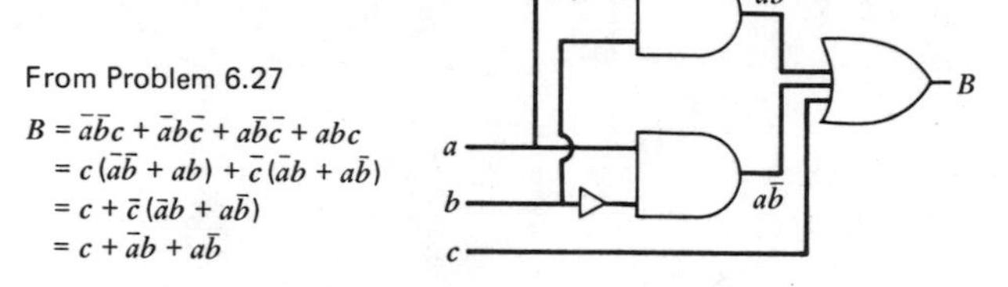

Note: This switch is not identical to the typical three-way switch in a house.

6.31

$$Cat = (a+b)(h+g) = \overline{\overline{(a+b)(h+g)}} = \overline{\overline{(a+b)} + \overline{(h+g)}}$$

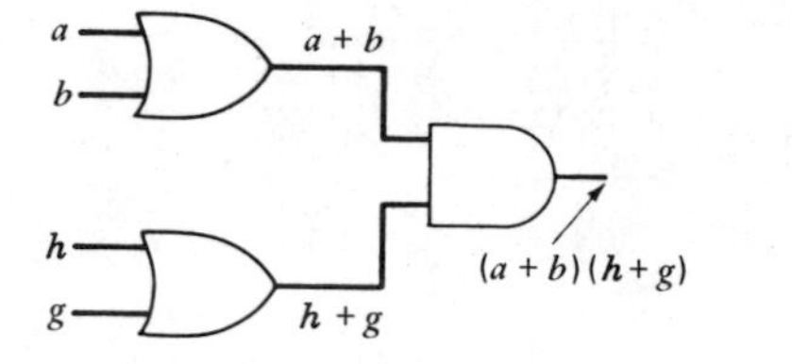

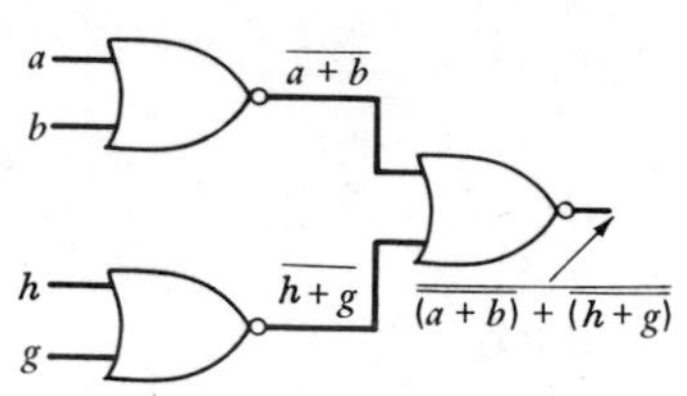

6.32

$$M = abc + a\bar{b}c + c\bar{a}b + b\bar{a}\bar{c}$$
$$= ac(b + \bar{b}) + \bar{a}b(c + \bar{c}) = ac + \bar{a}b$$

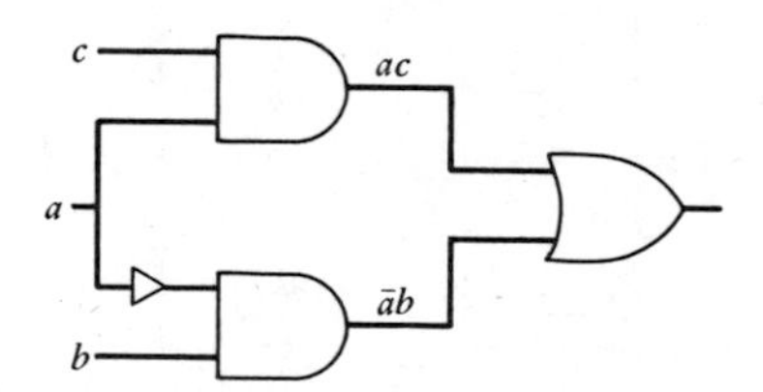

CHAPTER 7

7.1 (a) 407.08 in. (b) 762.5 mm (c) 10 339.75 mm
7.2 550 kPa
7.3 86.03 psi absolute
7.4 10 002 psi gage
7.5 94 380 lb, 18.34 ft up from bottom
7.6 3.521 kN, 0.257 m up from bottom
7.7 12 090 lb, 2.36 ft up from door bottom and centered
7.8 (a) 5880 kN at center (b) 588 kN up 0.66 m from bottom (c) 196 kN up 0.66 from bottom
7.9 94 572.8 kN
7.10 6.12 ft × 6.12 ft
7.11 $h = 2.1$ ft
7.12 920.7 psi
7.13 2244.6 ft·lb
7.14 353.6 psi
7.15 393.2 N
7.16 1.95 psi gage
7.17 $h = 346.1$ ft
7.18 $h = 4.7$ m
7.19 $P_A = 9.6$ psi gage
7.20 $V_2 = 108$ fps
7.21 $Q = 4.89$ ft^3/s
7.22 21.2 h
7.23 $d_2 = 3.8$ in.
7.24 $Q = 0.075$ m^3/s
7.25 $h = 14.53$ ft
7.26 $P_1 = 102.67$ psi
7.27 $Q = 0.0052$ m^3/s
7.28 $P_2 = 166.2$ psi
7.29 0.82 hp
7.30 70.9 hp
7.31 8.66 hp
7.32 2.88 hp
7.33 0.056 hp
7.34 172.7 hp
7.35 $V_2 = 22.3$ m/s
7.36 96.4 kW
7.37 34.35 hp

INDEX